# DEDICATION

For Ellen

R. H. F.

For Anna

B. S. B.

# ABOUT THE AUTHORS

## ROBERT H. FRANK

Professor Frank received his B.S. from Georgia Tech in 1966, then taught math and science for two years as a Peace Corps volunteer in rural Nepal. He received his M.A. in statistics and his Ph.D. in economics in 1972 from the University of California at Berkeley. He is the H. J. Louis Professor of Economics at the Johnson Graduate School of Management at Cornell University, where he has taught since 1972. During a leave of absence from Cornell he served as chief economist for the Civil Aeronautics Board (1978–1980), a Fellow at the Center for Advanced Study in the Behavioral Sciences (1992–1993), and Professor of American Civilization at l'École des Hautes Études en Sciences Sociales in Paris (2000–2001).

Professor Frank is the author of a best-selling intermediate economics textbook—*Microeconomics and Behavior,* Fifth Edition (McGraw-Hill/Irwin, 2003). He has published on a variety of subjects, including price and wage discrimination, public utility pricing, the measurement of unemployment spell lengths, and the distributional consequences of direct foreign investment. His research has focused on rivalry and cooperation in economic and social behavior. His books on these themes include *Choosing the Right Pond: Human Behavior and the Quest for Status* (Oxford University Press, 1985) and *Passions Within Reason: The Strategic Role of the Emotions* (W.W. Norton, 1988). He and Philip Cook are coauthors of *The Winner-Take-All Society* (The Free Press, 1995), which received a Critic's Choice Award and appeared on both the *New York Times* Notable Books list and *Business Week* Ten Best list for 1995. His most recent general-interest publication, *Luxury Fever* (The Free Press, 1999), was named to the Knight-Ridder Best Books list for 1999. He was awarded an Andrew W. Mellon Professorship (1987–1990), a Kenan Enterprise Award (1993), and a Merrill Scholars Program Outstanding Educator Citation (1991). Professor Frank's introductory microeconomics course has graduated more than 6,000 enthusiastic economic naturalists over the years.

## BEN S. BERNANKE

rofessor Bernanke received his B.A. in economics from Harvard University in 1975 and his Ph.D. in economics from MIT in 1979. He taught at the Stanford Graduate School of Business from 1979 to 1985 and moved to Princeton University in 1985, where he is the Howard Harrison and Gabrielle Snyder Beck Professor of Economics and Public Affairs, and where he served as Chairman of the Economics Department. He has consulted for the Board of Governors of the European Central Bank and other central banks, and he served on a U.S. State Department Committee that advises the Israeli government on economic policy. He is a member of the American Academy of Arts and Sciences, Fellow of the Econometrics Society, and a Research Associate for the National Bureau of Economic Research. He has been a visiting scholar at the Federal Reserve System in Boston, Philadelphia, and New York. And in August 2002, he was named to the Board of Governors of the Federal Reserve.

Professor Bernanke's intermediate textbook, with Andrew Abel, *Macroeconomics,* Fourth Edition (Addison-Wesley, 2001), is a best seller in its field. He has written more than 50 scholarly publications in macroeconomics, macroeconomic history, and finance. He has done significant research on the causes of the Great Depression, the role of financial markets and institutions in the business cycle, and measuring the effects of monetary policy on the economy. His two most recent books, both published by Princeton University Press, are *Inflation Targeting: Lessons from the International Experience* (with coauthors) and *Essays on the Great Depression.* He is the editor of the *American Economic Review* and has been the coeditor of the *NBER Macroeconomics Annual* and of *Economics Letters.* He has served as associate editor for the *Journal of Financial Intermediation,* the *Quarterly Journal of Economics,* the *Journal of Money, Credit, and Banking,* and the *Review of Economics and Statistics.* Professor Bernanke has taught principles of economics at both Stanford and Princeton.

# PREFACE

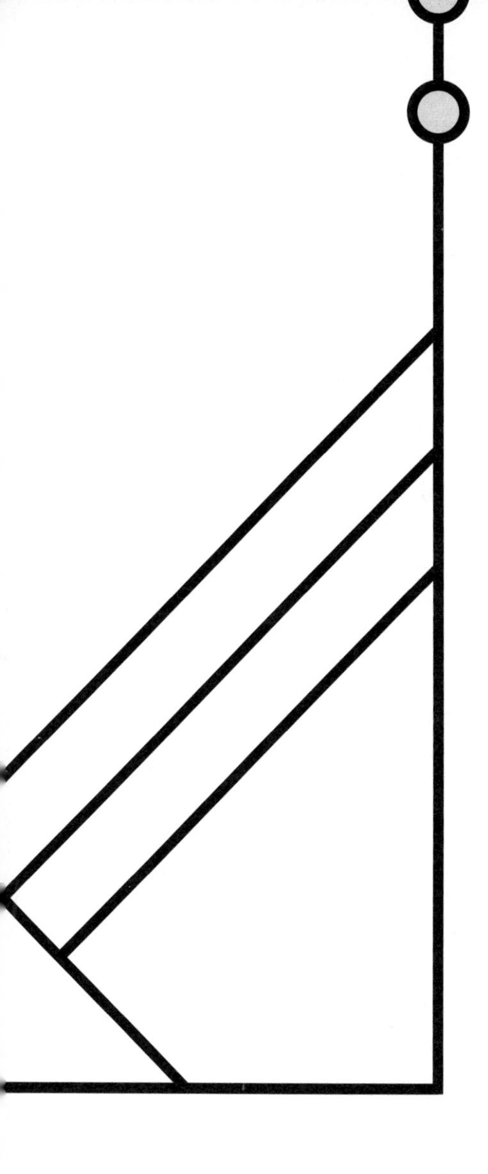

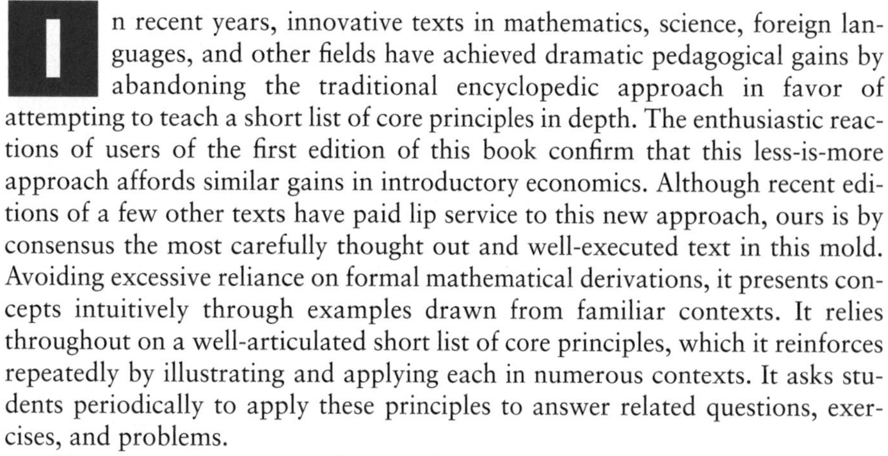

In recent years, innovative texts in mathematics, science, foreign languages, and other fields have achieved dramatic pedagogical gains by abandoning the traditional encyclopedic approach in favor of attempting to teach a short list of core principles in depth. The enthusiastic reactions of users of the first edition of this book confirm that this less-is-more approach affords similar gains in introductory economics. Although recent editions of a few other texts have paid lip service to this new approach, ours is by consensus the most carefully thought out and well-executed text in this mold. Avoiding excessive reliance on formal mathematical derivations, it presents concepts intuitively through examples drawn from familiar contexts. It relies throughout on a well-articulated short list of core principles, which it reinforces repeatedly by illustrating and applying each in numerous contexts. It asks students periodically to apply these principles to answer related questions, exercises, and problems.

The text encourages students to become "economic naturalists," people who employ basic economic principles to understand and explain what they observe in the world around them. An economic naturalist understands, for example, that infant safety seats are required in cars but not in airplanes because the marginal cost of space to accommodate these seats is typically zero in cars but often hundreds of dollars in airplanes. Such examples engage students' interest while teaching them to see each feature of their economic landscape as the reflection of an implicit or explicit cost-benefit calculation.

The second edition incorporates several significant pedagogical improvements. Based on extensive reviewer feedback, it offers (1) even more streamlined coverage of the cost-benefit approach in the introductory chapter; (2) exercises that are more closely tied to the examples; (3) for important or difficult concepts, expanded narrative explanations that are more accessible to average students; and (4) expanded coverage of several key topics (see below). The result is a revision that is even more clutter-free, engaging, and pedagogically effective than its predecessor.

## FEATURES

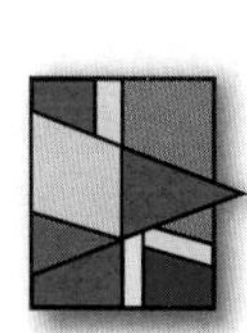

- **Core Principles Emphasized:** A few core principles do most of the work in economics. By focusing almost exclusively on these principles, the text ensures that students leave the course with a deep mastery of them. In contrast, traditional encyclopedic texts so overwhelm students with detail that they often leave the course with little useful working knowledge at all.

ECONOMIC NATURALIST

- **Economic Naturalism Expanded in Macro:** This feature, arguably more applied than in microeconomics, still involves an explicit or implicit cost-benefit calculation. In macro the economic naturalist might ask the following:
  - Why was the United States able to experience rapid growth and low inflation in the latter part of the 1990s?
  - What caused the 2001 recession in the United States?
  - How did the Fed react to recession and the terror attacks in 2001?

- **Active Learning Stressed:** The only way to learn to hit an overhead smash in tennis or to speak a foreign language is through repeated practice. The same is true for learning economics. Accordingly, we consistently introduce new ideas in the context of simple examples and then follow them with applications showing how they work in familiar settings. At frequent intervals, we pose exercises that both test and reinforce the understanding of these ideas. The end-of-chapter questions and problems are carefully crafted to help students internalize and extend core concepts. Experience with our first edition confirms that our text really does prepare students to apply basic economic principles to solve economic puzzles drawn from the real world.
- **Modern Macroeconomics:** Recent developments have renewed interest in cyclical fluctuations without challenging the importance of such long-run issues as growth, productivity, the evolution of real wages, and capital formation. Our treatment of these issues is organized as follows:
  - A five-chapter treatment of long-run issues is followed by an analysis of short-run fluctuations, followed by a modern treatment of short-term fluctuations and stabilization policy, emphasizing the important distinction between short- and long-run behavior of the economy.
  - Consistent with both media reporting and recent research on the central bank reaction function, we treat the interest rate rather than the money supply as the primary instrument of Fed policy.
  - The analysis of aggregate demand and aggregate supply relates output to inflation rather than to the price level, sidestepping the necessity of a separate derivation of the link between the output gap and inflation.
  - This book places a heavy emphasis on globalization, starting with an analysis of its effects on real wage inequality and progressing to such issues as the benefits of trade, the causes and effects of protectionism, the role of capital flows in domestic capital formation, the link between exchange rates and monetary policy, and the sources of speculative attacks on currencies.
- **Web site:** The site was developed by Scott Simkins of North Carolina A & T State University, an expert in the growing field of economics education on the World Wide Web. The ambitious web site contains a host of features that will enhance the principles classroom, including dynamic graphs, email updates, macroeconomic experiments, current news articles, information about the text, eLearning sessions, and more.

## IMPROVEMENTS

- **Introductory Material Shortened and Refined:** The material from the first edition's Chapters 1 and 2 has been reworked and condensed into one chapter in an effort to launch these important concepts as clearly and efficiently as possible. From the very beginning, the focus is on how rational people make choices among alternative courses of action.
- **Long-Run Coverage Expanded:** Chapter 11 of the first edition dealt with financial markets, money, and the Fed; capital flows was dealt with as part of a later, international chapter. Because of the importance of these topics, we have expanded this material into two long-run chapters in this edition. These chapters are now titled "Money, Prices, and the Federal Reserve" (Ch. 10) and "Financial Markets and International Capital Flows" (Ch. 11).

- **Short-Run Coverage More Accessible:** Economic concepts are more streamlined, providing a smoother flow of economic reasoning for students. There are new Economic Naturalists (ENs) that help students better understand how policymakers react to changes in economic conditions. These ENs allow students to bridge the gap between theory and practice more effectively. There is a complete graphical/verbal discussion of the key economic concepts in each chapter for instructors and students who find this material most accessible in a nonmathematical presentation. But for those instructors who wish to reinforce specific concepts with a more formal mathematical approach, chapter appendixes provide a full algebraic treatment of the corresponding chapter concepts.

  Students are provided with a clear, consistent framework for understanding modern economic theory and policymaking. They systematically develop a powerful model for understanding a wide range of short-run issues that are relevant in today's economy. Updated examples in each chapter are both topical and effective at illustrating key points. Each chapter provides students with examples and hands-on exercises to promote practice with key concepts and the workings of the short-run model. This hands-on approach to learning helps students better understand how economic policy affects interest rates, inflation, unemployment, exchange rates, and output and gets students more engaged in the learning process. Chapter-by-chapter benefits include the following:

  - The development of the Keynesian cross in Chapter 13 places more emphasis on graphs and verbal explanations, with an optional mathematical treatment in the appendix. The new presentation makes it easier for students to understand the basic economic concepts underlying the short-run model.

  - Chapter 14 continues to focus on the Federal Reserve's use of the interest rate as the primary policy tool, allowing students to more easily relate economic concepts to real-world macroeconomic policy decisions reported in the news.

  - The AD-AS model is developed systematically in Chapter 15 (based on concepts introduced in Chapters 13 and 14) using a graphical/verbal approach, allowing students to better understand the links among economic theory, real-world macroeconomic behavior, and macroeconomic policymaking. Again, an algebraic treatment of material in this chapter is provided in the appendix.

- **Modern International Finance:** Chapter 17 focuses on a particularly important variable in international economics, the exchange rate. The exchange rate plays a key role in determining patterns of trade. Moreover, the type of exchange rate systems a country adopts has important implications for the effectiveness of its macroeconomic policies.

## THE CHALLENGE

The world is a more competitive place now than it was when we started teaching in the 1970s. In arena after arena, business as usual is no longer good enough. Baseball players used to drink beer and go fishing during the off-season, but they now lift weights and ride exercise bicycles. Assistant professors used to work on their houses on weekends, but the current crop can be found most weekends at the office. The competition for student attention has grown similarly more intense.

There are many tempting courses in the typical college curriculum, and even more tempting diversions outside the classroom. Students are freer than ever to pick and choose. Yet many of us seem to operate under the illusion that most freshmen arrive with a burning desire to become economics majors. And many of us do not yet seem to have recognized that students' cognitive abilities and powers of concentration are scarce resources. To hold our ground we must become not only more selective in what we teach, but also more effective as advocates for our discipline. We must persuade students that we offer something of value.

A well-conceived and well-executed introductory course in economics can teach our students more about society and human behavior in a single term than virtually any other course in the university. This course can and should be an intellectual adventure of the first order. Not all students who take the kind of course we envisioned when writing this book will go on to become economics majors of course. But many will, and even those who do not will leave with a sense of admiration for the power of economic ideas.

A salesperson knows that he or she often gets only one chance to make a good first impression on a potential customer. Analogously, the principles course is often our only shot at persuading students to appreciate the value of economics. By trying to teach them everything we know—rather than teaching them the most important things we know—we too often squander this opportunity.

## SUPPLEMENTS

We continue to believe that an ancillary package is most useful if each element in it is part of a well-considered whole, and here, as with the first edition, we've tried hard to coordinate all the elements. We have also worked hard to improve each element within the package—for example, the PowerPoints are more extensive and more sophisticated, the Instructor's Manual now provides more help than before, the Test Bank is larger and the questions more closely tied to the textbook, the Study Guide sends students to the web site more often, the web site itself is more extensive and more accessible, and the DiscoverEcon Tutorial Software is more student-friendly than ever and now allows instructors e-submission capability and easy syllabus linking. Finally, we listened to you about what, in the first edition, needed work and tried to make it better.

### FOR THE INSTRUCTOR

**Instructor's Manual:** Prepared by Margaret Ray at Mary Washington College, this manual will be extremely useful for all teachers, but especially for those new to the job. In addition to such general topics as Using the Web Site, Economic Education Resources, Innovative Ideas, and Tips for Teaching, there will be, for each chapter, An Overview, An Outline, Core Principles, Important Concepts, Teaching Objectives, Web Site Applications, In-Class Activities, More Economic Naturalists, Answers to Textbook Problems, Sample Homework, and a Sample Reading Quiz.

**Test Bank:** Prepared by Nancy Jianakoplos at Colorado State University, this manual contains nearly 3,000 multiple-choice questions categorized by Teaching Objective (from the Study Guide); Learning Level (knowledge, comprehension, application, analysis); Type (graph, calculation, word problem); and Source (textbook, Study Guide, web site, unique).

**Computerized Test Bank:** The print test banks are also available in the latest Diploma test-generating software, ensuring maximum flexibility in test preparation, including the reconfiguring of graphing exercises. This Brownstone program is the gold standard of testing programs. It is available in both a Windows and Macintosh format.

**PowerPoints:** Prepared by Steve Smith and Jeff Caldwell at Rose State, these slides contain all of the illustrations in the textbook, along with a detailed, chapter-by-chapter review of the important ideas presented in the textbook. These teachers have done PowerPoints for many books at both the principles and intermediate level.

**Overhead Transparencies:** More than 150 four-color acetates contain all the illustrations presented in the textbook.

**Instructor's CD-ROM:** This remarkable Windows software program, which contains the complete Instructor's Manual, Computerized Test Bank, PowerPoints, and a full set of lecture notes for principles of microeconomics, prepared by Bob Frank for his successful introductory course at Cornell University, also allows the instructor to create presentations from any of the materials on the CD or from additional material that can be imported.

**Online Learning Center (www.mhhe.com/economics/frankbernanke2):** For teachers there are, among other things, an online newsletter called "Teaching Using the Web"; the Instructor's Manual; the PowerPoints; Economics on the Web, an annotated set of URLs/links to sites of interest to economists; a graphing library; and a description of what's on the student site and some optional material from the book.

## FOR THE STUDENT

**Study Guide:** Written by Jack Mogab and Bruce McClung at Southwest Texas State University, this book provides the following elements for each chapter: a Pretest; a Learning Objective Grid; a Key Point Review with Learning Tips; some Self-Tests (Key Term Matching, Multiple Choice, Problems) with answers; and an extension of the guide to the web site, where students may practice with graphing.

**Online Learning Center (www.mhhe.com/economics/frankbernanke2):** For students there are such useful and exciting features such as Interpreting the News—articles and summaries of relevant articles with analysis and discussion questions; a Math Tutor—help for those whose math skills are rusty; email updates—periodic sending of information/study tips; the Glossary from the textbook; and Economics on the Web—annotated URLs useful for economics students. Additionally, for each chapter there is an Electronic Learning Session, which opens with a brief recap of the chapter followed by a test with answers and analysis; this is followed by a set of study sessions based on Economic Naturalist Exercises; Graphing Exercises; PowerPoints; and Key Terms; and finally a second quiz, with answers and analysis.

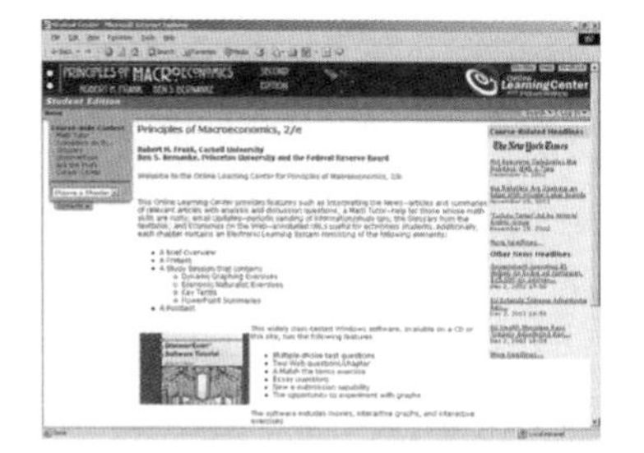

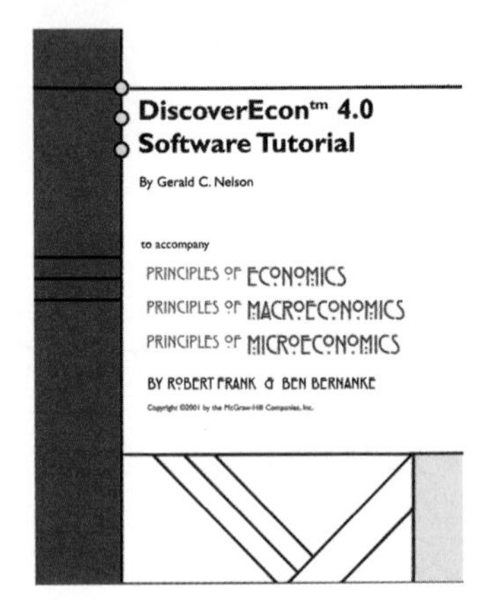


**DiscoverEcon CD-ROM:** Created by Gerald Nelson of the University of Illinois, DiscoverEcon is the best-selling academic economics software available. It is available as a CD or an online version. This widely class-tested software is Windows-based and text-specific. It is available either in a standard CD format or online by using a unique student pass code packaged free with each new book. The following features are contained in every chapter:

*Multiple-choice test questions:* The user chooses the number of questions in the test; the software selects randomly from the question bank for the chapter.

*Two web questions:* Students working online can access web addresses directly through their browsers. Students without web access can simply ignore these questions.

*Match the terms:* This exercise challenges the student to match randomly selected key glossary terms with their appropriate glossary explanation.

*Essay questions.*

**Additional features include:**

*New e-submission capability:* Instructors may now set up their courses for e-submission so that all exercise results for each student can be viewed and downloaded to other Windows applications (web-based version only).

*Easy syllabus linking:* For those instructors using a course management system, specific pages in DiscoverEcon can be linked to specific parts of the course web site. This makes navigation and therefore self-assessment easier than ever for both student and instructor.

*The opportunity to experiment with graphs,* those seen in the classroom and the textbook. The software includes *movies* (graphs are drawn step-by-step with explanatory text appearing as the graph is constructed); *interactive graphs* (students change a parameter and see how the graph changes); and *interactive exercises* (students interpret a graph based on concepts presented in the text and in the software).

**BusinessWeek Edition:** Your students can subscribe to 15 weeks of *BusinessWeek* for a specially priced rate. Students will receive a pass code card shrink-wrapped with their new text. The card directs students to a web site where they enter the code and then gain access to *BusinessWeek*'s registration page to enter address info and set up their print and online subscription as well.

**Wall Street Journal Edition:** Your students can subscribe to the *Wall Street Journal* for 15 weeks at a specially priced rate. Students will receive a "How To Use the *WSJ*" handbook plus a pass code card shrink-wrapped with the text. The card directs students to a web site where they enter the code and then gain access to the *WSJ* registration page to enter address info and set up their print and online subscription, and also set up their subscription to Dow Jones Interactive online for the span of the 15-week course.

## ACKNOWLEDGMENTS

We are greatly indebted to Professor Scott Simkins at North Carolina Agricultural and Technical State University for a number of important contributions to the book, most especially for providing Examples and Exercises in the short-run chapters.

Scott also updated or created Economic Naturalists throughout. Our thanks also go to our publisher, Gary Burke, for his unwavering faith in our project over the past several years. Without his support and encouragement, we never could have produced this book. Tom Thompson, our development editor, was enormously helpful as he guided us with intelligence, patience, and tact through three major revisions of the original manuscript, and further extensive revisions for the second edition. We also thank Paul Shensa, the sponsoring editor, and Marty Quinn, the marketing manager, whose considerable experience, insightful suggestions and extensive knowledge of the marketplace were of incalculable help. We are especially grateful to Betty Morgan, our superb manuscript editor. And we are also grateful to the production team, whose professionalism was outstanding: Kimberly Hooker, Project Manager; Melonie Salvati, Production Supervisor; Becky Szura, Supplements Coordinator, and Melissa Kansa, Media Tech Producer.

Finally, our sincere thanks to the following teachers and colleagues, whose thorough reviews and thoughtful suggestions led to innumerable substantive improvements to both editions:

Ehsan Ahmed
James Madison University

Ercument Aksoy
Los Angeles Valley College

Richard Anderson
Texas A&M University

Michael Balch
University of Iowa

Daniel Berkowitz
University of Pittsburgh

Guatam Bhattacharya
University of Kansas

Scott Bierman
Carleton College

Bruce Blonigen
University of Oregon

Beth Bogan
Princeton University

George Borts
Brown University

Isaac Brannon
University of Wisconsin–Oshkosh

Nancy Brooks
University of Vermont

Bruce Brown
California Polytechnic Institute–Pomona

Douglas Brown
Georgetown University

Marie Bussings-Burk
Southern Indiana University

David Carr
University of Colorado

Edward Castranova
California State University–Fullerton

Tom Cate
Northern Kentucky University

Jack Chambless
Valencia Community College

James Cover
University of Alabama

John D'Amico
Hilbert College

Carl Davidson
Michigan State University

Robert Deckle
University of Southern California

Lynn Pierson Doti
Chapman College

Donald Dutkowsky
Syracuse University

David Eaton
Murray State University

Nancy Fox
Saint Joseph's College

John Francis
University of Wisconsin–Oshkosh

Roger Franz
San Diego State University

Susan Gale
New York University

Johah Gelbach
University of Maryland

Linda Ghent
East Carolina University

Kirk Gifford
Ricks College

Robert Gillette
University of Kentucky

Stephen Gohman
University of Louisville

John Graham
Rutgers University–Newark

Refet Gurkaynak
Princeton University

Galina Hale
University of California–Berkeley

Russell Hardy
New Mexico State University

Mehdi Haririan
Bloomsburg University

Joseph Haslag
University of Missouri

Susan Hayes
Sonoma State University

Bruce Herrick
Washington and Lee University

William Hogan
University of Massachusetts–Dartmouth

Mary Jean Horney
Furman University

Frederick Inaba
Washington State University

Nancy Jianakoplos
Colorado State University

Robert Johnson
University of San Diego

Rogear Kaufman
Smith College

Elizabeth Kelley
University of Wisconsin–Madison

Herbert Kiesling
Indiana University

Bruce Kingma
State University of New York–Albany

Faik Koray
Louisiana State University

Maria Kula
Roger Williams University

Leonard Lardaro
University of Rhode Island

Mary Lesser
Iona College

Anthony Lima
California State University–Hayward

Bruce Linster
U.S. Air Force Academy

Tom Love
North Central University

Steven McCafferty
The Ohio State University

Edward McNertney
Texas Christian University

William Merrill
Iowa State University

Paul Nelson
Northeast Louisiana State University

Neil Niman
University of New Hampshire

Norman Obst
Michigan State University

Frank O'Connor
Eastern Kentucky University

Charles Okeke
Community College of Southern Nevada

Duane Oyen
University of Wisconsin–Eau Claire

Theodore Palivos
Louisiana State University

Jan Palmer
Ohio University

Hong Park
Saginaw Valley State University

Michael Potepan
San Francisco State University

Rahim Quazi
University of Georgia

Steve Robinson
University of North Carolina–Wilmington

Christina Romer
University of California–Berkeley

David Romer
University of California–Berkeley

Greg Rose
Sacramento City College

Jeffrey Rous
University of North Texas

Daniel Rubenson
Southern Oregon University

Richard Salvucci
Trinity University

Edward Scahill
University of Scranton

Pamela Schmitt
U.S. Naval Academy

Esther-Mirjam Sent
University of Notre Dame

Neil Skaggs
Illinois State University

John Solow
University of Iowa

Martin Spechler
Indiana University/Purdue University–Indianapolis

Dennis Starleaf
Iowa State University

Michael Stroup
Stephen F. Austin University

Helen Tauchen
University of North Carolina–Chapel Hill

Philip Taylor
Wesleyan University (Georgia)

Jennifer Tessendorf
Hobart & William Smith College

Nora Underwood
University of California–Davis

Kay Unger
University of Montana

Norm Van Cott
Ball State University

Stephan Weiler
Colorado State University

Charles Weise
Gettysburg College

Jeffrey Weiss
City University of New York–Baruch College

Richard Winkelman
Arizona State University

Mark Wohar
University of Nebraska–Omaha

Louise Wolitz
University of Texas–Austin

Darrel Young
University of Texas–Austin

Zenon Zygmont
Western Oregon University

# BRIEF CONTENTS

# CONTENTS

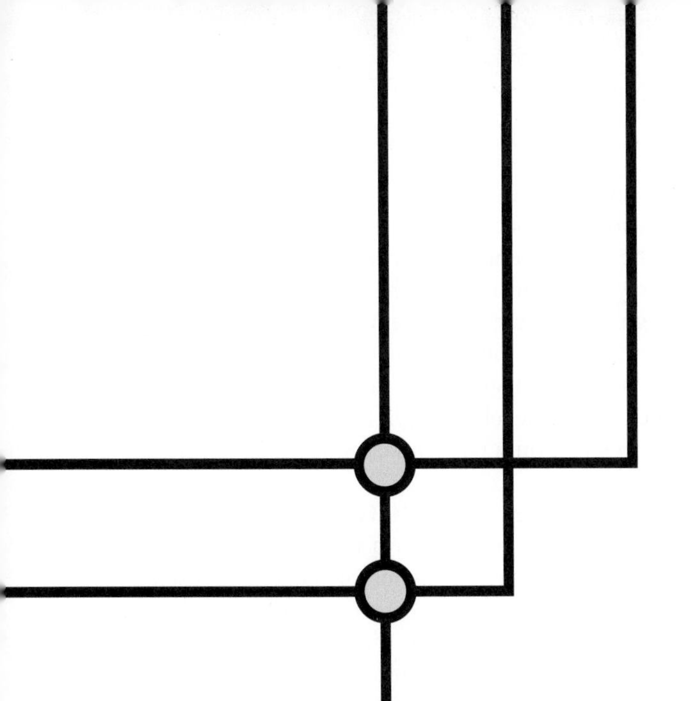

CHAPTER 4

# MACROECONOMICS: THE BIRD'S-EYE VIEW OF THE ECONOMY

In 1929 the economy of the United States slowed dramatically. Between August of 1929 and the end of 1930, the nation's factories and mines, facing sharp declines in sales, cut their production rates by a remarkable 31 percent. These cutbacks led in turn to mass layoffs: Between 1929 and 1930, the number of people without jobs almost tripled, from about 3 percent of the workforce to nearly 9 percent.[1] Financial markets were equally shaky. The stock market crashed in October 1929, and stocks lost nearly a third of their value in just three weeks.

At first, policymakers and the general public (except for those people who had put their life savings into the stock market) were concerned but not panic-stricken. Americans remembered that the nation had experienced a similar slowdown only eight years earlier, in 1921–1922. That episode had ended quickly, apparently on its own, and the decade that followed (popularly known as the roaring twenties) had been one of unparalleled prosperity. But the fall in production and the rise in unemployment that began in 1929 continued into 1931. In the spring of 1931 the economy seemed to stabilize briefly, and President Herbert Hoover optimistically proclaimed that "prosperity is just around

[1]The source for these and most other pre-1960 statistics cited in this chapter is the U.S. Bureau of the Census, *Historical Statistics of the United States: Colonial Times to 1970,* Washington, 1975.

the corner." But in mid-1931 the economy went into an even steeper dive. What historians now call the Great Depression had begun in earnest.

Labor statistics tell the story of the Great Depression from the worker's point of view. Unemployment was extremely high throughout the 1930s, despite government attempts to reduce it through large-scale public employment programs. At the worst point of the Depression, in 1933, one out of every four American workers was unemployed. Joblessness declined gradually to 17 percent of the workforce by 1936 but remained stuck at that level through 1939. Of those lucky enough to have jobs, many were able to work only part-time, while others worked for near-starvation wages.

In some other countries conditions were even worse. In Germany, which had never fully recovered from its defeat in World War I, nearly a third of all workers were without jobs, and many families lost their savings as major banks collapsed. Indeed, the desperate economic situation was a major reason for Adolf Hitler's election as chancellor of Germany in 1933. Introducing extensive government control over the economy, Hitler rearmed the country and ultimately launched what became the most destructive war in history, World War II.

How could such an economic catastrophe have happened? One often-heard hypothesis is that the Great Depression was caused by wild speculation on Wall Street, which provoked the stock market crash. But though stock prices may have been unrealistically high in 1929, there is little evidence to suggest that the fall in stock prices was a major cause of the Depression. A similar crash in October 1987, when stock prices fell a record 23 percent in one day—an event comparable in severity to the crash of October 1929—did not slow the economy significantly. Another reason to doubt that the 1929 stock market crash caused the Great Depression is that, far from being confined to the United States, the Depression was a worldwide event, affecting countries that did not have well-developed stock markets at the time. The more reasonable conclusion is that the onset of the Depression probably caused the stock market crash, rather than the other way round.

Another explanation for the Depression, suggested by some economists in the 1930s, was that free-market economies like those of the United States and Germany are "naturally" unstable, prone to long periods of low production and high unemployment. But this idea too has fallen out of favor, since the period after World War II has generally been one of prosperity and economic growth throughout the industrialized world.

What *did* cause the Great Depression, then? Today most economists who have studied the period blame *poor economic policymaking* both in the United States and in other major industrialized countries. Of course, policymakers did not set out to create an economic catastrophe. Rather, they fell prey to misconceptions of the time about how the economy worked. In other words, the Great Depression, far from being inevitable, *might have been avoided*—if only the state of economic knowledge had been better. From today's perspective, the Great Depression was to economic policymaking what the voyage of the *Titanic* was to ocean navigation.

Could better economic policies have prevented the Great Depression?

One of the few benefits of the Great Depression was that it forced economists and policymakers of the 1930s to recognize that there were major gaps in their understanding of how the economy works. This recognition led to the development of a new subfield within economics, called macroeconomics. Recall from Chapter 1 that *macroeconomics* is the study of the performance of national economies and the policies governments use to try to improve that performance.

This chapter will introduce the subject matter and some of the tools of macroeconomics. Although understanding episodes like the Great Depression remains an important concern of macroeconomists, the field has expanded to include the analysis of many other aspects of national economies. Among the issues macroeconomists study are the sources of long-run economic growth and development, the causes of high unemployment, and the factors that determine the rate of inflation. Appropriately enough in a world in which economic "globalization" preoccupies businesspeople and policymakers, macroeconomists also study how national economies interact. Since the performance of the national economy has an important bearing on the availability of jobs, the wages workers earn, the prices they pay, and the rates of return they receive on their saving, it's clear that macroeconomics addresses bread-and-butter issues that affect virtually everyone.

In light of the nation's experience during the Great Depression, macroeconomists are particularly concerned with understanding how *macroeconomic policies* work and how they should be applied. **Macroeconomic policies** are government actions designed to affect the performance of the economy as a whole (as opposed to policies intended to affect the performance of the market for a particular good or service, such as sugar or haircuts). The hope is that by understanding more fully how government policies affect the economy, economists can help policymakers do a better job—and avoid serious mistakes, such as those that were made during the Great Depression. On an individual level, educating people about macroeconomic policies and their effects will make for a better-informed citizenry, capable of making well-reasoned decisions in the voting booth.

***macroeconomic policies*** government actions designed to affect the performance of the economy as a whole

## THE MAJOR MACROECONOMIC ISSUES

We defined macroeconomics as the study of the performance of the national economy as well as the policies used to improve that performance. Let's now take a closer look at some of the major economic issues that macroeconomists study.

### ECONOMIC GROWTH AND LIVING STANDARDS

Although the wealthy industrialized countries (such as the United States, Canada, Japan, and the countries of western Europe) are certainly not free from poverty, hunger, and homelessness, the typical person in those countries enjoys a standard of living better than at any previous time or place in history. By *standard of living* we mean the degree to which people have access to goods and services that make their lives easier, healthier, safer, and more enjoyable. People with a high living standard enjoy more and better consumer goods: sports utility vehicles, camcorders, cellular phones, and the like. But they also benefit from a longer life expectancy and better general health (the result of high-quality medical care, good nutrition, and good sanitation), from higher literacy rates (the result of greater access to education), from more time and opportunity for cultural enrichment and recreation, from more interesting and fulfilling career options, and from better working conditions. Of course, the *scarcity principle* will always apply—even for the citizen of a rich country, having more of one good thing means having less of another. But higher incomes make these choices much less painful than they would be otherwise. Choosing between a larger apartment and a nicer car is much easier than choosing between feeding your children adequately and sending them to school, the kind of hard choice people in the poorest nations face.

Americans sometimes take their standard of living for granted, or even as a "right." As a Paul Simon lyric proclaims, "God bless our standard of living—let's keep it that way!" But we should realize that the way we live today is radically different from the way people have lived throughout most of history. The current standard of living in the United States is the result of several centuries of sustained *economic growth,* a process of steady increase in the quantity and quality of the goods and services the economy can produce. The basic equation is simple: The more we can produce, the more we can consume. Though not everyone in a society shares equally in the fruits of economic growth, in most cases growth brings an improvement in the average person's standard of living.

To get a sense of the extent of economic growth over time, examine Figure 4.1, which shows how the output of the U.S. economy has increased since 1900. (We discuss the measure of output used here, real gross domestic product, in Chapter 5.) Although output fluctuates at times, the overall trend has been unmistakably upward. Indeed, in 2001 the output of the U.S. economy was more than 25 times what it was in 1900 and more than 5 times its level in 1950. What caused this remarkable economic growth? Can it continue? Should it? These are some of the questions macroeconomists try to answer.

**FIGURE 4.1**
**Output of the U.S. Economy, 1900–2001.**
The output of the U.S. economy has increased by more than 25 times since 1900 and by more than 5 times since 1950.

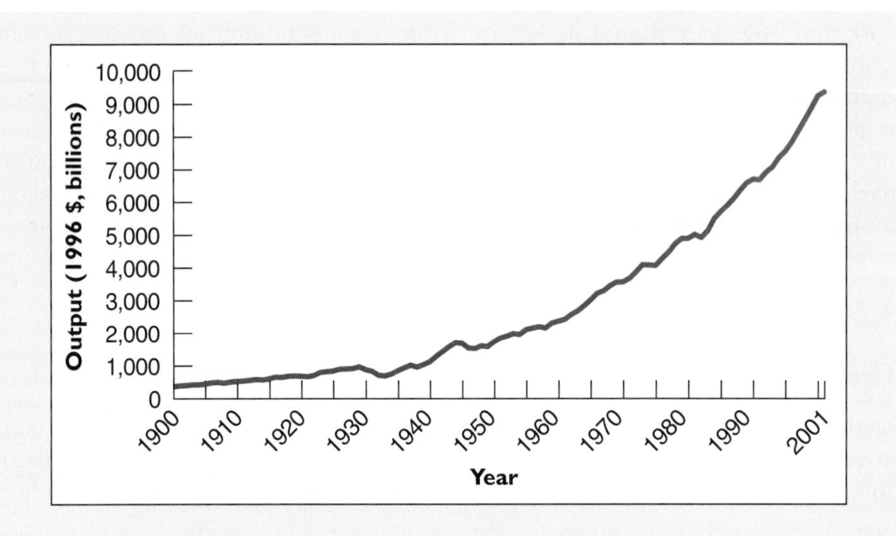


One reason for the growth in U.S. output over the last century has been the rapid growth of the U.S. population, and hence the number of workers available. Because of population growth, increases in *total* output cannot be equated with improvements in the general standard of living. Although increased output means that more goods and services are available, increased population implies that more people are sharing those goods and services. Because the population changes over time, output *per person* is a better indicator of the average living standard than total output.

Figure 4.2 shows output per person in the United States since 1900 (the blue line). Note that the long-term increase in output per person is smaller than the increase in total output shown in Figure 4.1 because of population growth. Nevertheless, the gains made over this long period are still impressive: In 2001 a typical U.S. resident consumed nearly seven times the quantity of goods and services available to a typical resident at the beginning of the century. To put this increase into perspective, according to recent estimates nearly 60 million U.S. households now own two or more automobiles; only about 8 million U.S. households (many of them located in cities, with good access to public transportation) have no car. A remarkable 98 percent of U.S. households own a television—95 percent a color television—and two-thirds subscribe to cable. And as of 2002, nearly 60 percent of adults in the United States were regular users of the Internet.

Nearly 60 million American households own two or more automobiles.

Nor has the rise in output been reflected entirely in increased availability of consumer goods. For example, as late as 1960, only 41 percent of U.S. adults over age 25 had completed high school, and less than 8 percent had completed four years of college. Today, over 80 percent of the adult population have at least a high school diploma, and about 25 percent have a college degree. Over half the students currently leaving high school will go on to college. Higher incomes, which allow young people to continue their schooling rather than work to support themselves and their families, are a major reason for these increases in educational levels.

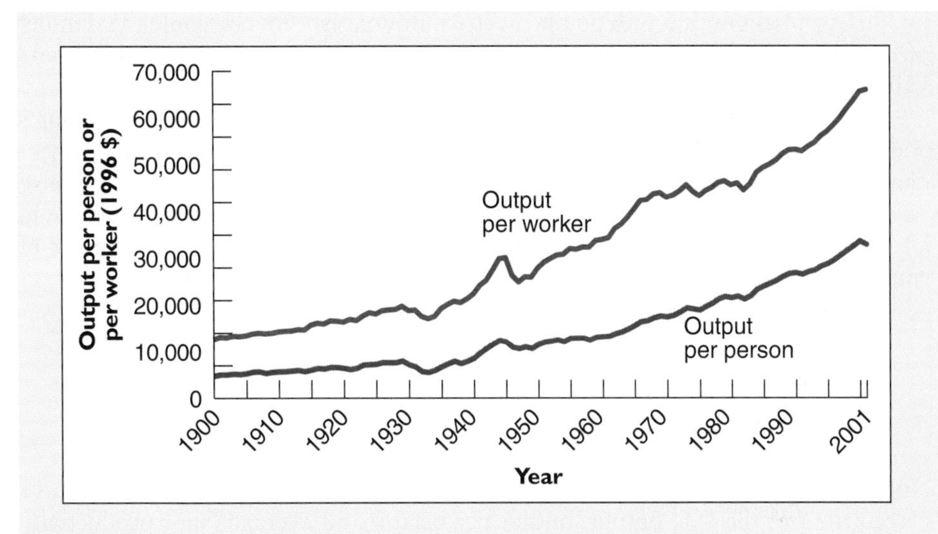


**FIGURE 4.2**
**Output per Person and per Worker in the U.S. Economy, 1900–2001.**
The red line shows the output per worker in the U.S. economy since 1900, and the blue line shows output per person. Both have risen substantially. Relative to 1900, output per person today is nearly seven times greater, and output per worker is more than five times greater.

## PRODUCTIVITY

While growth in output per person is closely linked to changes in what the typical person can *consume,* macroeconomists are also interested in changes in what the average worker can *produce*. Figure 4.2 shows how output per employed worker (that is, total output divided by the number of people working) has changed since 1900 (red line). The figure shows that in 2001 a U.S. worker could produce more than five times the quantity of goods and services produced by a worker at the beginning of the twentieth century, despite the fact that the workweek is now much shorter than it was 100 years ago.

Economists refer to output per employed worker as **average labor productivity.** As Figure 4.2 shows, average labor productivity and output per person are closely related. This relationship makes sense—as we noted earlier, the more we can produce, the more we can consume. Because of this close link to the average living standard, average labor productivity and the factors that cause it to increase over time are of major concern to macroeconomists.

***average labor productivity*** output per employed worker

Although the long-term improvement in output per worker is impressive, the *rate* of improvement has slowed somewhat since the 1970s. Between 1950 and 1973 in the United States, output per employed worker increased by 2.1 percent per year. But from 1973 to 1995 the average rate of increase in output per worker was less than 1 percent per year. Since 1995 the pace of productivity growth seems to have picked up again, however, to nearly 2 percent per year. Slowing productivity growth leads to less rapid improvement in living standards, since the supply of goods and services cannot grow as quickly as it does during periods of rapid growth in productivity. Identifying the causes of productivity slowdowns and speedups is thus an important challenge for macroeconomists.

The current standard of living in the United States is not only much higher than in the past but also much higher than in many other nations today. Why have many of the world's countries, including both the developing nations of Asia, Africa, and Latin America and some formerly communist countries of eastern Europe not enjoyed the same rates of economic growth as the industrialized countries? How can the rate of economic growth be improved in these countries? Once again, these are questions of keen interest to macroeconomists.

**EXAMPLE 4.1**

### Productivity and living standards in China and the United States

In 2001 the value of the output of the U.S. economy was about $10,200 billion. In the same year, the estimated value of the output of the People's Republic of China was $1,160 billion (U.S). The populations of the United States and China in 2001 were about 285 million and 1,262 million, respectively, while the number of employed workers in the two countries were approximately 135 million and 710 million.

Find output per person and average labor productivity for the United States and China in 2001. What do the results suggest about comparative living standards in the two countries?

Output per person is simply total output divided by the number of people in an economy, and average labor productivity is output divided by the number of employed workers. Doing the math we get the following results for 2001:

| | United States | China |
|---|---|---|
| Output per person | $35,790 | $ 919 |
| Average labor productivity | $75,556 | $1,634 |

Note that, although the total output of the Chinese economy is more than 11 percent that of the U.S. output, output per person and average labor productivity in China are each only about 2.5 and 2 percent, respectively, of what they are in the United States. Thus, though the Chinese economy may someday rival the U.S. economy in total output, for the time being there remains a large gap in productivity. This gap translates into striking differences in the living standard between the two countries—in access to consumer goods, health care, transportation, education, and other benefits of affluence.

## RECESSIONS AND EXPANSIONS

Economies do not always grow steadily; sometimes they go through periods of unusual strength or weakness. A look back at Figure 4.1 shows that although output generally grows over time, it does not always grow smoothly. Particularly striking is the decline in output during the Great Depression of the 1930s, followed by the sharp increase in output during World War II (1941–1945). But the figure shows many more moderate fluctuations in output as well.

Slowdowns in economic growth are called *recessions;* particularly severe economic slowdowns, like the one that began in 1929, are called *depressions.* In the United States, major recessions occurred in 1973–1975 and 1981–1982 (find those recessions in Figure 4.1). More modest downturns occurred in 1990–1991 and 2001. During recessions economic opportunities decline: Jobs are harder to find, people with jobs are less likely to get wage increases, profits are lower, and more companies go out of business. Recessions are particularly hard on economically disadvantaged people, who are most likely to be thrown out of work and have the hardest time finding new jobs.

Sometimes the economy grows unusually quickly. These periods of rapid economic growth are called *expansions,* and particularly strong expansions are called *booms.* During an expansion, jobs are easier to find, more people get raises and promotions, and most businesses thrive.

The alternating cycle of recessions and expansions raises some questions that are central to macroeconomics. What causes these short-term fluctuations in the rate of economic growth? Can government policymakers do anything about them? Should they try?

## UNEMPLOYMENT

The *unemployment rate,* the fraction of people who would like to be employed but can't find work, is a key indicator of the state of the labor market. When the unemployment rate is high, work is hard to find, and people who do have jobs typically find it harder to get promotions or wage increases.

Figure 4.3 shows the unemployment rate in the United States since 1900. Unemployment rises during recessions—note the dramatic spike in unemployment during the Great Depression, as well as the increases in unemployment during the 1973–1975 and 1981–1982 recessions. But even in the so-called good times, such as the 1960s and the 1990s, some people are unemployed. Why does unemployment rise so sharply during periods of recession? And why are there always unemployed people, even when the economy is booming?

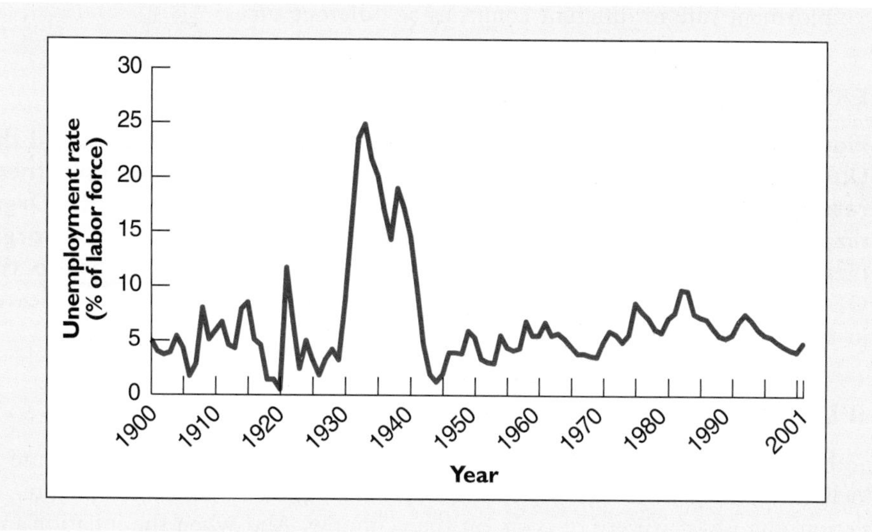


**FIGURE 4.3**
**The U.S. Unemployment Rate, 1900–2001.**
The unemployment rate is the percentage of the labor force that is out of work. Unemployment spikes upward during recessions and depressions, but the unemployment rate is always above zero, even in good times.

**EXAMPLE 4.2**

**Increases in unemployment during recessions**

Using monthly data on the national civilian unemployment rate, find the increase in the unemployment rate between the onset of recession in November 1973, January 1980, and July 1990 and the peak unemployment rate in the following three years. Compare these increases in unemployment to the increase during the Great Depression.

Unemployment data are collected by the U.S. Bureau of Labor Statistics (BLS) and can be obtained from the BLS home page (http://stats.bls.gov/datahome.htm). Hard copy sources include the *Survey of Current Business*, the *Federal Reserve Bulletin*, and *Economic Indicators*. Monthly data from the BLS home page yield the following comparisons:

| Unemployment rate at beginning of recession (%) | Peak unemployment rate (%) | Increase in unemployment rate (%) |
|---|---|---|
| 4.8 (Nov. 1973) | 9.0 (May 1975) | +4.2 |
| 6.3 (Jan. 1980) | 10.8 (Nov./Dec. 1982) | +4.5 |
| 5.5 (July 1990) | 7.8 (June 1992) | +2.3 |

Unemployment increased significantly following the onset of each recession, although the impact of the 1990 recession on the labor market was clearly less serious than that of the 1973 and the 1980 recessions. (Actually, the 1980 recession was a "double dip"—a short recession in 1980, followed by a longer one in 1981–82.) In comparison, during the Great Depression the unemployment rate rose from about 3 percent in 1929 to about 25 percent in 1933, as we mentioned in the introduction to this chapter. Clearly, the 22 percentage point change in the unemployment rate that Americans experienced in the Great Depression dwarfs the effects of these three postwar recessions.

One question of great interest to macroeconomists is why unemployment rates sometimes differ markedly from country to country. For the past two decades unemployment rates in western Europe have more often than not been measured in the "double digits." On average, more than 10 percent of the European workforce has been out of a job during this period, a rate roughly double that in the United States. The high unemployment is particularly puzzling, because during the 1950s and 1960s, European unemployment rates were generally much lower than those in the United States. What explains these differences in the unemployment rate in different countries at different times?

**EXERCISE 4.1**

**Find the most recent unemployment rates for France, Germany, and the United Kingdom, and compare them to the most recent unemployment rate for the United States. A useful source is the home page of the Organization for Economic Cooperation and Development (OECD), an organization of industrialized countries (http://www.oecd.org/). See also the OECD's publication *Main Economic Indicators*. Is unemployment still lower in the United States than in western Europe?**

## INFLATION

Another important economic statistic is the rate of *inflation,* which is the rate at which prices in general are increasing over time. As we discuss in Chapter 6, inflation imposes a variety of costs on the economy. And when the inflation rate is high, people on fixed incomes, such as pensioners who receive a fixed dollar payment each month, can't keep up with the rising cost of living.

In recent years inflation has been relatively low in the United States, but that has not always been the case (see Figure 4.4 for data on U.S. inflation since 1900). During the 1970s, inflation was a major problem; in fact, many people told poll takers that inflation was "public enemy number one." Why was inflation high in the 1970s, and why is it relatively low today? What difference does it make to the average person?

**FIGURE 4.4**
**The U.S. Inflation Rate, 1900–2001.**
The U.S. inflation rate has fluctuated over time. Inflation was high in the 1970s but has been quite low recently.

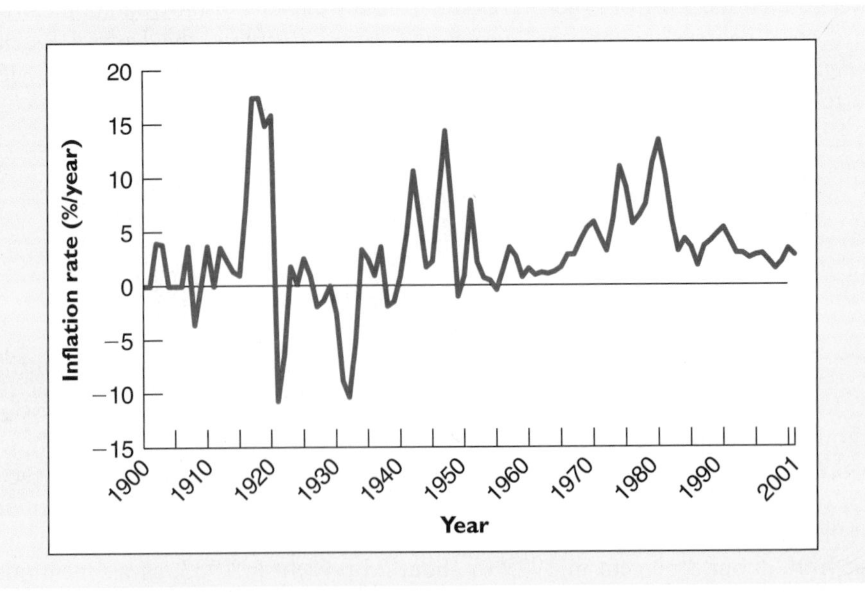


As with unemployment rates, the rate of inflation can differ markedly from country to country. For example, in 2001 the inflation rate was less than 3 percent in the United States, but the nation of Ukraine averaged over 400 percent annual inflation for the whole decade of the 1990s. What accounts for such large differences in inflation rates between countries?

Inflation and unemployment are often linked in policy discussions. One reason for this linkage is the oft-heard argument that unemployment can be reduced only at the cost of higher inflation and that inflation can be reduced only at the cost of higher unemployment. Must the government accept a higher rate of inflation to bring down unemployment, and vice versa?

## ECONOMIC INTERDEPENDENCE AMONG NATIONS

National economies do not exist in isolation but are increasingly interdependent. The United States, because of its size and the wide variety of goods and services it produces, is one of the most self-sufficient economies on the planet. Even so, in 1999 the United States exported about 10.8 percent of all the goods and services it produced and imported from abroad 13.3 percent of the goods and services that Americans used.

Sometimes international flows of goods and services become a matter of political and economic concern. For example, congressional representatives of states producing steel or textiles repeatedly complain that low-priced imports of these goods threaten the jobs of their constituents. Ross Perot, the Texas businessman and presidential candidate, predicted such problems when he opposed the adoption of the North American Free Trade Agreement (NAFTA) and similar agreements designed to promote international trade in goods and services. (Perot made famous the phrase "giant sucking sound" to describe what he thought free trade would do to American jobs.) Are free trade agreements, in which countries agree not to tax or otherwise block the international flow of goods and services, a good or bad thing?

*"I don't know what the hell happened—one minute I'm at work in Flint, Michigan, then there's a giant sucking sound and suddenly here I am in Mexico."*

A related issue is the phenomenon of *trade imbalances,* which occur when the quantity of goods and services that a country sells abroad (its *exports*) differs significantly from the quantity of goods and services its citizens buy from abroad (its *imports*). Figure 4.5 shows U.S. exports and imports (of goods only to allow for a longer data series) since 1900, measured as a percentage of the economy's total output. Prior to the 1970s the United States generally exported more than

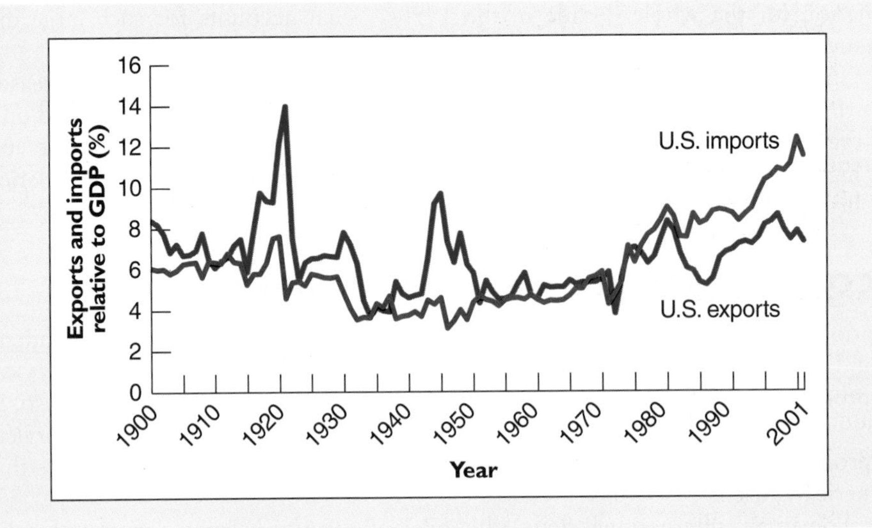


**FIGURE 4.5**
**Exports and Imports as a Share of U.S. Output, 1900–2001.**
The blue line shows U.S. exports of goods as a percentage of U.S. output. The red line shows U.S. imports of goods relative to U.S. output. For much of its history the United States has exported more than it imported, but over the past two decades imports have greatly outstripped exports.

it imported. (Notice the major export booms that occurred after both world wars, when the United States was helping to reconstruct Europe.) Since the 1970s, however, imports to the United States have outstripped exports, creating a situation called a *trade deficit*. Other countries—Japan, for example—export much more than they import. A country such as Japan is said to have a *trade surplus*. What causes trade deficits and surpluses? Are they harmful or helpful?

## RECAP THE MAJOR MACROECONOMIC ISSUES

- *Economic growth and living standards.* Over the last century the industrialized nations have experienced remarkable economic growth and improvements in living standards. Macroeconomists study the reasons for this extraordinary growth and try to understand why growth rates vary markedly among nations.
- *Productivity.* Average labor productivity, or output per employed worker, is a crucial determinant of living standards. Macroeconomists ask, What causes slowdowns and speedups in the rate of productivity growth?
- *Recessions and expansions.* Economies experience periods of slower growth (recessions) and more rapid growth (expansions). Macroeconomists examine the sources of these fluctuations and the government policies that attempt to moderate them.
- *Unemployment.* The unemployment rate is the fraction of people who would like to be employed but can't find work. Unemployment rises during recessions, but there are always unemployed people even during good times. Macroeconomists study the causes of unemployment, including the reasons why it sometimes differs markedly across countries.
- *Inflation.* The inflation rate is the rate at which prices in general are increasing over time. Questions macroeconomists ask about inflation include, Why does inflation vary over time and across countries? Must a reduction in inflation be accompanied by an increase in unemployment, or vice versa?
- *Economic interdependence among nations.* Modern economies are highly interdependent. Related issues studied by macroeconomists include the desirability of free trade agreements and the causes and effects of trade imbalances.

## MACROECONOMIC POLICY

We have seen that macroeconomists are interested in why different countries' economies perform differently and why a particular economy may perform well in some periods and poorly in others. Although many factors contribute to economic performance, government policy is surely among the most important. Understanding the effects of various policies and helping government officials develop better policies are important objectives of macroeconomists.

### TYPES OF MACROECONOMIC POLICY

We defined macroeconomic policies as government policies that affect the performance of the economy as a whole, as opposed to the market for a particular good or service. There are three major types of macroeconomic policy: *monetary policy, fiscal policy,* and *structural policy.*

The term **monetary policy** refers to the determination of the nation's money supply. (Cash and coin are the basic forms of money, although as we will see, modern economies have other forms of money as well.) For reasons that we will discuss in later chapters, most economists agree that changes in the money supply affect important macroeconomic variables, including national output, employment, interest rates, inflation, stock prices, and the international value of the dollar. In virtually all countries, monetary policy is controlled by a government institution called the *central bank.* The Federal Reserve System, often called the Fed for short, is the central bank of the United States.

***monetary policy*** determination of the nation's money supply

**Fiscal policy** refers to decisions that determine the government's budget, including the amount and composition of government expenditures and government revenues. The balance between government spending and taxes is a particularly important aspect of fiscal policy. When government officials spend more than they collect in taxes, the government runs a *deficit,* and when they spend less, the government's budget is in *surplus.* As with monetary policy, economists generally agree that fiscal policy can have important effects on the overall performance of the economy. For example, many economists believe that the large deficits run by the federal government during the 1980s were harmful to the nation's economy. Likewise, many would say that the balancing of the federal budget that occurred during the 1990s contributed to the nation's strong economic performance during that decade. Most recently, the federal budget has moved once again into deficit.

***fiscal policy*** decisions that determine the government's budget, including the amount and composition of government expenditures and government revenues

#### EXERCISE 4.2

**The Congressional Budget Office (CBO) is the government agency that is charged with projecting the federal government's surpluses or deficits. From the CBO's home page (http://www.cbo.gov/), find the most recent value of the federal government's surplus or deficit and the CBO's projected values for the next five years. How do you think these projections are likely to affect congressional deliberations on taxation and government spending?**

Finally, the term **structural policy** includes government policies aimed at changing the underlying structure, or institutions, of the nation's economy. Structural policies come in many forms, from minor tinkering to ambitious overhauls of the entire economic system. The move away from government control of the economy and toward a more market-oriented approach in many formerly communist countries, such as Poland, the Czech Republic, and Hungary, is a large-scale example of structural policy. Many developing countries have tried similar structural reforms. Supporters of structural policy hope that, by changing the basic characteristics of the economy or by remaking its institutions, they can stimulate economic growth and improve living standards.

***structural policy*** government policies aimed at changing the underlying structure, or institutions, of the nation's economy

## POSITIVE VERSUS NORMATIVE ANALYSES OF MACROECONOMIC POLICY

Macroeconomists are frequently called upon to analyze the effects of a proposed policy. For example, if Congress is debating a tax cut, economists in the Congressional Budget Office or the Treasury may be asked to prepare an analysis of the likely effects of the tax cut on the overall economy, as well as on specific industries, regions, or income groups. An objective analysis aimed at determining only the economic consequences of a particular policy—not whether those consequences are desirable—is called a **positive analysis.** In contrast, a **normative analysis** includes recommendations on whether a particular policy *should* be implemented. While a positive analysis is supposed to be objective and scientific, a normative analysis involves the *values* of the person or organization doing the analysis—conservative, liberal, or middle-of-the-road.

***positive analysis*** addresses the economic consequences of a particular event or policy, not whether those consequences are desirable

***normative analysis*** addresses the question of whether a policy *should* be used; normative analysis inevitably involves the values of the person doing the analysis

While pundits often joke that economists cannot agree among themselves, the tendency for economists to disagree is exaggerated. When economists do disagree, the controversy often centers on normative judgments (which relate to economists' personal values) rather than on positive analysis (which reflects objective knowledge of the economy). For example, liberal and conservative economists might agree that a particular tax cut would increase the incomes of the relatively wealthy (positive analysis). But they might vehemently disagree on whether the policy *should* be enacted, reflecting their personal views about whether wealthy people deserve a tax break (normative analysis).

The next time you hear or read about a debate over economic issues, try to determine whether the differences between the two positions are primarily *positive* or *normative*. If the debate focuses on the actual effects of the event or policy under discussion, then the disagreement is over positive issues. But if the main question has to do with conflicting personal opinions about the *desirability* of those effects, the debate is normative. The distinction between positive and normative analyses is important, because objective economic research can help to resolve differences over positive issues. When people differ for normative reasons, however, economic analysis is of less use.

### EXERCISE 4.3

**Which of the following statements are positive and which are normative? How can you tell?**

a. **A tax increase is likely to lead to lower interest rates.**

b. **Congress should increase taxes to reduce the inappropriately high level of interest rates.**

c. **A tax increase would be acceptable if most of the burden fell on those with incomes over $100,000.**

d. **Higher tariffs (taxes on imports) are needed to protect American jobs.**

e. **An increase in the tariff on imported steel would increase employment of American steelworkers.**

### RECAP MACROECONOMIC POLICY

Macroeconomic policies affect the performance of the economy as a whole. The three types of macroeconomic policy are monetary policy, fiscal policy, and structural policy. *Monetary policy,* which in the United States is under the control of the Federal Reserve System, refers to the determination of the nation's money supply. *Fiscal policy* involves decisions about the government budget, including its expenditures and tax collections. *Structural policy* refers to government actions to change the underlying structure or institutions of the economy. Structural policy can range from minor tinkering to a major overhaul of the economic system, as with the formerly communist countries that are attempting to convert to market-oriented systems.

The analysis of a proposed policy can be positive or normative. A *positive analysis* addresses the policy's likely economic consequences, but not whether those consequences are desirable. A *normative analysis* addresses the question of whether a proposed policy *should* be used. Debates about normative conclusions inevitably involve personal values and thus generally cannot be resolved by objective economic analysis alone.

## AGGREGATION

In Chapter 1 we discussed the difference between macroeconomics, the study of national economies, and microeconomics, the study of individual economic entities, such as households and firms, and the markets for specific goods and services. The main difference between the fields is one of perspective: Macroeconomists take a "bird's-eye view" of the economy, ignoring the fine details to understand how the system works as a whole. Microeconomists work instead at "ground level," studying the economic behavior of individual households, firms, and markets. Both perspectives are useful—indeed essential—to understand what makes an economy work.

Although macroeconomics and microeconomics take different perspectives on the economy, the basic tools of analysis are much the same. In the chapters to come you will see that macroeconomists apply the same core principles as microeconomists in their efforts to understand and predict economic behavior. Even though a national economy is a much bigger entity than a household or even a large firm, the choices and actions of individual decision makers ultimately determine the performance of the economy as a whole. So, for example, to understand saving behavior at the national level, the macroeconomist must first consider what motivates an individual family or household to save. The core principles introduced in Part 1 prove very useful for attacking such questions.

### EXERCISE 4.4

**Which of the following questions would be studied primarily by macroeconomists? By microeconomists? Explain.**

**a. Does increased government spending lower the unemployment rate?**

**b. Does Microsoft Corporation's dominance of the software industry harm consumers?**

**c. Would a school voucher program improve the quality of education in the United States? (Under a voucher program, parents are given a fixed amount of government aid, which they may use to send their children to any school, public or private.)**

**d. Should government policymakers aim to reduce inflation still further?**

**e. Why is the average rate of household saving low in the United States?**

**f. Does the increase in the number of consumer products being sold over the Internet threaten the profits of conventional retailers?**

While macroeconomists use the core principles of economics to understand and predict individual economic decisions, they need a way to relate millions of individual decisions to the behavior of the economy as a whole. One important tool they use to link individual behavior to national economic performance is **aggregation,** the adding up of individual economic variables to obtain economy-wide totals.

***aggregation*** the adding up of individual economic variables to obtain economywide totals

For example, macroeconomists don't care whether consumers drink Pepsi or Coke, go to the movie theater or rent videos, drive a convertible or a sports utility vehicle. These individual economic decisions are the province of microeconomics. Instead, macroeconomists add up consumer expenditures on all goods and services during a given period to obtain *aggregate,* or total, consumer expenditure. Similarly, a macroeconomist would not focus on plumbers' wages versus electricians' but would concentrate instead on the average wage of all workers. By focusing on aggregate variables, like total consumer expenditures or the average wage, macroeconomists suppress the mind-boggling details of a complex modern economy to see broad economic trends.

**EXAMPLE 4.3**

## Aggregation (1): A national crime index

To illustrate not only why aggregation is needed but also some of the problems associated with it, consider an issue that is only partly economic: crime. Suppose policymakers want to know whether *in general* the problem of crime in the United States is getting better or worse. How could an analyst obtain a statistical answer to that question?

Police keep detailed records of the crimes reported in their jurisdictions, so in principle a researcher could determine precisely how many purse snatchings occurred last year on New York City subways. But data on the number of crimes of each type in each jurisdiction would produce stacks of computer output. Is there a way to add up, or aggregate, all the crime data to get some sense of the national trend?

Law enforcement agencies such as the FBI use aggregation to obtain national *crime rates,* which are typically expressed as the number of "serious" crimes committed per 100,000 population. For example, the FBI reported that in 2000 some 11.6 million serious crimes (both violent crimes and property crimes) occurred in the United States (http://www.fbi.gov). Dividing the number of crimes by the U.S. population in 2000, which was about 280 million, and multiplying by 100,000 yields the crime rate for 2000, equal to about 4,100 crimes per 100,000 people. This rate represented a substantial drop from the crime rate in 1992, which was nearly 5,700 crimes per 100,000 people. So aggregation (the adding up of many different crimes into a national index) indicates that, in general, serious crime decreased in the United States between 1992 and 2000.

Although aggregation of crime statistics reveals the "big picture," it may obscure important details. The FBI crime index lumps together relatively minor crimes such as petty theft with very serious crimes such as murder and rape. Most people would agree that murder and rape do far more damage than a typical theft, so adding together these two very different types of crimes might give a false picture of crime in the United States. For example, although the U.S. crime rate fell 28 percent between 1992 and 2000, the murder rate fell by more than 40 percent. Since murder is the most serious of crimes, the reduction in crime between 1992 and 2000 was probably more significant than the change in the overall crime rate indicates. The aggregate crime rate glosses over other important details, such as the fact that the most dramatic reductions in crime occurred in urban areas. This loss of detail is a cost of aggregation, the price analysts pay for the ability to look at broad economic or social trends.

**EXAMPLE 4.4**

## Aggregation (2): U.S. exports

The United States exports a wide variety of products and services to many different countries. Kansas farmers sell grain to Russia, Silicon Valley programmers sell software to France, and Hollywood movie studios sell entertainment the world over. Suppose macroeconomists want to compare the total quantities of American-made goods sold to various regions of the world. How could such a comparison be made?

Economists can't add bushels of grain, lines of code, and movie tickets—the units aren't comparable. But they can add the *dollar values* of each—the revenue farmers earned from foreign grain sales, the royalties programmers received for their exported software, and the revenues studios reaped from films shown abroad. By comparing the dollar values of U.S. exports to Europe, Asia, Africa, and other regions in a particular year, economists are able to determine which regions are the biggest customers for American-made goods.

**RECAP AGGREGATION**

Macroeconomics, the study of national economies, differs from microeconomics, the study of individual economic entities (such as households and firms) and the markets for specific goods and services. Macroeconomists take a "bird's-eye view" of the economy. To study the economy as a whole, macroeconomists make frequent use of aggregation, the adding up of individual economic variables to obtain economywide totals. For example, a macroeconomist is more interested in the determinants of total U.S. exports, as measured by total dollar value, than in the factors that determine the exports of specific goods. A cost of aggregation is that the fine details of the economic situation are often obscured.

## STUDYING MACROECONOMICS: A PREVIEW

This chapter introduced many of the key issues of macroeconomics. In the chapters to come we will look at each of these issues in more detail. The next two chapters (Chapters 5 and 6) cover the *measurement* of economic performance, including key variables like the level of economic activity, the extent of unemployment, and the rate of inflation. Obtaining quantitative measurements of the economy, against which theories can be tested, is the crucial first step in answering basic macroeconomic questions like those raised in this chapter.

In Part 3 we will study economic behavior over relatively long periods of time. Chapter 7 examines economic growth and productivity improvement, the fundamental determinants of the average standard of living in the long run. Chapter 8 discusses the long-run determination of employment, unemployment, and wages. In Chapter 9 we study saving and its link to the creation of new capital goods, such as factories and machines. The role played in the economy by money, and its relation to the rate of inflation, is covered in Chapter 10, which also introduces the Federal Reserve, the central bank of the United States, and discusses

some of its policy tools. Chapter 11 looks at both domestic and international financial markets and their role in allocating saving to productive uses, in particular their role in promoting international capital flows.

John Maynard Keynes, a celebrated British economist, once wrote that "In the long run, we are all dead." Keynes's statement was intended as an ironic comment on the tendency of economists to downplay short-run economic problems on the grounds that "in the long run," the operation of the free market will always restore economic stability. Keynes, who was particularly active and influential during the Great Depression, correctly viewed the problem of massive unemployment, whether "short run" or not, as the most pressing economic issue of the time.

So why start our study of macroeconomics with the long run? Keynes's comment notwithstanding, long-run economic performance is extremely important, accounting for most of the substantial differences in living standards and economic well-being the world over. Furthermore, studying long-run economic behavior provides important background for understanding short-term fluctuations in the economy.

We turn to those short-term fluctuations in Part 4. Chapter 12 provides background on what happens during recessions and expansions, as well as some historical perspective. Chapter 13 discusses one important source of short-term economic fluctuations, variations in aggregate spending. The chapter also shows how, by influencing aggregate spending, fiscal policy may be able to moderate economic fluctuations. The second major policy tool for stabilizing the economy, monetary policy, is the subject of Chapter 14. Chapter 15 brings inflation into the analysis and discusses the circumstances under which macroeconomic policymakers may face a short-term trade-off between inflation and unemployment.

The international dimension of macroeconomics is the focus of Part 5. Chapter 16 focuses on the issue of international trade and the costs and benefits of unrestricted trade. Chapter 17 introduces exchange rates between national currencies. We will discuss how exchange rates are determined and how they affect the workings of the economy and macroeconomic policy.

## SUMMARY

- Macroeconomics is the study of the performance of national economies and of the policies governments use to try to improve that performance. Some of the broad issues macroeconomists study are:

  Sources of economic growth and improved living standards
  Trends in *average labor productivity,* or output per employed worker
  Short-term fluctuations in the pace of economic growth (recessions and expansions)
  Causes and cures of unemployment and inflation
  Economic interdependence among nations

- To help explain differences in economic performance among countries, or in economic performance in the same country at different times, macroeconomists study the implementation and effects of macroeconomic policies. *Macroeconomic policies* are government actions designed to affect the performance of the economy as a whole. Macroeconomic policies include *monetary policy* (the determination of the nation's money supply), *fiscal policy* (relating to decisions about the government's budget), and *structural policy* (aimed at affecting the basic structure and institutions of the economy).

- In studying economic policies, economists apply both *positive analysis* (an objective attempt to determine the consequences of a proposed policy) and *normative analysis* (which addresses whether a particular policy *should* be adopted). Normative analysis involves the values of the person doing the analysis.

- Macroeconomics is distinct from microeconomics, which focuses on the behavior of individual economic entities and specific markets. Macroeconomists make heavy use of *aggregation,* which is the adding up of individual economic variables into economywide totals. Aggregation allows macroeconomists to study the "big picture" of the economy, while ignoring fine details about individual households, firms, and markets.

## KEY TERMS

aggregation (104)
average labor productivity (95)
fiscal policy (101)
macroeconomic policies (93)
monetary policy (101)
normative analysis (102)
positive analysis (102)
structural policy (101)

## REVIEW QUESTIONS

1. How did the experience of the Great Depression motivate the development of the field of macroeconomics?
2. Generally, how does the standard of living in the United States today compare to the standard of living in other countries? To the standard of living in the United States a century ago?
3. Why is average labor productivity a particularly important economic variable?
4. True or false: Economic growth within a particular country generally proceeds at a constant rate. Explain.
5. True or false: Differences of opinion about economic policy recommendations can always be resolved by objective analysis of the issues. Explain.
6. Baseball statistics, such as batting averages, are calculated and reported for each individual player, for each team, and for the league as a whole. What purposes are served by doing this? Relate to the idea of aggregation in macroeconomics.
7. What type of macroeconomic policy (monetary, fiscal, structural) might include each of the following actions?
   a. A broad government initiative to reduce the country's reliance on agriculture and promote high-technology industries
   b. A reduction in income tax rates
   c. Provision of additional cash to the banking system
   d. An attempt to reduce the government budget deficit by reducing spending
   e. A decision by a developing country to reduce government control of the economy and to become more market-oriented

## PROBLEMS

1. Over the next 50 years the Japanese population is expected to decline, while the fraction of the population that is retired is expected to increase sharply. What are the implications of these population changes for total output and average living standards in Japan, assuming that average labor productivity continues to grow? What if average labor productivity stagnates?
2. Is it possible for average living standards to rise during a period in which average labor productivity is falling? Discuss, using a numerical example for illustration.
3. The Bureau of Economic Analysis, or BEA, is a government agency that collects a wide variety of statistics about the U.S. economy. From the BEA's home page (http://www.bea.doc.gov) find data for the most recent year available on U.S. exports and imports of goods and services. Is the United States running a trade surplus or deficit? Calculate the ratio of the surplus or deficit to U.S. exports.
4. Which of the following statements are positive and which are normative?
   a. If the Federal Reserve raises interest rates, demand for housing is likely to fall.
   b. The Federal Reserve should raise interest rates to keep inflation at an acceptably low level.
   c. Stock prices are likely to fall over the next year as the economy slows.
   d. A reduction in the capital gains tax (the tax on profits made in the stock market) would lead to a 10 to 20 percent increase in stock prices.
   e. Congress should not reduce capital gains taxes without also providing tax breaks for lower-income people.
5. Which of the following would be studied by a macroeconomist? By a microeconomist?
   a. The worldwide operations of General Motors
   b. The effect of government subsidies on sugar prices
   c. Factors affecting average wages in the U.S. economy
   d. Inflation in developing countries
   e. The effects of tax cuts on consumer spending

## ANSWERS TO IN-CHAPTER EXERCISES

4.1 Your answer will depend upon the current unemployment rate available at the OECD web site.

4.2 Your answer will depend upon the current CBO budget data.

4.3 a. Positive. This is a prediction of the effect of a policy, not a value judgment on whether the policy should be used.
b. Normative. Words like *should* and *inappropriately* express value judgments about the policy.
c. Normative. The statement is about the desirability of certain types of policies, not their likely effects.
d. Normative. The statement is about desirability of a policy.
e. Positive. The statement is a prediction of the likely effects of a policy, not a recommendation on whether the policy should be used.

4.4 a. Macroeconomists. Government spending and the unemployment rate are aggregate concepts pertaining to the national economy.
b. Microeconomists. Microsoft, though large, is an individual firm.
c. Microeconomists. The issue relates to the supply and demand for a specific service, education.
d. Macroeconomists. Inflation is an aggregate, economywide concept.
e. Macroeconomists. Average saving is an aggregate concept.
f. Microeconomists. The focus is on a relatively narrow set of markets and products rather than on the economy as a whole.

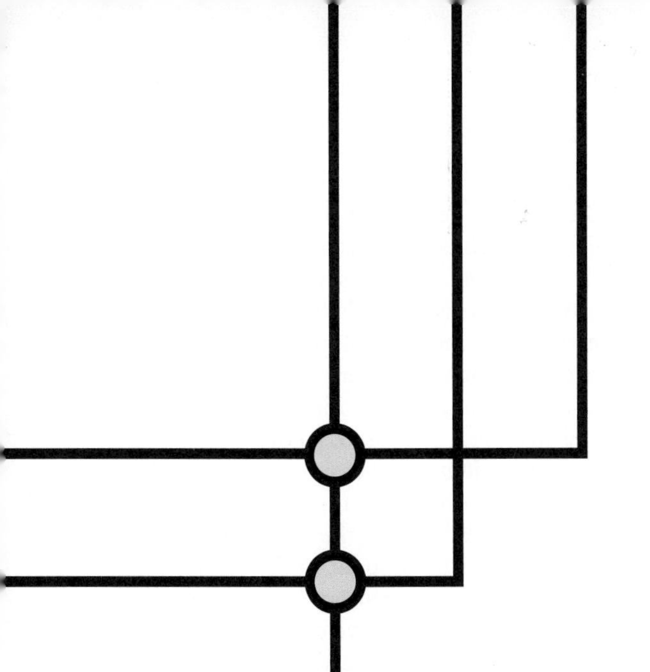

CHAPTER

5

# MEASURING ECONOMIC ACTIVITY: GDP AND UNEMPLOYMENT

"Nonfarm payrolls grew at a 2 percent rate in the third quarter . . ."

"The Dow Jones stock market index closed up 93 points yesterday in moderate trading . . ."

"Inflation appears subdued as the consumer price index registered an increase of only 0.2 percent last month . . ."

"The unemployment rate last month rose to 6.1 percent, its highest level since . . ."

News reports like these fill the airwaves—some TV and radio stations carry nothing else. In fact, all kinds of people are interested in economic data. The average person hopes to learn something that will be useful in a business decision, a financial investment, or a career move. The professional economist depends on economic data in much the same way that a doctor depends on a patient's vital signs—pulse, blood pressure, and temperature—to make an accurate diagnosis. To understand economic developments and to be able to give useful advice to policymakers, businesspeople, and financial investors, an economist simply must have up-to-date, accurate data. Political leaders and policymakers also need economic data to help them in their decisions and planning.

Interest in measuring the economy, and attempts to do so, date back as far as the mid-seventeenth century, when Sir William Petty (1623–1687) conducted a detailed survey of the land and wealth of Ireland. The British government's purpose in commissioning the survey was to determine the capacity of the Irish people to pay taxes to the Crown. But Petty used the opportunity to measure a variety of social and economic variables and went on to conduct pioneering studies of wealth, production, and population in several other countries. A firm believer in the idea that scientific progress depends first and foremost on accurate measurement, he once interrupted a meeting of the British Royal Society (a distinguished association of scientists, of which Petty was a founding member) to correct a speaker who had used the phrase "considerably bigger." A rule should be passed barring such vague terms, Petty proposed, so that "no word might be used but what marks either number, weight, or measure."[1]

Not until the twentieth century, though, did economic measurement come into its own. World War II was an important catalyst for the development of accurate economic statistics, since its very outcome was thought to depend on the mobilization of economic resources. Two economists, Simon Kuznets in the United States and Richard Stone in the United Kingdom, developed comprehensive systems for measuring a nation's output of goods and services, which were of great help to Allied leaders in their wartime planning. Kuznets and Stone each received a Nobel Prize in economics for their work, which became the basis for the economic accounts used today by almost all the world's countries. The governments of the United States and many other countries now collect and publish a wealth of statistics covering all aspects of their economies.

In this chapter and the next we will discuss how economists measure three basic macroeconomic variables that arise frequently in analyses of the state of the economy: the *gross domestic product,* or *GDP,* the *rate of unemployment,* and the *rate of inflation.* The focus of this chapter is on the first two of these statistics, GDP and the unemployment rate, which both measure the overall level of economic activity in a country.

Measuring economic activity might sound like a straightforward and uncontroversial task, but that is not the case. Indeed, the basic measure of a nation's output of goods and services, the gross domestic product, or GDP, has been criticized on many grounds. Some critics have complained that GDP does not adequately reflect factors such as the effect of economic growth on the environment or the rate of resource depletion. Because of problems like these, they charge, policies based on GDP statistics are likely to be flawed. Unemployment statistics have also been the subject of some controversy. By the end of this chapter you will understand how official measures of output and unemployment are constructed and used and will have gained some insight into these debates over their accuracy. Understanding the strengths and limitations of economic data is the first critical step toward becoming an intelligent user of economic statistics, as well as a necessary background for the economic analysis in the chapters to come.

## GROSS DOMESTIC PRODUCT: MEASURING THE NATION'S OUTPUT

Chapter 4 emphasized the link between an economy's output of goods and services and its living standard. We noted that high levels of output per person, and per worker, are typically associated with a high standard of living. But what, exactly, does "output" mean? To study economic growth and productivity scientifically, we need to be more precise about how economists define and measure an economy's output.

The most frequently used measure of an economy's output is called the *gross domestic product,* or *GDP.* GDP is intended to measure how much an economy

[1]This story is reported by Charles H. Hull, "Petty's Place in the History of Economic Theory," *Quarterly Journal of Economics,* May 14, 1900, pp. 307–340.

produces in a given period, such as a quarter (three months) or a year. More precisely, **gross domestic product (GDP)** is the market value of the final goods and services produced in a country during a given period. To understand this definition, let's take it apart and examine each of its parts separately. The first key phrase in the definition is "market value."

***gross domestic product (GDP)*** the market value of the final goods and services produced in a country during a given period

## MARKET VALUE

A modern economy produces many different goods and services, from dental floss (a good) to acupuncture (a service). Macroeconomists are not interested in this kind of detail, however; rather, their goal is to understand the behavior of the economy as a whole. For example, a macroeconomist might ask, Has the overall capacity of the economy to produce goods and services increased over time? If so, by how much?

To be able to talk about concepts like the "total output" or "total production"—as opposed to the production of specific items like dental floss—economists need to *aggregate* the quantities of the many different goods and services into a single number. They do so by adding up the *market values* of the different goods and services the economy produces. A simple example will illustrate the process. In the imaginary economy of Orchardia, total production is 4 apples and 6 bananas. To find the total output of Orchardia, we could add the number of apples to the number of bananas and conclude that total output is 10 pieces of fruit. But what if this economy also produced 3 pairs of shoes? There really is no sensible way to add apples and bananas to shoes.

Suppose though that we know that apples sell for \$0.25 each, bananas for \$0.50 each, and shoes for \$20.00 a pair. Then the market value of this economy's production, or its GDP, is equal to

$$(4 \text{ apples} \times \$0.25/\text{apple}) + (6 \text{ bananas} \times \$0.50/\text{banana}) + (3 \text{ pairs of shoes} \times \$20.00/\text{pair}) = \$64.00.$$

Notice that when we calculate total output this way, the more expensive items (the shoes) receive a higher weighting than the cheaper items (the apples and bananas). In general, the amount people are willing to pay for an item is an indication of the economic benefit they expect to receive from it (see Chapter 3). For this reason higher-priced items should count for more in a measure of aggregate output.

### EXAMPLE 5.1

### Orchardia's GDP

Suppose Orchardia were to produce 3 apples, 3 bananas, and 4 pairs of shoes at the same prices as in the preceding text. What is its GDP now?

Now the Orchardian GDP is equal to

$$(3 \text{ apples} \times \$0.25/\text{apple}) + (3 \text{ bananas} \times \$0.50/\text{banana}) + (4 \text{ pairs of shoes} \times \$20.00/\text{pair}) = \$82.25.$$

Notice that Orchardian GDP is higher in Example 5.1 than in the previous text, even though two of the three goods (apples and bananas) are being produced in smaller quantities than before. The reason is that the good whose production has increased (shoes) is much more valuable than the goods whose production has decreased (apples and bananas).

### EXERCISE 5.1

**Suppose Orchardia produces the same quantities of the three goods as originally at the same prices (see text preceding Example 5.1). In addition, it produces 5 oranges at \$0.30 each. What is the GDP of Orchardia now?**

## EXERCISE 5.2

**Following are data for a recent month on U.S. production of passenger cars and other light vehicles (a category that includes minivans, light trucks, and sports utility vehicles). The data are broken down into two categories: U.S. auto producers (GM, Ford, and Chrysler, now DaimlerChrysler) and foreign-owned plants (such as Honda, Toyota, and BMW). The average selling price is $17,000 for passenger cars and $25,000 for other light vehicles.**

| | Passenger cars | Other light vehicles |
|---|---|---|
| U.S. producers | 471,000 | 714,000 |
| Foreign-owned plants | 227,000 | 63,000 |

**Compare the output of U.S. producers to that of foreign-owned plants in terms of both the total number of vehicles produced and their market values (contribution to GDP). Explain why the two measures give different impressions of the relative importance of production by U.S.-owned and foreign-owned plants.**

Market values provide a convenient way to add together, or aggregate, the many different goods and services produced in a modern economy. A drawback of using market values, however, is that not all economically valuable goods and services are bought and sold in markets. For example, the unpaid work of a homemaker, although it is of economic value, is not sold in markets and so isn't counted in GDP. But paid housekeeping and child care services, which are sold in markets, do count. This distinction can create some pitfalls, as Example 5.2 shows.

## EXAMPLE 5.2 Women's labor force participation and GDP measurement

The percentage of adult American women working outside the home has increased dramatically in the past four decades, from less than 40 percent in 1960 to about 60 percent today (see Figure 5.1). This trend has led to a substantial increase in the demand for paid day care and housekeeping services, as working wives and mothers require more help at home. How have these changes affected measured GDP?

**FIGURE 5.1**
**Percentages of American Men and Women over Age 16 Working Outside the Home, 1960–2001.**
The fraction of American women working outside the home has risen by about 20 percentage points since 1960, while the fraction of men working outside the home has declined slightly.

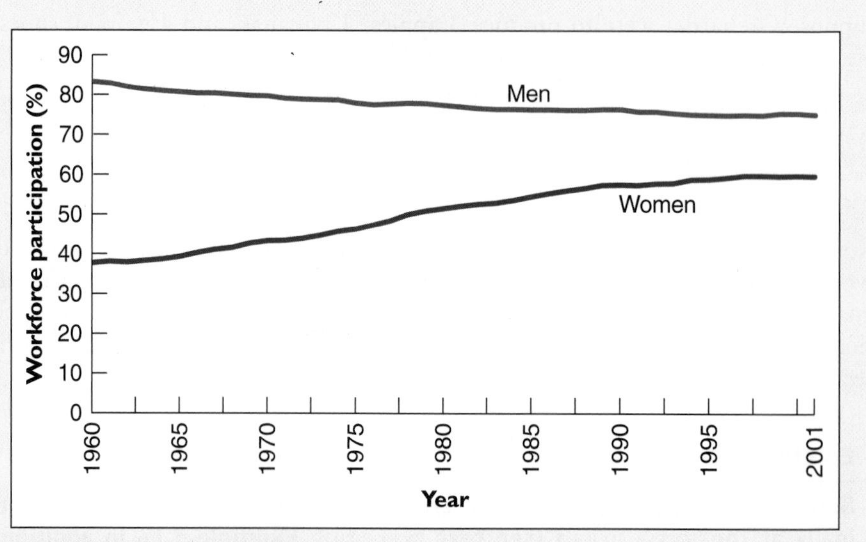


SOURCE: *Economic Report of the President,* February 2002 (http://w3.access.gpo.gov/eop/index.html).

The entry of many women into the labor market has raised measured GDP in two ways. First, the goods and services that women produce in their new jobs have contributed directly to increasing GDP. Second, the fact that paid workers took over previously unpaid housework and child care duties has increased measured GDP by the amount paid to those workers. The first of these two changes represents a genuine increase in economic activity, but the second reflects a transfer of existing economic activities from the unpaid sector to the market sector. Overall, then, the increase in measured GDP associated with increased participation in the labor force by women probably overstates the actual increase in economic activity.

**Why has female participation in the labor market increased by so much? What explains the trends illustrated in Figure 5.1?**

In a world governed only by economic principles—without social conventions, customs, or traditions—homemaking tasks like cleaning, cooking, and child rearing would be jobs like any other. As such, they would be subject to the principle of comparative advantage: Those people (either men or women) whose comparative advantage lay in performing homemaking tasks would specialize in them, freeing people whose comparative advantage lies elsewhere to work outside the home. In other words, homemaking tasks would be done by those with the lowest opportunity cost in those tasks. In such a world, to see a woman with a medical degree doing housework would be very unusual—her opportunity cost of doing housework would be too high.

But of course we don't live in a world driven only by economic considerations. Traditionally, social custom has severely limited the economic opportunities of women (and in some societies still does). However, social restrictions on women have weakened considerably over the past century, particularly in the industrialized countries, as a result of the increased educational attainment of women, the rise of the feminist movement, and other factors. As traditional social restraints on women have loosened, domestic arrangements have moved in the direction dictated by comparative advantage—to an increasing degree, homemaking tasks are now performed by paid specialists, while the majority of women (and men) work outside the home.

Although homemaking activities are excluded from measured GDP, in a few cases goods and services that are not sold in markets are included in GDP. By far the most important are the goods and services provided by federal, state, and local governments. The protection provided by the army and navy, the transportation convenience of the interstate highway system, and the education provided by the public school system are examples of publicly provided goods and services that are not sold in markets. As market prices for publicly provided goods and services do not exist, economic statisticians add to the GDP the *costs* of providing those goods and services as rough measures of their economic value. For example, to include public education in the GDP, the statisticians add to GDP the salaries of teachers and administrators, the costs of textbooks and supplies, and the like. Similarly, the economic value of the national defense establishment is approximated, for the purposes of measuring GDP, by the *costs* of defense: the pay earned by soldiers and sailors, the costs of acquiring and maintaining weapons, and so on.

With a few exceptions, like publicly provided goods and services, GDP is calculated by adding up market values. However, not all goods and services that have a market value are counted in GDP. As we will see next, GDP includes only those goods and services that are the end products of the production process, called *final goods and services*. Goods and services that are used up in the production process are not counted in GDP.

Why is the female labor force participation rate now more than 50 percent greater than in the 1960s?

## FINAL GOODS AND SERVICES

Many goods are used in the production process. Before a baker can produce a loaf of bread, grain must be grown and harvested, then the grain must be ground into flour, and, together with other ingredients, baked into bread. Of the three major goods that are produced during this process—the grain, the flour, and the bread—only the bread is used by consumers. Because producing the bread is the ultimate purpose of the process, the bread is called a *final good.* In general, a **final good or service** is the end product of a process, the product or service that consumers actually use. The goods or services produced on the way toward making the final product—here, the grain and the flour—are called **intermediate goods or services.**

***final goods or services*** goods or services consumed by the ultimate user; because they are the end products of the production process, they are counted as part of GDP

***intermediate goods or services*** goods or services used up in the production of final goods and services and therefore not counted as part of GDP

Since we are interested in measuring only those items that are of direct economic value, *only final goods and services are included in GDP.* Intermediate goods and services are *not* included. To illustrate, suppose that the grain from the previous example has a market value of $0.50 (the price the milling company paid for the grain). The grain is then ground into flour, which has a market value of $1.20 (the price the baker paid for the flour). Finally, the flour is made into a loaf of fine French bread, worth $2.00 at the local store. In calculating the contribution of these activities to GDP, would we want to add together the values of the grain, the flour, and the bread? No, because the grain and flour are intermediate goods, valuable only because they can be used to make bread. So in this example, the total contribution to GDP is $2.00, the value of the loaf of bread, the final product.

Example 5.3 illustrates the same distinction but this time with a focus on services.

**EXAMPLE 5.3**

### The barber and his assistant

Your barber charges $10 for a haircut. In turn, the barber pays his assistant $2 per haircut in return for sharpening the scissors, sweeping the floor, and other chores. For each haircut given, what is the total contribution of the barber and his assistant, taken together, to GDP?

The answer to this problem is $10, the price, or market value, of the haircut. The haircut is counted in GDP because it is the final service, the one that actually has value to the final user. The services provided by the assistant have value only because they contribute to the production of the haircut; thus they are not counted in GDP.

Example 5.4 illustrates that the same good can be either intermediate or final, depending on how it is used.

**EXAMPLE 5.4**

### A good that can be either intermediate or final

Farmer Brown produces $100 worth of milk. He sells $40 worth of milk to his neighbors and uses the rest to feed his pigs, which he sells to his neighbors for $120. What is Farmer Brown's contribution to the GDP?

The final goods in this example are the $40 worth of milk and the $120 worth of pigs sold to the neighbors. Adding $40 and $120, we get $160, which is Farmer Brown's contribution to the GDP. Note that part of the milk Farmer Brown produced serves as an intermediate good and part as a final good. The $60 worth of milk that is fed to the pigs is an intermediate good, and so it is not counted in GDP. The $40 worth of milk sold to the neighbors is a final good, and so it is counted.

***capital good*** a long-lived good, which is itself produced and used to produce other goods and services

A special type of good that is difficult to classify as intermediate or final is a capital good. A **capital good** is a long-lived good, which is itself produced and used to produce other goods and services. Factories and machines are examples of capital goods. Capital goods do not fit the definition of final goods, since their purpose is to produce other goods. On the other hand, they are not used up during the production process, except over a very long period, so they are not exactly intermediate goods either. For purposes of measuring GDP, economists have agreed to classify newly produced capital goods as final goods. Otherwise, a country that invested in its future by building modern factories and buying new machines would be counted as having a lower GDP than a country that devoted all its resources to producing consumer goods.

We have established the rule that only final goods and services (including newly produced capital goods) are counted in GDP. Intermediate goods and services, which are used up in the production of final goods and services, are not counted. In practice, however, this rule is not easy to apply, because the production process often stretches over several periods. To illustrate, recall the earlier example of the grain that was milled into flour, which in turn was baked into a loaf of French bread. The contribution of the whole process to GDP is $2, the value of the bread (the final product). Suppose, though, that the grain and the flour were produced near the end of the year 2002 and the bread was baked early the next year in 2003. In this case, should we attribute the $2 value of the bread to the GDP for the year 2002 or to the GDP for the year 2003?

Neither choice seems quite right, since part of the bread's production process occurred in each year. Part of the value of the bread should probably be counted in the year 2002 GDP and part in the year 2003 GDP. But how should we make the split? To deal with this problem, economists determine the market value of final goods and services indirectly, by adding up the *value added* by each firm in the production process. The **value added** by any firm equals the market value of its product or service minus the cost of inputs purchased from other firms. As we'll see, summing the value added by all firms (including producers of both intermediate and final goods and services) gives the same answer as simply adding together the value of final goods and services. But the value-added method eliminates the problem of dividing the value of a final good or service between two periods.

***value added*** for any firm, the market value of its product or service minus the cost of inputs purchased from other firms

To illustrate this method, let's revisit the example of the French bread, which is the result of multiple stages of production. We have already determined that the total contribution of this production process to GDP is $2, the value of the bread. Let's show now that we can get the same answer by summing value added. Suppose that the bread is the ultimate product of three corporations: ABC Grain Company, Inc., produces grain; General Flour produces flour; and Hot'n'Fresh Baking produces the bread. If we make the same assumptions as before about the market value of the grain, the flour, and the bread, what is the value added by each of these three companies?

ABC Grain Company produces $0.50 worth of grain, with no inputs from other companies, so ABC's value added is $0.50. General Flour uses $0.50 worth of grain from ABC to produce $1.20 worth of flour. The value added by General Flour is thus the value of its product ($1.20) less the cost of purchased inputs ($0.50), or $0.70. Finally, Hot'n'Fresh Baking buys $1.20 worth of flour from General Flour and uses it to produce $2.00 worth of bread. So the value added by Hot'n'Fresh is $0.80. These calculations are summarized in Table 5.1.

**TABLE 5.1**
**Value Added in Bread Production**

| Company | Revenues − | Cost of purchased inputs = | Value added |
|---|---|---|---|
| ABC Grain | $0.50 | $0.00 | $0.50 |
| General Flour | $1.20 | $0.50 | $0.70 |
| Hot'n'Fresh | $2.00 | $1.20 | $0.80 |
| Total | | | $2.00 |

You can see that summing the value added by each company gives the same contribution to GDP, $2.00, as the method based on counting final goods and services only. Basically, the value added by each firm represents the portion of the value of the final good or service that the firm creates in its stage of production. Summing the value added by all firms in the economy yields the total value of final goods and services, or GDP.

You can also see now how the value-added method solves the problem of production processes that bridge two or more periods. Suppose that the grain and flour are produced during the year 2002 but the bread is not baked until 2003. Using the value-added method, the contribution of this production process to the year 2002 GDP is the value added by the grain company plus the value added by the flour company, or $1.20. The contribution of the production process to the year 2003 GDP is the value added by the baker, which is $0.80. Thus part of the value of the final product, the bread, is counted in the GDP for each year, reflecting the fact that part of the production of the bread took place in each year.

### EXERCISE 5.3

**Amy's card shop receives a shipment of Valentine's Day cards in December 2002. Amy pays the wholesale distributor of the cards a total of $500. In February 2003 she sells the cards for a total of $700. What are the contributions of these transactions to GDP in the years 2002 and 2003?**

We have now established that GDP is equal to the market value of final goods and services. Let's look at the last part of the definition, "produced within a country during a given period."

## PRODUCED WITHIN A COUNTRY DURING A GIVEN PERIOD

The word *domestic* in the term *gross domestic product* tells us that GDP is a measure of economic activity within a given country. Thus, only production that takes place within the country's borders is counted. For example, the GDP of the United States includes the market value of *all* cars produced within U.S. borders, even if they are made in foreign-owned plants (recall Exercise 5.2). However, cars produced in Mexico by a U.S.-based company like General Motors are *not* counted.

We have seen that GDP is intended to measure the amount of production that occurs during a given period, such as the calendar year. For this reason, only goods and services that are actually produced during a particular year are included in the GDP for that year. Example 5.5 and Exercise 5.4 illustrate.

### EXAMPLE 5.5 The sale of a house and GDP

A 20-year-old house is sold to a young family for $200,000. The family pays the real estate agent a 6 percent commission, or $12,000. What is the contribution of this transaction to GDP?

Because the house was not produced during the current year, its value is *not* counted in this year's GDP. (The value of the house was included in the GDP 20 years earlier, the year the house was built.) In general, purchases and sales of existing assets, such as old houses or used cars, do not contribute to the current year's GDP. However, the $12,000 fee paid to the real estate agent represents the market value of the agent's services in helping the family find the house and make the purchase. Since those services were provided during the current year, the agent's fee *is* counted in current-year GDP.

**EXERCISE 5.4**

**Lotta Doe sells 100 shares of stock in Benson Buggywhip for $50 per share. She pays her broker a 2 percent commission for executing the sale. How does Lotta's transaction affect the current-year GDP?**

**MEASURING GDP**

Gross domestic product (GDP) equals

**the market value**

GDP is an aggregate of the market values of the many goods and services produced in the economy.

Goods and services that are not sold in markets, such as unpaid housework, are not counted in GDP. An important exception is goods and services provided by the government, which are included in GDP at the government's cost of providing them.

**of final goods and services**

Final goods and services (which include capital goods, such as factories and machines) are counted in GDP. Intermediate goods and services, which are used up in the production of final goods and services, are not counted.

In practice, the value of final goods and services is determined by the value-added method. The value added by any firm equals the firm's revenue from selling its product minus the cost of inputs purchased from other firms. Summing the value added by all firms in the production process yields the value of the final good or service.

**produced in a country during a given period.**

Only goods and services produced within a nation's borders are included in GDP.

Only goods and services produced during the current year (or the portion of the value produced during the current year) are counted as part of the current-year GDP.

## THE EXPENDITURE METHOD FOR MEASURING GDP

GDP is a measure of the quantity of goods and services *produced* by an economy. But any good or service that is produced will also be *purchased* and used by some economic agent—a consumer buying Christmas gifts or a firm investing in new machinery, for example. For many purposes, knowing not only how much is produced, but who uses it and how, is important.

Economic statisticians divide the users of the final goods and services that make up the GDP for any given year into four categories: *households, firms, governments,* and the *foreign sector* (that is, foreign purchasers of domestic products). They assume that all the final goods and services that are produced in a country in a given year will be purchased and used by members of one or more of these four groups. Furthermore, the amounts that purchasers spend on various goods and services should be equal to the market values of those goods and services. As a result, GDP can be measured with equal accuracy by either of two methods: (1) adding up the market values of all the final goods and services that are produced domestically, or (2) adding up the total amount spent by each of the four groups on final goods and services and subtracting spending on imported goods and services. The values obtained by the two methods will be the same.

**TABLE 5.2**
**Expenditure Components of U.S. GDP, 2001 (billions of dollars)**

| | | |
|---|---|---|
| **Consumption** | | **6,987.0** |
| Durable goods | 835.9 | |
| Nondurable goods | 2,041.3 | |
| Services | 4,109.9 | |
| **Investment** | | **1,586.0** |
| Business fixed investment | 1,201.6 | |
| Residential investment | 444.8 | |
| Inventory investment | −60.3 | |
| **Government purchases** | | **1,858.0** |
| **Net exports** | | **−348.9** |
| Exports | 1,034.1 | |
| Imports | 1,383.0 | |
| **Total:** Gross domestic product | | **10,082.2** |

SOURCE: Bureau of Economic Analysis (http://www.bea.doc).

Corresponding to the four groups of final users are four components of expenditure: consumption, investment, government purchases, and net exports. That is, households consume, firms invest, governments make government purchases, and the foreign sector buys the nation's exports. Table 5.2 gives the dollar values for each of these components for the U.S. economy in 2001. As the table shows, GDP for the United States in 2001 was about $10.1 trillion, roughly $35,400 per person. Detailed definitions of the components of expenditure, and their principal subcomponents, follow. As you read through them, refer to Table 5.2 to get a sense of the relative importance of each type of spending.

***consumption expenditure, or consumption*** spending by households on goods and services, such as food, clothing, and entertainment

**Consumption expenditure,** or simply **consumption,** is spending by households on goods and services such as food, clothing, and entertainment. Consumption expenditure is subdivided into three subcategories:

- *Consumer durables* are long-lived consumer goods such as cars and furniture. Note that new houses are not treated as consumer durables but as part of investment.
- *Consumer nondurables* are shorter-lived goods like food and clothing.
- *Services,* a large component of consumer spending, include everything from haircuts and taxi rides to legal, financial, and educational services.

***investment*** spending by firms on final goods and services, primarily capital goods and housing

**Investment** is spending by firms on final goods and services, primarily capital goods and housing. Investment is divided into three subcategories:

- *Business fixed investment* is the purchase by firms of new capital goods such as machinery, factories, and office buildings. (Remember that for the purposes of calculating GDP, long-lived capital goods are treated as final goods rather than as intermediate goods.) Firms buy capital goods to increase their capacity to produce.
- *Residential investment* is construction of new homes and apartment buildings. For GDP accounting purposes, residential investment is treated as an investment by the business sector, which then sells the homes to households.
- *Inventory investment* is the addition of unsold goods to company inventories. In other words, the goods that a firm produces but doesn't sell during

the current period are treated, for accounting purposes, as if the firm had bought those goods from itself. (This convention guarantees that production equals expenditure.) Inventory investment can take a negative value, as it did in 2001, if the value of inventories on hand falls over the course of the year.

People often refer to purchases of financial assets, such as stocks or bonds, as "investments." That use of the term is different from the definition we give here. A person who buys a share of a company's stock acquires partial ownership of the *existing* physical and financial assets controlled by the company. A stock purchase does not usually correspond to the creation of *new* physical capital, however, and so is not investment in the sense we are using the term in this chapter. We will generally refer to purchases of financial assets, such as stocks and bonds, as "financial investments," to distinguish them from a firm's investment in new capital goods, such as factories and machines.

**Government purchases** are purchases by federal, state, and local governments of final goods, such as fighter planes, and services, such as teaching in public schools. Government purchases do *not* include *transfer payments,* which are payments made by the government in return for which no current goods or services are received. Examples of transfer payments (which, again, are *not* included in government purchases) are Social Security benefits, unemployment benefits, pensions paid to government workers, and welfare payments. Interest paid on the government debt is also excluded from government purchases.

***government purchases*** purchases by federal, state, and local governments of final goods and services; government purchases do *not* include *transfer payments,* which are payments made by the government in return for which no current goods or services are received, nor do they include interest paid on the government debt

**Net exports** equal exports minus imports.

- *Exports* are domestically produced final goods and services that are sold abroad.
- *Imports* are purchases by domestic buyers of goods and services that were produced abroad. Imports are subtracted from exports to find the net amount of spending on domestically produced goods and services.

***net exports*** exports minus imports

A country's net exports reflect the net demand by the rest of the world for its goods and services. Net exports can be negative, since imports can exceed exports in any given year. As Table 5.2 shows, the United States had significantly greater imports than exports in 2001.

The relationship between GDP and expenditures on goods and services can be summarized by an equation. Let

$Y$ = gross domestic product, or output
$C$ = consumption expenditure
$I$ = investment
$G$ = government purchases
$NX$ = net exports.

Using these symbols, we can write that GDP equals the sum of the four types of expenditure algebraically as

$$Y = C + I + G + NX.$$

**EXAMPLE 5.6**

**Measuring GDP by production and by expenditure**

An economy produces 1,000,000 automobiles valued at $15,000 each. Of these, 700,000 are sold to consumers, 200,000 are sold to businesses, 50,000 are sold to the government, and 25,000 are sold abroad. No automobiles are imported. The automobiles left unsold at the end of the year are held in inventory by the auto producers. Find GDP in terms of (a) the market value of production and (b) the components of expenditure. You should get the same answer both ways.

The market value of the production of final goods and services in this economy is 1,000,000 autos times $15,000 per auto, or $15 billion.

To measure GDP in terms of expenditure, we must add spending on consumption, investment, government purchases, and net exports. Consumption is 700,000 autos times $15,000, or $10.5 billion. Government purchases are 50,000 autos times $15,000, or $0.75 billion. Net exports are equal to exports (25,000 autos at $15,000, or $0.375 billion) minus imports (zero), so net exports are $0.375 billion.

But what about investment? Here we must be careful. The 200,000 autos that are sold to businesses, worth $3 billion, count as investment. But notice too that the auto companies produced 1,000,000 automobiles but sold only 975,000 (700,000 + 200,000 + 50,000 + 25,000). Hence 25,000 autos were unsold at the end of the year and were added to the automobile producers' inventories. This addition to producer inventories (25,000 autos at $15,000, or $0.375 billion) counts as inventory investment, which is part of total investment. Thus total investment spending equals the $3 billion worth of autos sold to businesses plus the $0.375 billion in inventory investment, or $3.375 billion.

Recapitulating, in this economy consumption is $10.5 billion, investment (including inventory investment) is $3.375 billion, government purchases equal $0.75 billion, and net exports are $0.375 billion. Summing these four components of expenditure yields $15 billion—the same value for GDP that we got by calculating the market value of production.

**EXERCISE 5.5**

**Extending Example 5.6, suppose that 25,000 of the automobiles purchased by households are imported rather than domestically produced. Domestic production remains at 1,000,000 autos valued at $15,000 each. Once again, find GDP in terms of (a) the market value of production and (b) the components of expenditure.**

**RECAP EXPENDITURE COMPONENTS OF GDP**

GDP can be expressed as the sum of expenditures on domestically produced final goods and services. The four types of expenditure that are counted in the GDP, and the economic groups that make each type of expenditure, are as follows:

| Who makes the expenditure? | Type of expenditure | Examples |
|---|---|---|
| Households | Consumption | Food, clothes, haircuts, new cars |
| Business firms | Investment | New factories and equipment, new houses, increases in inventory stocks |
| Governments | Government purchases | New school buildings, new military hardware, salaries of soldiers and government officials |
| Foreign sector | Net exports, or exports minus imports | Exported manufactured goods, legal or financial services provided by domestic residents to foreigners |

## GDP AND THE INCOMES OF CAPITAL AND LABOR

The GDP can be thought of equally well as a measure of total production or as a measure of total expenditure—either method of calculating the GDP gives the same final answer. There is yet a third way to think of the GDP, which is as the *incomes of capital and labor.*

Whenever a good or service is produced or sold, the revenue from the sale is distributed to the workers and the owners of the capital involved in the production of the good or service. Thus, except for some technical adjustments that we will ignore, GDP also equals labor income plus capital income. *Labor income* (equal to about 75 percent of GDP) comprises wages, salaries, and the incomes of the self-employed. *Capital income* (about 25 percent of GDP) is made up of payments to owners of physical capital (such as factories, machines, and office buildings) and intangible capital (such as copyrights and patents). The components of capital income include items such as profits earned by businessowners, the rents paid to owners of land or buildings, interest received by bondholders, and the royalties received by the holders of copyrights or patents. Both labor income and capital income are to be understood as measured prior to payment of taxes; ultimately, of course, a portion of both types of income is captured by the government in the form of tax collections.

Figure 5.2 may help you visualize the three equivalent ways of thinking about GDP: the market value of production, the total value of expenditure, and the sum of labor income and capital income. The figure also roughly captures the relative importance of the expenditure and income components. About 65 percent of expenditure is consumption spending, about 20 percent is government purchases, and the rest is investment spending and net exports. (Actually, as Table 5.2 or Figure 4.5 shows, net exports have been negative in recent years, reflecting the U.S. trade deficit.) As we mentioned, labor income is about 75 percent of total income, with capital income making up the rest.

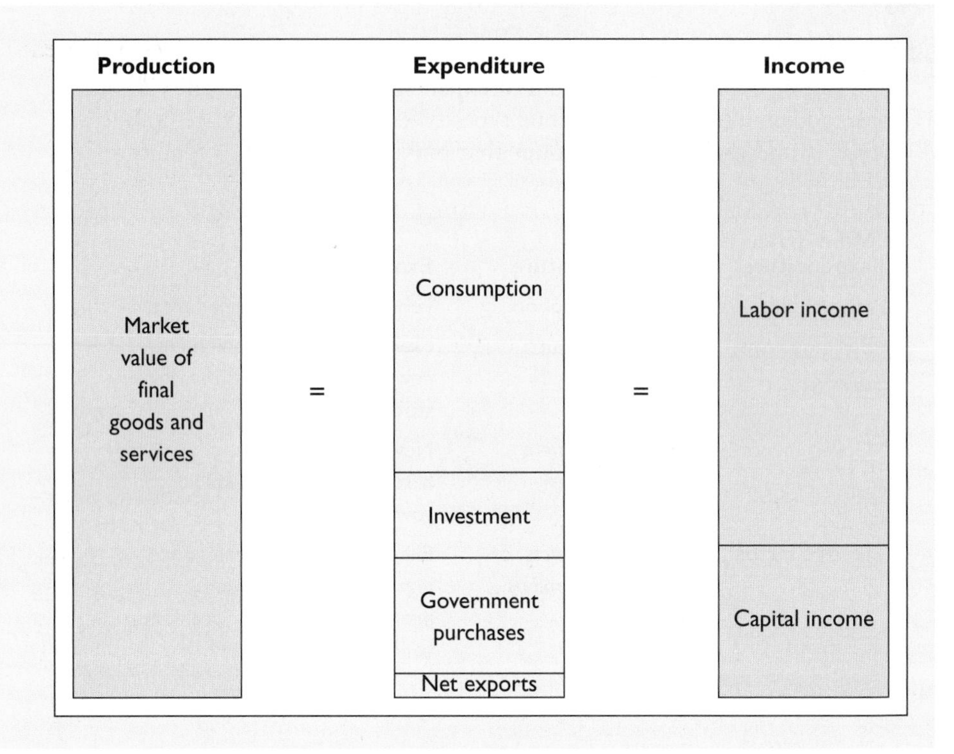


**FIGURE 5.2**
**The Three Faces of GDP.** The GDP can be expressed equally well as (1) the market value of production, (2) total expenditure (consumption, investment, government purchases, net exports), or (3) total income (labor income and capital income).

## NOMINAL GDP VERSUS REAL GDP

As a measure of the total production of an economy over a given period, such as a particular year, GDP is useful in comparisons of economic activity in different places. For example, GDP data for the year 2002, broken down state by state, could be used to compare aggregate production in New York and California during that year. However, economists are interested in comparing levels of economic activity not only in different *locations* but *over time* as well. For example, a president who is running for reelection on the basis of successful economic policies might want to know by how much output in the U.S. economy had increased during his term.

Using GDP to compare economic activity at two different points in time may give misleading answers, however, as the following example shows. Suppose for the sake of illustration that the economy produces only pizzas and calzones. The prices and quantities of the two goods in the years 2000 and 2004, the beginning and end of the president's term, are shown in Table 5.3. If we calculate GDP in each year as the market value of production, we find that the GDP for 2000 is (10 pizzas × $10/pizza) + (15 calzones × $5/calzone) = $175. The GDP for 2004 is (20 pizzas × $12/pizza) + (30 calzones × $6/calzone) = $420. Comparing the GDP for the year 2004 to the GDP for the year 2000, we might conclude that it is 2.4 times greater ($420/$175).

**TABLE 5.3**
**Prices and Quantities in 2000 and 2004**

| | Quantity of pizzas | Price of pizzas | Quantity of calzones | Price of calzones |
|---|---|---|---|---|
| 2000 | 10 | $10 | 15 | $5 |
| 2004 | 20 | $12 | 30 | $6 |

But look more closely at the data given in Table 5.3. Can you see what is wrong with this conclusion? The quantities of both pizzas and calzones produced in the year 2004 are exactly twice the quantities produced in the year 2000. If economic activity, as measured by actual production of both goods, exactly doubled over the four years, why do the calculated values of GDP show a greater increase?

The answer, as you also can see from the table, is that prices as well as quantities rose between 2000 and 2004. Because of the increase in prices, the *market value* of production grew more over those four years than the *physical volume* of production. So in this case, GDP is a misleading gauge of economic growth during the president's term, since the physical quantities of the goods and services produced in any given year, not the dollar values, are what determine people's economic well-being. Indeed, if the prices of pizzas and calzones had risen 2.4 times between 2000 and 2004, GDP would have risen 2.4 times as well, with no increase in physical production! In that case, the claim that the economy's (physical) output had more than doubled during the president's term would obviously be wrong.

As this example shows, if we want to use GDP to compare economic activity at different points in time, we need some method of excluding the effects of price changes. In other words, we need to adjust for inflation. To do so, economists use a common set of prices to value quantities produced in different years. The standard approach is to pick a particular year, called the *base year,* and use the prices from that year to calculate the market value of output. When GDP is calculated using the prices from a base year, rather than the current year's prices, it is called **real GDP,** to indicate that it is a measure of real physical production. Real GDP is GDP adjusted for inflation. To distinguish real GDP, in which quantities produced are valued at base-year prices, from GDP valued at current-year prices, economists refer to the latter measure as **nominal GDP.**

***real GDP*** a measure of GDP in which the quantities produced are valued at the prices in a base year rather than at current prices; real GDP measures the actual *physical volume* of production

***nominal GDP*** a measure of GDP in which the quantities produced are valued at current-year prices; nominal GDP measures the *current dollar value* of production

**EXAMPLE 5.7**

#### Calculating the change in real GDP over the president's term

Using data from Table 5.3 and assuming that 2000 is the base year, find real GDP for the years 2004 and 2000. By how much did real output grow between 2000 and 2004?

To find real GDP for the year 2004, we must value the quantities produced that year using the prices in the base year, 2000. Using the data in Table 5.3:

Year 2004 real GDP = (year 2004 quantity of pizzas × year 2000 price of pizzas) + (year 2004 quantity of calzones × year 2000 price of calzones)
$= (20 \times \$10) + (30 \times \$5)$
$= \$350.$

The real GDP of this economy in the year 2004 is $350. What is the real GDP for 2000?

By definition, the real GDP for 2000 equals 2000 quantities valued at base-year prices. The base year in this example happens to be 2000, so real GDP for 2000 equals 2000 quantities valued at 2000 prices, which is the same as nominal GDP for 2000. In general, in the base year, real GDP and nominal GDP are the same. We already found nominal GDP for 2000, $175, so that is also the real GDP for 2000.

We can now determine how much real production has actually grown over the four-year period. Since real GDP was $175 in 2000 and $350 in 2004, the physical volume of production doubled between 2000 and 2004. This conclusion makes good sense, since Table 5.3 shows that the production of both pizzas and calzones exactly doubled over the period. By using real GDP, we have eliminated the effects of price changes and obtained a reasonable measure of the actual change in physical production over the four-year span.

Of course, the production of all goods will not necessarily grow in equal proportion, as in Example 5.7. Exercise 5.6 asks you to find real GDP when pizza and calzone production grow at different rates.

### EXERCISE 5.6

**Suppose production and prices of pizza and calzone in 2000 and 2004 are as follows:**

| | Quantity of pizzas | Price of pizzas | Quantity of calzones | Price of calzones |
|---|---|---|---|---|
| 2000 | 10 | $10 | 15 | $5 |
| 2004 | 30 | $12 | 30 | $6 |

**These data are the same as those in Table 5.3, except that pizza production has tripled rather than doubled between 2000 and 2004. Find real GDP in 2004 and 2000, and calculate the growth in real output over the four-year period. (Continue to assume that 2000 is the base year.)**

If you complete Exercise 5.6, you will find that the growth in real GDP between 2000 and 2004 reflects a sort of average of the growth in physical production of pizzas and calzones. Real GDP therefore remains a useful measure of overall physical production, even when the production of different goods and services grows at different rates.

The method of calculating real GDP just described was followed for many decades by the Bureau of Economic Analysis (BEA), the U.S. government agency responsible for GDP statistics. However, in recent years the BEA has adopted a more complicated procedure of determining real GDP, called *chain weighting*. The new procedure makes the official real GDP data less sensitive to the particular base year chosen. However, the chain-weighting and traditional approaches share the basic idea of valuing output in terms of base-year prices, and the results obtained by the two methods are generally similar.

**NOMINAL GDP VERSUS REAL GDP**

Real GDP is calculated using the prices of goods and services that prevailed in a base year rather than in the current year. Nominal GDP is calculated using current-year prices. Real GDP is GDP adjusted for inflation; it may be thought of as measuring the physical volume of production. Comparisons of economic activity at different times should always be done using real GDP, not nominal GDP.

## REAL GDP IS NOT THE SAME AS ECONOMIC WELL-BEING

Government policymakers pay close attention to real GDP, often behaving as if the greater the GDP, the better. However, real GDP is *not* the same as economic well-being. At best, it is an imperfect measure of economic well-being because, for the most part, it captures only those goods and services that are priced and sold in markets. Many factors that contribute to people's economic well-being are not priced and sold in markets and thus are largely or even entirely omitted from GDP. Maximizing real GDP is not, therefore, always the right goal for government policymakers. Whether or not policies that increase GDP will also make people better off has to be determined on a case-by-case basis.

To understand why an increase in real GDP does not always promote economic well-being, let's look at some factors that are not included in GDP but do affect whether people are better off.

## LEISURE TIME

Most Americans (and most people in other industrialized countries as well) work many fewer hours than their great-grandparents did 100 years ago. Early in the twentieth century some industrial workers—steelworkers, for example—worked as many as 12 hours a day, 7 days a week. Today, the 40-hour workweek is typical. Today, Americans also tend to start working later in life (after college or graduate school), and, in many cases, they are able to retire earlier. The increased leisure time available to workers in the United States and other industrialized countries—which allows them to pursue many worthwhile activities, including being with family and friends, participating in sports and hobbies, and pursuing cultural and educational activities—is a major benefit of living in a wealthy society. These extra hours of leisure are not priced in markets, however, and therefore are not reflected in GDP.

**Why do people work fewer hours today than their great-grandparents did?**

Americans start work later in life, retire earlier, and in many cases work fewer hours per week than people of 50 or 100 years ago.

The *opportunity cost* of working less—retiring earlier, for example, or working fewer hours per week—is the earnings you forgo by not working. If you can make $400 per week at a summer job in a department store, for example, then leaving the job two weeks early to take a trip with some friends has an opportunity cost of $800. The fact that people are working fewer hours today suggests that their opportunity cost of forgone earnings is lower than their grandparents' and great-grandparents' opportunity cost. Why this difference?

Over the past century, rapid economic growth in the United States and other industrialized countries has greatly increased the purchasing power of the average worker's wages (see Chapter 8). In other words, the typical worker today can buy more goods and services with his or her hourly earnings than ever before. This fact would seem to suggest that the opportunity cost of forgone earnings (measured in terms of what those earnings can buy) is greater, not smaller, today than in earlier times. But because the buying power of wages is so much higher today than in the past, Americans can achieve a reasonable standard of living by working fewer hours than they did in the past. Thus, while your grandparents may have had to work long hours to pay the rent or put food on the table, today the extra income from working long hours is more likely to buy relative luxuries, like nicer clothes or a fancier car. Because such discretionary purchases are easier to give up than basic food and shelter, the true opportunity cost of forgone earnings is lower today than it was 50 years ago. As the opportunity cost of leisure has fallen, Americans have chosen to enjoy more of it.

## NONMARKET ECONOMIC ACTIVITIES

Not all economically important activities are bought and sold in markets; with a few exceptions, such as government services, nonmarket economic activities are omitted from GDP. We mentioned earlier the example of unpaid housekeeping services. Another example is volunteer services, such as the volunteer fire and rescue squads that serve many small towns. The fact that these unpaid services are left out of GDP does *not* mean that they are unimportant. The problem is that, because there are no market prices and quantities for unpaid services, estimating their market values is very difficult.

How far do economists go wrong by leaving nonmarket economic activities out of GDP? The answer depends on the type of economy being studied. Although nonmarket economic activities exist in all economies, they are particularly important in poor economies. For example, in rural villages of developing countries, people commonly trade services with each other or cooperate on various tasks without exchanging any money. Families in these communities also tend to be relatively self-sufficient, growing their own food and providing many of their own basic services (recall the many skills of the Nepalese cook Birkhaman, described in Chapter 2). Because such nonmarket economic activities are not counted in official statistics, GDP data may substantially understate the true amount of economic activity in the poorest countries. In 1999, according to the United Nations,[2] the official GDP per person in Nepal was about $218, an amount that seems impossibly low. Part of the explanation for this figure is that because the Nepalese seldom use formal markets, many economic activities that would ordinarily be included in GDP are excluded from it in Nepal.

Closely related to nonmarket activities is what is called the *underground economy,* which includes transactions that are never reported to government officials and data collectors. The underground economy encompasses both legal and illegal activities, from informal babysitting jobs to organized crime. For instance, some people pay temporary or part-time workers like housecleaners and painters in cash, which allows these workers to avoid paying taxes on their income. Economists who have tried to estimate the value of such services by studying how much cash the public holds have concluded that these sorts of transactions are quite important, even in advanced industrial economies.

## ENVIRONMENTAL QUALITY AND RESOURCE DEPLETION

China has recently experienced tremendous growth in real GDP. But in expanding its manufacturing base, it has also suffered a severe decline in air and water quality. Increased pollution certainly detracts from the quality of life, but because air and water quality are not bought and sold in markets, the Chinese GDP does not reflect this downside of their economic growth.

The exploitation of finite natural resources also tends to be overlooked in GDP. When an oil company pumps and sells a barrel of oil, GDP increases by the value of the oil. But the fact that there is one less barrel of oil in the ground, waiting to be pumped sometime in the future, is not reflected in GDP.

A number of efforts have been made to incorporate factors like air quality and resource depletion into a comprehensive measure of GDP. Doing so is difficult, since it often involves placing a dollar value on intangibles, like having a clean river to swim in instead of a dirty one. But the fact that the benefits of environmental quality and resource conservation are hard to measure in dollars and cents does not mean that they are unimportant.

## QUALITY OF LIFE

What makes a particular town or city an attractive place in which to live? Some desirable features you might think of are reflected in GDP: spacious, well-constructed homes, good restaurants and stores, a variety of entertainment, and high-quality medical services. However, other indicators of the good life are not sold in markets and so may be omitted from GDP. Examples include a low crime rate, minimal traffic congestion, active civic organizations, and open space. Thus

[2]See p. 110 for *first* footnote.

citizens of a rural area may be justified in opposing the construction of a new shopping center because of its presumed negative effect on the quality of life—even though the new center may increase GDP.

### POVERTY AND ECONOMIC INEQUALITY

GDP measures the *total* quantity of goods and services produced and sold in an economy, but it conveys no information about who gets to enjoy those goods and services. Two countries may have identical GDPs but differ radically in the distribution of economic welfare across the population. Suppose, for example, that in one country—call it Equalia—most people have a comfortable middle-class existence; both extreme poverty and extreme wealth are rare. But in another country, Inequalia—which has the same real GDP as Equalia—a few wealthy families control the economy, and the majority of the population lives in poverty. While most people would say that Equalia has a better economic situation overall, that judgment would not be reflected in the GDPs of the two countries, which are the same.

In the United States absolute poverty has been declining. Today, many families whose income is below today's official "poverty line" (in 2001, $17,960 for a family of four) own a television, a car, and in some cases their own home. Some economists have argued that people who are considered poor today live as well as many middle-class people did in the 1950s.

But, though absolute poverty seems to be decreasing in the United States, inequality of income has generally been rising. The chief executive officer of a large U.S. corporation may earn hundreds of times what the typical worker in the same firm receives. Psychologists tell us that people's economic satisfaction depends not only on their absolute economic position—the quantity and quality of food, clothing, and shelter they have—but on what they have compared to what others have. If you own an old, beat-up car but are the only person in your neighborhood to have a car, you may feel privileged. But if everyone else in the neighborhood owns a luxury car, you are likely to be less satisfied. To the extent that such comparisons affect people's well-being, inequality matters as well as absolute poverty. Again, because GDP focuses on total production rather than on the distribution of output, it does not capture the effects of inequality.

## BUT GDP IS RELATED TO ECONOMIC WELL-BEING

You might conclude from the list of important factors omitted from the official figures that GDP is useless as a measure of economic welfare. Indeed, numerous critics have made that claim. Clearly, in evaluating the effects of a proposed economic policy, considering only the likely effects on GDP is not sufficient. Planners must also ask whether the policy will affect aspects of economic well-being that are not captured in GDP. Environmental regulations may reduce production of steel, for example, which reduces the GDP. But that fact is not a sufficient basis on which to decide whether such regulations are good or bad. The right way to decide such questions is to apply the *cost-benefit principle* (see Chapter 1). Are the benefits of cleaner air worth more to people than the costs the regulations impose in terms of lost output and lost jobs? If so, then the regulations should be adopted; otherwise, they should not.

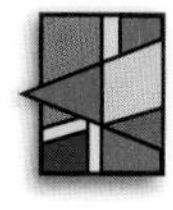

Although looking at the effects of a proposed policy on real GDP is not a good enough basis on which to evaluate a policy, real GDP per person *does* tend to be positively associated with many things people value, including a high material standard of living, better health and life expectancies, and better education. We discuss next some of the ways in which a higher real GDP implies greater economic well-being.

## AVAILABILITY OF GOODS AND SERVICES

Obviously, citizens of a country with a high GDP are likely to possess more and better goods and services (after all, that is what GDP measures). On average, people in high-GDP countries enjoy larger, better-constructed, and more comfortable homes, higher-quality food and clothing, a greater variety of entertainment and cultural opportunities, better access to transportation and travel, better communications and sanitation, and other advantages. While social commentators may question the value of material consumption—and we agree that riches do not necessarily bring happiness or peace of mind—the majority of people in the world place great importance on achieving material prosperity. Throughout history people have made tremendous sacrifices and taken great risks to secure a higher standard of living for themselves and their families. In fact, to a great extent the United States was built by people who were willing to leave their native lands, often at great personal hardship, in hopes of bettering their economic condition.

## HEALTH AND EDUCATION

Beyond an abundance of consumer goods, a high GDP brings other more basic advantages. Table 5.4 shows the differences between rich and poor countries with regard to some important indicators of well-being, including life

**TABLE 5.4**
**GDP and Basic Indicators of Well-Being**

| Indicator | All developing countries | Least developed countries | Industrialized countries |
|---|---|---|---|
| GDP per person (U.S. dollars) | 3,530 | 1,170 | 25,860 |
| Life expectancy at birth (years) | 64.5 | 51.7 | 78.0 |
| Infant mortality rate (per1,000 live births) | 61 | 100 | 6 |
| Under-5 mortality rate (per 1,000 live births) | 89 | 159 | 6 |
| Doctors (per 100,000 people) | 78 | 30 | 252 |
| Incidence of HIV/AIDS (% in 15–49 age group) | 1.3 | 4.3 | 0.3 |
| Undernourished people (%) | 18 | 38 | Negligible |
| Primary enrollment rate (as % of age group) | 85.7 | 60.4 | 99.9 |
| Secondary enrollment rate (as % of age group) | 60.4 | 31.2 | 96.2 |
| Adult literacy rate (%) | 72.9 | 51.7 | 98.6 |

SOURCE: United Nations, *Human Development Report,* 2000–2001, available at http://hdr.undp.org/. All data are for 1999, except doctors per 100,000 people (1992–1995), undernourished people (1996–1998), and enrollment rates (1997). GDP data are adjusted to account for local differences in prices of basic commodities and services (adjusted for purchasing power parity).

expectancy, infant and child mortality rates, number of doctors, measures of nutrition, and educational opportunity. Three groups of countries are compared: (1) developing countries as a group (total population, 4.6 billion); (2) the least developed countries (25 countries with a total population of about 600 million); and (3) the industrialized countries (25 countries, including the United States, Canada, the western European countries, and Japan, with a total population of 850 million). As the first row of Table 5.4 shows, these three groups of countries have radically different levels of GDP per person. Most notably, GDP per person in the industrialized countries is more than 20 times that of the least developed countries.[3]

How do these large differences in GDP relate to other measures of well-being? Table 5.4 shows that on some of the most basic measures of human welfare, the developing countries fare much worse than the industrial countries. A child born in one of the least developed countries has a 10 percent (100/1,000) chance of dying before its first birthday and about a 16 percent (159/1,000) chance of dying before its fifth birthday. The corresponding figures for the industrialized countries are 0.6 percent (6/1,000) and 0.6 percent (6/1,000), respectively. A child born in an industrialized country has a life expectancy of about 78 years, compared to about 52 years for a child born in one of the least developed countries. Superior nutrition, sanitation, and medical services in the richer countries account for these large discrepancies in basic welfare. For treatment of the sick, industrialized countries have about 252 doctors per 100,000 people, compared to 30 doctors per 100,000 people in the least-developed countries. Doctors in the poorest countries, despite their fewer numbers, must contend with much higher rates of illness. For example, the incidence of HIV/AIDS in the least developed countries is 4.3 percent of the population aged 15–49, about 14 times the rate in industrialized countries.

A child born in one of the least developed countries has a 16 percent chance of dying before its fifth birthday.

On another important dimension of human well-being, literacy and education rates, high-GDP countries also have the advantage. As Table 5.4 shows, in the industrialized countries the percentage of adults who can read and write exceeds 98 percent, almost twice the percentage (52 percent) in the poorest developing countries. The percentage of children of primary-school age who are enrolled in school is virtually 100 percent in industrialized countries, compared to about 60 percent in the least developed countries. At the secondary (high school) level the difference is even greater, with about 96 percent of children enrolled in industrialized countries and 31 percent enrolled in the poorest countries. Furthermore, enrollment rates do not capture important differences in the quality of education available in rich and poor countries, as measured by indicators such as the educational backgrounds of teachers and student–teacher ratios. Once again, the average person in an industrialized country seems to be better off than the average person in a poor developing country.

**Why do far fewer children complete high school in poor countries than in rich countries?**

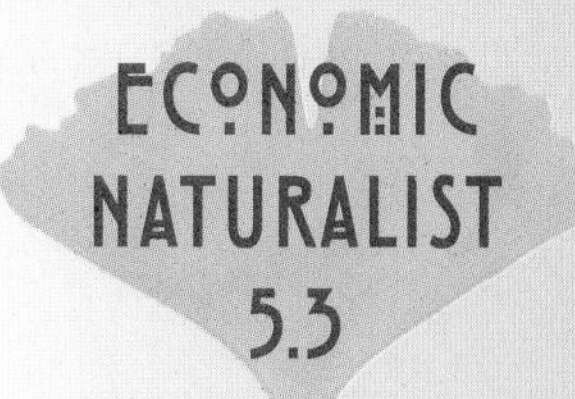


One possible explanation is that people in poor countries place a lower priority on getting an education than people in rich countries. But immigrants from poor countries often put a heavy emphasis on education—though it may be that people who emigrate from poor countries are unrepresentative of the population as a whole.

An economic naturalist's explanation for the lower schooling rates in poor countries would rely not on cultural differences but on differences in *opportunity*

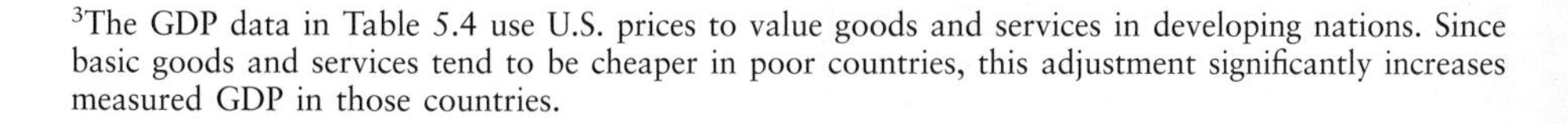

[3]The GDP data in Table 5.4 use U.S. prices to value goods and services in developing nations. Since basic goods and services tend to be cheaper in poor countries, this adjustment significantly increases measured GDP in those countries.

*cost.* In poor societies, most of which are heavily agricultural, children are an important source of labor. Beyond a certain age, sending children to school imposes a high opportunity cost on the family. Children who are in school are not available to help with planting, harvesting, and other tasks that must be done if the family is to survive. In addition, the cost of books and school supplies imposes a major hardship on poor families. In rich, nonagricultural countries, school-age children have few work opportunities, and their potential earnings are small relative to other sources of family income. The low opportunity cost of sending children to school in rich countries is an important reason for the higher enrollment rates in those countries.

In Chapter 7 we will discuss the costs and benefits of economic growth—which in practice means growth in real GDP per person—in greater depth. In that context we will return to the question of whether a growing real GDP is necessarily equated with greater economic well-being.

### REAL GDP AND ECONOMIC WELL-BEING

Real GDP is at best an imperfect measure of economic well-being. Among the factors affecting well-being omitted from real GDP are the availability of leisure time, nonmarket services such as unpaid homemaking and volunteer services, environmental quality and resource conservation, and quality-of-life indicators such as a low crime rate. The GDP also does not reflect the degree of economic inequality in a country. Because real GDP is not the same as economic well-being, proposed policies should not be evaluated strictly in terms of whether or not they increase the GDP.

Although GDP is not the same as economic well-being, it is positively associated with many things that people value, including a higher material standard of living, better health, longer life expectancies, and higher rates of literacy and educational attainment. This relationship between real GDP and economic well-being has led many people to emigrate from poor nations in search of a better life and has motivated policymakers in developing countries to try to increase their nations' rates of economic growth.

## THE UNEMPLOYMENT RATE

In assessing the level of economic activity in a country, economists look at a variety of statistics. Besides real GDP, one statistic that receives a great deal of attention, both from economists and from the general public, is the rate of unemployment. The unemployment rate is a sensitive indicator of conditions in the labor market. When the unemployment rate is low, jobs are secure and relatively easier to find. Low unemployment is often associated with improving wages and working conditions as well, as employers compete to attract and retain workers.

We will discuss labor markets and unemployment in detail in Chapter 8. This chapter will explain how the unemployment rate and some related statistics are defined and measured. It will close with a discussion of the costs of unemployment, both to the unemployed and to the economy as a whole.

### MEASURING UNEMPLOYMENT

In the United States, defining and measuring unemployment is the responsibility of the Bureau of Labor Statistics, or BLS. Each month the BLS surveys about

60,000 randomly selected households. Each person in those households who is 16 years or older is placed in one of three categories:

1. *Employed.* A person is employed if he or she worked full-time or part-time (even for a few hours) during the past week or is on vacation or sick leave from a regular job.
2. *Unemployed.* A person is unemployed if he or she did not work during the preceding week but made some effort to find work (for example, by going to a job interview) in the past four weeks.
3. *Out of the labor force.* A person is considered to be out of the labor force if he or she did not work in the past week and did not look for work in the past four weeks. In other words, people who are neither employed nor unemployed (in the sense of looking for work but not being able to find it) are "out of the labor force." Full-time students, unpaid homemakers, retirees, and people unable to work because of disabilities are examples of people who are out of the labor force.

Based on the results of the survey, the BLS estimates how many people in the whole country fit into each of the three categories.

To find the unemployment rate, the BLS must first calculate the size of the *labor force.* The **labor force** is defined as the total number of employed and unemployed people in the economy (the first two categories of respondents to the BLS survey). The **unemployment rate** is then defined as the number of unemployed people divided by the labor force. Notice that people who are out of the labor force (because they are in school, have retired, or are disabled, for example) are not counted as unemployed and thus do not affect the unemployment rate. In general, a high rate of unemployment indicates that the economy is performing poorly.

Another useful statistic is the **participation rate,** or the percentage of the working-age population in the labor force (that is, the percentage that is either employed or looking for work). Figure 5.1 showed participation rates for American women and men since 1960. The participation rate is calculated by dividing the labor force by the working-age (16+) population.

Table 5.5 illustrates the calculation of key labor market statistics, using data based on the BLS survey for July 2002. In that month unemployment was 5.9 percent of the labor force. The participation rate was 66.5 percent; that is, about two out of every three adults had a job or were looking for work. Figure 5.3 shows the U.S. unemployment rate since 1960. Unemployment rates were exceptionally

***labor force*** the total number of employed and unemployed people in the economy

***unemployment rate*** the number of unemployed people divided by the labor force

***participation rate*** the percentage of the working-age population in the labor force (that is, the percentage that is either employed or looking for work)

**TABLE 5.5**
**U.S. Employment Data, July 2002 (in millions)**

| | |
|---|---|
| Employed | 134.04 |
| Plus: | |
| Unemployed | 8.35 |
| Equals: Labor force | 142.39 |
| Plus: | |
| Not in labor force | 71.63 |
| Equals: | |
| Working-age (over 16) population | 214.02 |

Unemployment rate = unemployed/labor force = 8.35/142.39 = 5.9%

Participation rate = labor force/working-age population = 142.39/214.02 = 66.5%

SOURCE: Bureau of Labor Statistics (http://stats.bls.gov).

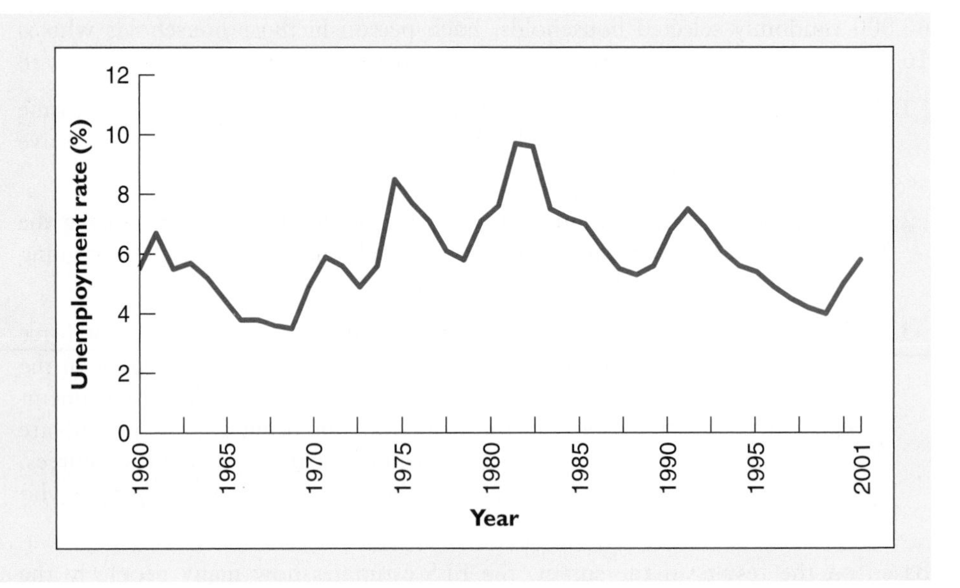


**FIGURE 5.3**
**The U.S. Unemployment Rate since 1960.**
The unemployment rate—the fraction of the U.S. labor force that is unemployed—was just above 4 percent in the late 1990s, the lowest recorded rate since the latter part of the 1960s. Unemployment rose to about 6 percent in 2002 due to a recession.

low—just above 4 percent—in the late 1960s and the late 1990s. By this measure, the latter part of the 1990s was an exceptionally good time for American workers. However, unemployment rose in 2001–2002 as the nation fell into recession.

**EXERCISE 5.7**

**Following are July 2002 BLS U.S. employment data for African-Americans.**

| | |
|---|---|
| Employed | 14.976 million |
| Unemployed | 1.642 million |
| Not in the labor force | 9.343 million |

**Find the labor force, the working-age population, the unemployment rate, and the participation rate for African-Americans and compare your results to those in Table 5.5.**

## THE COSTS OF UNEMPLOYMENT

Unemployment imposes *economic, psychological,* and *social* costs on a nation. From an economic perspective, the main cost of unemployment is the output that is lost because the workforce is not fully utilized. Much of the burden of the reduced output is borne by the unemployed themselves, whose incomes fall when they are not working and whose skills may deteriorate from lack of use. However, society at large also bears part of the economic cost of unemployment. For example, workers who become unemployed are liable to stop paying taxes and start receiving government support payments, such as unemployment benefits. This net drain on the government's budget is a cost to all taxpayers.

The *psychological* costs of unemployment are felt primarily by unemployed workers and their families. Studies show that lengthy periods of unemployment can lead to a loss of self-esteem, feelings of loss of control over one's life, depression, and even suicidal behavior.[4] The unemployed worker's family is likely to feel increased psychological stress, compounded by the economic difficulties created by the loss of income.

[4]For a survey of the literature on the psychological effects of unemployment, see William Darity Jr., and Arthur H. Goldsmith, "Social Psychology, Unemployment and Macroeconomics," *Journal of Economic Perspectives,* 10:121–140, Winter 1996.

The *social* costs of unemployment are a result of the economic and psychological effects. People who have been unemployed for a while tend not only to face severe financial difficulties but also to feel anger, frustration, and despair. Not surprisingly, increases in unemployment tend to be associated with increases in crime, domestic violence, alcoholism, drug abuse, and other social problems. The costs created by these problems are borne not only by the unemployed but by society in general, as more public resources must be spent to counteract these problems—for example, by hiring more police to control crime.

## THE DURATION OF UNEMPLOYMENT

In assessing the impact of unemployment on jobless people, economists must know how long individual workers have been without work. Generally, the longer a person has been out of work, the more severe are the economic and psychological costs that person will face. People who are unemployed for only a few weeks, for example, are not likely to suffer a serious reduction in their standard of living, since for a short period they can draw upon their savings and perhaps on government benefits. Nor would we expect someone who is unemployed for only a short time to experience psychological problems such as depression or loss of self-esteem, at least not to the same extent as someone who has been out of work for months or years.

In its surveys, therefore, the BLS asks respondents how long they have been unemployed. A period during which an individual is continuously unemployed is called an **unemployment spell**; it begins when the worker becomes unemployed and ends when the worker either finds a job or leaves the labor force. (Remember, people outside the labor force are not counted as unemployed.) The length of an unemployment spell is called its **duration.** The duration of unemployment rises during recessions, reflecting the greater difficulty of finding work during those periods.

***unemployment spell*** a period during which an individual is continuously unemployed

***duration*** the length of an unemployment spell

At any given time a substantial fraction of unemployed workers has been unemployed for six months or more; we will refer to this group as the *long-term unemployed.* Long-term unemployment creates the highest economic, psychological, and social costs, both for the unemployed themselves and for society as a whole.

Although long-term unemployment is a serious problem, many unemployment spells are quite short. For example, in July 2002, despite a weak labor market, nearly 35 percent of the unemployed had been out of work for just 5 weeks or less, and another 30 percent had been unemployed for 5 to 14 weeks. In other words, only about 35 percent of the unemployed had been without a job for as long as 14 weeks (about 3 months). These statistics are a bit deceptive, however, because short unemployment spells can arise from two very different patterns of labor-market experience. Some people have short unemployment spells that end in their finding a stable long-term job. For the most part, these workers, whom we will refer to as the *short-term unemployed,* do not bear a high cost of unemployment. But other workers have short unemployment spells that typically end either in their withdrawal from the labor force or in a short-term or temporary job that soon leaves the worker unemployed again. Workers whose unemployment spells are broken up by brief periods of employment or withdrawal from the labor force are referred to as the *chronically unemployed.* In terms of the costs of unemployment, the experience of these workers is similar to that of the long-term unemployed.

## THE UNEMPLOYMENT RATE VERSUS "TRUE" UNEMPLOYMENT

Like GDP measurement, unemployment measurement has its critics. Most of them argue that the official unemployment rate understates the true extent of unemployment. They point in particular to two groups of people who are not counted among the unemployed: so-called *discouraged workers* and *involuntary part-time workers.*

***discouraged workers*** people who say they would like to have a job but have not made an effort to find one in the past four weeks

**Discouraged workers** are people who say they would like to have a job but have not made an effort to find one in the past four weeks. Often, discouraged workers tell the survey takers that they have not searched for work because they have tried without success in the past, or because they are convinced that labor-market conditions are such that they will not be able to find a job. Because they have not sought work in the past four weeks, discouraged workers are counted as being out of the labor force rather than unemployed. Some observers have suggested that treating discouraged workers as unemployed would provide a more accurate picture of the labor market.

*Involuntary part-time workers* are people who say they would like to work full-time but are able to find only part-time work. Because they do have jobs, involuntary part-time workers are counted as employed rather than unemployed. Some economists have suggested that these workers should be counted as partially unemployed.

In response to these criticisms, in recent years the BLS has released special unemployment rates that include estimates of the number of discouraged workers and involuntary part-time workers. In July 2002, when the official unemployment rate was 5.9 percent (see Table 5.5), the BLS calculated that if both discouraged workers and involuntary part-time workers were counted as unemployed, the unemployment rate would have been 9.4 percent. So the problem of discouraged and underemployed workers appears to be fairly significant.

Whether in an official or adjusted version, the unemployment rate is a good overall indicator of labor-market conditions. A high unemployment rate tends to be bad news even for those people who are employed, since raises and promotions are hard to come by in a "slack" labor market. We will discuss the causes and cures of unemployment at some length in Chapter 8 and subsequent chapters.

## SUMMARY


- The basic measure of an economy's output is *gross domestic product (GDP),* the market value of the final goods and services produced in a country during a given period. Expressing output in terms of market values allows economists to aggregate the millions of goods and services produced in a modern economy.

- Only *final goods and services* (which include *capital goods*) are counted in GDP, since they are the only goods and services that directly benefit final users. *Intermediate goods and services,* which are used up in the production of final goods and services, are not counted in GDP, nor are sales of existing assets, such as a 20-year-old house. Summing the value added by each firm in the production process is a useful method of determining the value of final goods and services.

- GDP can also be expressed as the sum of four types of expenditure: *consumption, investment, government purchases,* and *net exports.* These four types of expenditure correspond to the spending of households, firms, the government, and the foreign sector, respectively.

- To compare levels of GDP over time, economists must eliminate the effects of inflation. They do so by measuring the market value of goods and services in terms of the prices in a base year. GDP measured in this way is called *real GDP,* while GDP measured in terms of current-year prices is called *nominal GDP.* Real GDP should always be used in making comparisons of economic activity over time.

- Real GDP per person is an imperfect measure of economic well-being. With a few exceptions, notably government purchases of goods and services (which are included in GDP at their cost of production), GDP includes only those goods and services sold in markets. It excludes important factors that affect people's well-being, such as the amount of leisure time available to them, the value of unpaid or volunteer services, the quality of the environment, quality of life indicators such as the crime rate, and the degree of economic inequality.

- Real GDP is still a useful indicator of economic well-being, however. Countries with a high real GDP per person not only enjoy high average standards of living; they also tend to have higher life expectancies, low rates of infant and child mortality, and high rates of school enrollment and literacy.

- The unemployment rate, perhaps the best-known indicator of the state of the labor market, is based on surveys conducted by the Bureau of Labor Statistics. The surveys classify all respondents over age 16 as employed, unemployed, or not in the labor force. The *labor force* is the sum of employed and unemployed workers—that is, people who have a job or are looking for one. The *unemployment rate* is calculated as the number of unemployed workers divided by the labor force. The *participation rate* is the percentage of the working-age population that is in the labor force.

- The costs of unemployment include the economic cost of lost output, the psychological costs borne by unemployed workers and their families, and the social costs associated with problems like increased crime and violence. The greatest costs are imposed by long *unemployment spells* (periods of unemployment). Critics of the official unemployment rate argue that it understates "true" unemployment by excluding *discouraged workers* and involuntary part-time workers.

## KEY TERMS

capital good (114)
consumption expenditure (118)
discouraged workers (134)
duration (of an unemployment spell) (133)
final goods or services (114)
government purchases (119)
gross domestic product (GDP) (111)
intermediate goods or services (114)
investment (118)
labor force (131)
net exports (119)
nominal GDP (123)
participation rate (131)
real GDP (123)
unemployment rate (131)
unemployment spell (133)
value added (115)

## REVIEW QUESTIONS

1. Why do economists use market values when calculating GDP? What is the economic rationale for giving high-value items more weight in GDP than low-value items?
2. A large part of the agricultural sector in developing countries is subsistence farming, in which much of the food that is produced is consumed by the farmer and the farmer's family. Discuss the implications of this fact for the measurement of GDP in poor countries.
3. Give examples of each of the four types of aggregate expenditure. Which of the four represents the largest share of GDP in the United States? Can an expenditure component be negative? Explain.
4. Al's Shoeshine Stand shined 1,000 pairs of shoes last year and 1,200 pairs this year. He charged $4 for a shine last year and $5 this year. If last year is taken as the base year, find Al's contribution to both nominal GDP and real GDP in both years. Which measure would be better to use if you were trying to measure the change in Al's productivity over the past year? Why?
5. Would you say that real GDP per person is a useful measure of economic well-being? Defend your answer.
6. True or false: A high participation rate in an economy implies a low unemployment rate. Explain.
7. What are the costs of a high unemployment rate? Do you think providing more generous government benefits to the unemployed would increase these costs, reduce these costs, or leave them unchanged? Discuss.

## PROBLEMS

1. George and Al, stranded on an island, use clamshells for money. Last year George caught 300 fish and 5 wild boars. Al grew 200 bunches of bananas. In the two-person economy that George and Al set up, fish sell for 1 clamshell each, boars sell for 10 clamshells each, and bananas go for 5 clamshells a bunch. George paid Al a total of 30 clamshells for helping him to dig bait for fishing, and he also purchased five of Al's mature banana trees for 30 clamshells each. What is the GDP of George's and Al's island in terms of clamshells?
2. How would each of the following transactions affect the GDP of the United States?
   a. The U.S. government pays $1 billion in salaries for government workers.
   b. The U.S. government pays $1 billion to Social Security recipients.
   c. The U.S. government pays a U.S. firm $1 billion for newly produced airplane parts.
   d. The U.S. government pays $1 billion in interest to holders of U.S. government bonds.
   e. The U.S. government pays $1 billion to Saudi Arabia for crude oil to add to U.S. official oil reserves.

3. Intelligence Incorporated produces 100 computer chips and sells them for $200 each to Bell Computers. Using the chips and other labor and materials, Bell produces 100 personal computers. Bell sells the computers, bundled with software that Bell licenses from Macrosoft at $50 per computer, to PC Charlie's for $800 each. PC Charlie's sells the computers to the public for $1,000 each. Calculate the total contribution to GDP using the value-added method. Do you get the same answer by summing up the market values of final goods and services?

4. For each of the following transactions, state the effect both on U.S. GDP and on the four components of aggregate expenditure.
   a. Your mother-in-law buys a new car from a U.S. producer.
   b. Your mother-in-law buys a new car imported from Sweden.
   c. Your mother-in-law's car rental business buys a new car from a U.S. producer.
   d. Your mother-in-law's car rental business buys a new car imported from Sweden.
   e. The U.S. government buys a new, domestically produced car for the use of your mother-in-law, who has been appointed the ambassador to Sweden.

5. Here are some data for an economy. Find its GDP. Explain your calculation.

| | |
|---|---|
| Consumption expenditures | $600 |
| Exports | 75 |
| Government purchases of goods and services | 200 |
| Construction of new homes and apartments | 100 |
| Sales of existing homes and apartments | 200 |
| Imports | 50 |
| Beginning-of-year inventory stocks | 100 |
| End-of-year inventory stocks | 125 |
| Business fixed investment | 100 |
| Government payments to retirees | 100 |
| Household purchases of durable goods | 150 |

6. The nation of Potchatoonie produces hockey pucks, cases of root beer, and back rubs. Here are data on prices and quantities of the three goods in the years 2000 and 2005.

| | Pucks | | Root beer | | Back rubs | |
|---|---|---|---|---|---|---|
| **Year** | **Quantity** | **Price** | **Quantity** | **Price** | **Quantity** | **Price** |
| 2000 | 100 | $5 | 300 | $20 | 100 | $20 |
| 2005 | 125 | $7 | 250 | $20 | 110 | $25 |

   Assume that 2000 is the base year. Find nominal GDP and real GDP for both years.

7. The government is considering a policy to reduce air pollution by restricting the use of "dirty" fuels by factories. In deciding whether to implement the policy, how, if at all, should the likely effects of the policy on real GDP be taken into account? Discuss.

8. Here is a report from a not-very-efficient BLS survey taker: "There were 65 people in the houses I visited, 10 of them children under 16 and 10 retired; 25 people had full-time jobs, and 5 had part-time jobs. There were 5 full-time homemakers, 5 full-time students over age 16, and 2 people who were disabled and cannot work. The remaining people did not have jobs but all said they would like one. One of these people had not looked actively for work for 3 months, however."

   Find the labor force, the unemployment rate, and the participation rate implied by the survey taker's report.

9. Ellen is downloading labor market data for the most recent month, but her connection is slow and so far this is all she has been able to get:

| Unemployment rate | 5.0% |
|---|---|
| Participation rate | 62.5% |
| Not in the labor force | 60 million |

Find the labor force, the working-age population, the number of employed workers, and the number of unemployed workers.

10. The towns of Sawyer and Thatcher each have a labor force of 1,200 people. In Sawyer, 100 people were unemployed for the entire year, while the rest of the labor force was employed continuously. In Thatcher every member of the labor force was unemployed for 1 month and employed for 11 months.
    a. What is the average unemployment rate over the year in each of the two towns?
    b. What is the average duration of unemployment spells in each of the two towns?
    c. In which town do you think the costs of unemployment are higher? Explain.

## ANSWERS TO IN-CHAPTER EXERCISES

5.1 In the text, GDP was calculated to be $64.00. If in addition Orchardia produces 5 oranges at $0.30 each, GDP is increased by $1.50 to $65.50.

5.2 Plants owned by U.S. companies produced a total of 1,185,000 vehicles, or 4.09 times the 290,000 vehicles produced by foreign-owned plants. In market value terms, with passenger cars valued at $17,000 and other light vehicles at $25,000, plants owned by U.S. companies produced (471,000 × $17,000) + (714,000 × $25,000), = $25,857,000 worth of vehicles. Foreign-owned plants produced (227,000 × $17,000) + (63,000 × $25,000) = $5,434,000 worth of vehicles. In market value terms U.S.-owned plants outproduced the foreign-owned plants by a ratio of 4.76 to 1. The U.S. producers have a greater advantage when output is compared in market value terms instead of in terms of number of vehicles because the U.S. companies produce relatively more of the higher-value types of vehicles than the foreign companies do.

5.3 The value added of the wholesale distributor together with the ultimate producers of the cards is $500. Amy's value added—her revenue less her payments to other firms—is $200. Since the cards were produced and purchased by Amy during the year 2002 (we assume), the $500 counts toward year 2001 GDP. The $200 in value added originating in Amy's card shop counts in year 2003 GDP, since Amy actually sold the cards in that year.

5.4 The sale of stock represents a transfer of ownership of part of the assets of Benson Buggywhip, not the production of new goods or services. Hence the stock sale itself does not contribute to GDP. However, the broker's commission of $100 (2 percent of the stock sale proceeds) represents payment for a current service and is counted in GDP.

5.5 As in Example 5.6, the market value of domestic production is 1,000,000 autos times $15,000 per auto, or $15 billion.

Also as in Example 5.6, consumption is $10.5 billion and government purchases are $0.75 billion. However, because 25,000 of the autos that are purchased are imported rather than domestic, the domestic producers have unsold inventories at the end of the year of 50,000 (rather than 25,000 as in Example 5.6). Thus inventory investment is 50,000 autos times $15,000, or $0.75 billion, and total investment (autos purchased by businesses plus inventory investment) is $3.75 billion. Since exports and imports are equal (both are 25,000 autos), net exports (equal to exports minus imports) are zero. Notice that since we subtract imports to get net exports, it is unnecessary also to subtract imports from consumption. Consumption is defined as total purchases by households, not just purchases of domestically produced goods.

Total expenditure is $C + I + G + NX$ = $10.5 billion + $3.75 billion + $0.75 billion + 0 = $15 billion, the same as the market value of production.

5.6 Real GDP in the year 2004 equals the quantities of pizzas and calzones produced in the year 2004, valued at the market prices that prevailed in the base year 2000. So real GDP in 2004 = (30 pizzas × \$10/pizza) + (30 calzones × \$5/calzone) = \$450.

Real GDP in 2000 equals the quantities of pizzas and calzones produced in 2000, valued at 2000 prices, which is \$175. Notice that since 2000 is the base year, real GDP and nominal GDP are the same for that year.

The real GDP in the year 2004 is \$450/\$175, or about 2.6 times what it was in 2000. Hence the expansion of real GDP lies between the threefold increase in pizza production and the doubling in calzone production that occurred between 2000 and 2004.

5.7

$$\text{Labor force} = \text{Employed} + \text{Unemployed}$$
$$= 14.976 \text{ million} + 1.642 \text{ million} = 16.618 \text{ million}$$

$$\text{Working-age population} = \text{Labor force} + \text{Not in labor force}$$
$$= 16.618 \text{ million} + 9.343 \text{ million} = 25.961 \text{ million}$$

$$\text{Unemployment rate} = \frac{\text{Unemployed}}{\text{Labor force}}$$
$$= \frac{1.642 \text{ million}}{16.618 \text{ million}} = 9.9\%$$

$$\text{Participation rate} = \frac{\text{Labor force}}{\text{Working-age population}}$$
$$= \frac{16.618 \text{ million}}{25.961 \text{ million}} = 64.0\%$$

In 2002 African-Americans represented between 10% and 11% of the U.S. labor force and approximately 12% of the working-age population. Note that while the participation rate for African-Americans is similar to that of the overall population the unemployment rate for African-Americans is substantially higher.

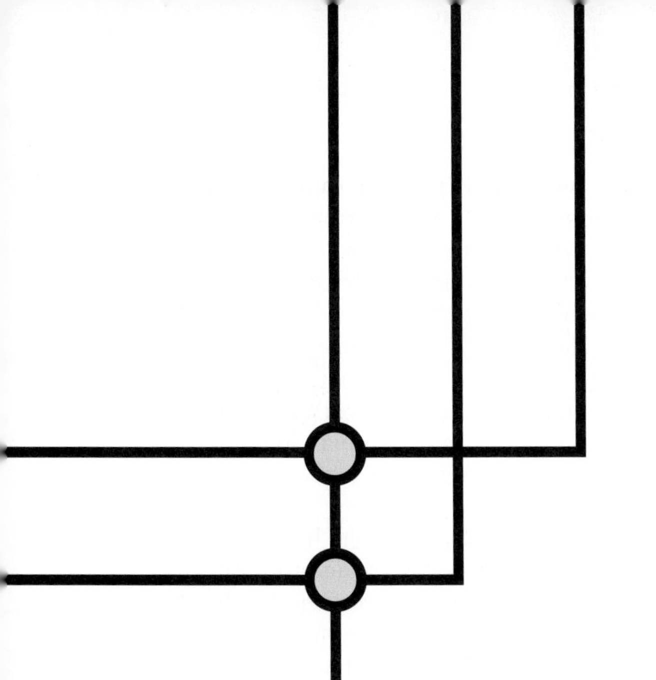

CHAPTER

6

# MEASURING THE PRICE LEVEL AND INFLATION

There is a story about a Wall Street investor, call her Lotta Doe, who was the first volunteer for an experiment in suspended animation. A team of scientists put Lotta into a deep sleep. Thirty years later she awoke, not having aged a day. Lotta's first action upon awakening was to run to a pay phone to call her broker. When the now-aged broker picked up the phone, Lotta asked how much her portfolio of financial investments, left to grow these 30 years, was now worth. "As of this morning," the broker replied, "your net worth is approximately $100 million."

Lotta was, of course, thrilled with this news. But at that moment the operator's voice broke into the call, and Lotta heard, "Please deposit $5 million for the next 3 minutes."

The story illustrates a simple but very important point, which is that the value of money depends entirely on the prices of the goods and services one wants to buy. A $100 million nest egg is a substantial fortune at the prices prevailing in the United States today, but it is only a pittance if a phone call costs $5 million. Likewise, high and sustained inflation—a rapid and ongoing increase in the prices of most goods and services—can radically reduce the buying power of a given amount of money. Apparently, there had been quite a lot of inflation during the 30 years Lotta Doe was in suspended animation.

Inflation can make a comparison of economic conditions at different points in time quite difficult. We remember being able to buy both a comic book and a chocolate sundae for a quarter. Today the same two items might cost $4 or $5. You might conclude from this fact that kids were much better off in "the good old days," but were they really? Without more information, we can't tell, for though the prices of comic books and sundaes have gone up, so have allowances. The real question is whether young people's spending money has increased as much as or more than the prices of the things they want to buy. If so, then they are no worse off today than we were when we were young and candy bars cost a nickel.

Inflation also creates uncertainty when we try to look into the future, to ask questions such as, "How much should I plan to save for retirement?" The answer to this question depends on how much inflation is likely to occur before one retires (and thus how much heating oil, food, and clothing will cost). Inflation can pose similar problems for policymakers. For example, to plan long-term government spending programs they must estimate how much the government's purchases will cost several years in the future.

An important benefit of studying macroeconomics is learning how to avoid the confusion inflation interjects into comparisons of economic conditions over time or projections for the future. In this chapter, a continuation of our study of the construction and interpretation of economic data, we will see how both prices and inflation are measured and how dollar amounts, such as the price of a comic book, can be "adjusted" to eliminate the effects of inflation. Quantities that are measured in dollars (or other currency units) and then adjusted for inflation are called *real* quantities (recall, for example, the concept of real GDP in Chapter 5). By working with real quantities, economists can avoid the kind of confusion that beset Lotta Doe.

More important than the complications inflation creates for economic measurement are the costs that it imposes on the economy. In this chapter we will see why high inflation can significantly impair an economy's performance, to the extent that economic policymakers claim a low and stable rate of inflation as one of their chief objectives. We will conclude the chapter by showing how inflation is linked to another key economic variable, the rate of interest on financial assets.

## THE CONSUMER PRICE INDEX: MEASURING THE PRICE LEVEL

The basic tool economists use to measure the price level and inflation in the U.S. economy is the *consumer price index,* or CPI for short. The CPI is a measure of the "cost of living" during a particular period. Specifically, the **consumer price index (CPI)** for any period measures the cost in that period of a standard set, or basket, of goods and services *relative* to the cost of the same basket of goods and services in a fixed year, called the *base year.*

***consumer price index (CPI)*** for any period, measures the cost in that period of a standard basket of goods and services relative to the cost of the same basket of goods and services in a fixed year, called the *base year*

To illustrate how the CPI is constructed, suppose the government has designated 1995 as the base year. Assume for the sake of simplicity that in 1995 a typical American family's monthly household budget consisted of spending on just three items: rent on a two-bedroom apartment, hamburgers, and movie tickets. In reality, of course, families purchase hundreds of different items each month, but the basic principles of constructing the CPI are the same no matter how many items are included. Suppose too that the family's average monthly expenditures in 1995, the base year, were as shown in Table 6.1.

**TABLE 6.1**
**Monthly Household Budget of the Typical Family in 1995 (Base Year)**

| Item | Cost (in 1995) |
|---|---|
| Rent, two-bedroom apartment | $500 |
| Hamburgers (60 at $2.00 each) | 120 |
| Movie tickets (10 at $6.00 each) | 60 |
| Total expenditure | $680 |

Now let's fast-forward to the year 2000. Over that period, the prices of various goods and services are likely to have changed; some will have risen and some fallen. Let's suppose that by the year 2000 the rent that our family pays for their two-bedroom apartment has risen to $630. Hamburgers now cost $2.50 each, and the price of movie tickets has risen to $7.00 each. So, in general, prices have been rising.

By how much did the family's cost of living increase between 1995 and 2000? Table 6.2 shows that if the typical family wanted to consume the *same basket of goods and services* in the year 2000 as they did in the year 1995, they would have to spend $850 per month, or $170 more than the $680 per month they spent in 1995. In other words, to live the same way in the year 2000 as they did in the year 1995, the family would have to spend 25 percent more ($170/$680) each month. So, in this example, the cost of living for the typical family rose 25 percent between 1995 and 2000.

**TABLE 6.2**
**Cost of Reproducing the 1995 (Base-Year) Basket of Goods and Services in Year 2000**

| Item | Cost (in 2000) | Cost (in 1995) |
|---|---|---|
| Rent, two-bedroom apartment | $630 | $500 |
| Hamburgers (60 at $2.50 each) | 150 | 120 |
| Movie tickets (10 at $7.00 each) | 70 | 60 |
| Total expenditure | $850 | $680 |

The government—actually, the Bureau of Labor Statistics (BLS), the same agency that is responsible for determining the unemployment rate—calculates the official consumer price index (CPI) using essentially the same method. The first step in deriving the CPI is to pick a base year and determine the basket of goods and services that were consumed by the typical family during that year. In practice, the government learns how consumers allocate their spending through a detailed survey, called the Consumer Expenditure Survey, in which randomly selected families record every purchase they make and the price they paid over a given month. (Quite a task!) Let's call the basket of goods and services that results the *base-year basket*. Then, each month BLS employees visit thousands of stores and conduct numerous interviews to determine the current prices of the goods and services in the base-year basket. The CPI in any given year is computed using this formula:

$$\text{CPI} = \frac{\text{Cost of base-year basket of goods and services in current year}}{\text{Cost of base-year basket of goods and services in base year}}.$$

Returning to the example of the typical family that consumes three goods, we can calculate the CPI in the year 2000 as

$$\text{CPI in year 2000} = \frac{\$850}{\$680} = 1.25.$$

In other words, in this example the cost of living in the year 2000 is 25 percent higher than it was in 1995, the base year. Notice that the base-year CPI is always equal to 1.00, since in that year the numerator and the denominator of the CPI formula are the same. The CPI for a given period (such as a month or year) measures the cost of living in that period *relative* to what it was in the base year.

Often news reporters multiply the CPI by 100 to get rid of the decimal point. If we were to do that here, the year 2000 CPI would be expressed as 125 rather than 1.25, and the base-year CPI would be expressed as 100 rather than 1.00. However, some calculations we will do later in the chapter are simplified if the CPI is stated in decimal form, so we will not adopt the convention of multiplying it by 100.

EXAMPLE 6.1

**Measuring the typical family's cost of living**

Suppose that in addition to the three goods and services the typical family consumed in 1995 they also bought four sweaters at $30 each. In the year 2000 the same sweaters cost $50 each. The prices of the other goods and services in 1995 and 2000 were the same as in Table 6.2. Find the change in the family's cost of living between 1995 and 2000.

In the example in the text, the cost of the base-year (1995) basket was $680. Adding four sweaters at $30 each raises the cost of the base-year basket to $800. What does this same basket (including the four sweaters) cost in 2000? The cost of the apartment, the hamburgers, and the movie tickets is $850, as before. Adding the cost of the four sweaters at $50 each raises the total cost of the basket to $1,050. The CPI equals the cost of the basket in 2000 divided by the cost of the basket in 1995 (the base year), or $1,050/$800 = 1.31. We conclude that the family's cost of living rose 31 percent between 1995 and 2000.

EXERCISE 6.1

**Returning to the three-good example in Tables 6.1 and 6.2, find the year 2000 CPI if the rent on the apartment falls from $500 in 1995 to $400 in 2000. The prices for hamburgers and movie tickets in the two years remain the same as in the two tables.**

***price index*** a measure of the average price of a given class of goods or services relative to the price of the same goods and services in a base year

The CPI is not itself the price of a specific good or service; it is a *price index*. A **price index** measures the average price of a class of goods or services relative to the price of those same goods or services in a base year. The CPI is an especially well-known price index, one of many economists use to assess economic trends. For example, because manufacturers tend to pass on increases in the prices of raw materials to their customers, economists use indexes of raw materials' prices to try to forecast changes in the prices of manufactured goods. Other indexes are used to study the rate of price change in energy, food, health care, and other major sectors.

EXERCISE 6.2

**The consumer price index captures the cost of living for the "typical" or average family. Suppose you were to construct a personal price index to measure changes in your own cost of living over time. In general, how would you go about constructing such an index? Why might changes in your personal price index differ from changes in the CPI?**

## INFLATION

***rate of inflation*** the annual percentage rate of change in the price level, as measured, for example, by the CPI

The CPI provides a measure of the average *level* of prices relative to prices in the base year. *Inflation,* in contrast, is a measure of how fast the average price level is *changing* over time. The **rate of inflation** is defined as the annual percentage rate of change in the price level, as measured, for example, by the CPI. Suppose, for example, that the CPI has a value of 1.25 in the year 2002 and a value of 1.30 in the year 2003. The rate of inflation between 2002 and 2003 is the percentage increase in the price level, or the increase in the price level (0.05) divided by the initial price level (1.25), which is equal to 4 percent.

**EXAMPLE 6.2**

### Calculating inflation rates: 1972–1976

CPI values for the years 1972 through 1976 are shown below. Find the rates of inflation between 1972 and 1973, 1973 and 1974, 1974 and 1975, and 1975 and 1976.

| Year | CPI |
|---|---|
| 1972 | 0.418 |
| 1973 | 0.444 |
| 1974 | 0.493 |
| 1975 | 0.538 |
| 1976 | 0.569 |

The inflation rate between 1972 and 1973 is the percentage increase in the price level between those years, or (0.444 – 0.418)/0.418 = 0.026/0.418 = 0.062 = 6.2 percent. Do the calculations on your own to confirm that inflation during each of the next three years was 11.0, 9.1, and 5.8 percent, respectively. During the 1970s, inflation rates were much higher than the 2 to 3 percent inflation rates that have prevailed in recent years.

**EXERCISE 6.3**

**Below are CPI values for the years 1929 through 1933. Find the rates of inflation between 1929 and 1930, 1930 and 1931, 1931 and 1932, and 1932 and 1933.**

| Year | CPI |
|---|---|
| 1929 | 0.171 |
| 1930 | 0.167 |
| 1931 | 0.152 |
| 1932 | 0.137 |
| 1933 | 0.130 |

**How did inflation rates in the 1930s differ from those of the 1970s?**

The results of the calculations for Exercise 6.3 include some examples of *negative* inflation rates. A situation in which the prices of most goods and services are falling over time so that inflation is negative is called **deflation.** The early 1930s was the last time the United States experienced significant deflation. Japan has experienced relatively mild deflation during the past decade.

***deflation*** a situation in which the prices of most goods and services are falling over time so that inflation is negative

## ADJUSTING FOR INFLATION

The CPI is an extremely useful tool. Not only does it allow us to measure changes in the cost of living; it can also be used to adjust economic data to eliminate the effects of inflation. In this section we will see how the CPI can be used to convert quantities measured at current dollar values into real terms, a process called *deflating*. We will also see that the CPI can be used to convert real quantities into current-dollar terms, a procedure called *indexing*. Both procedures are useful not only to economists but to anyone who needs to adjust payments, accounting measures, or other economic quantities for the effects of inflation.

## DEFLATING A NOMINAL QUANTITY

***nominal quantity*** a quantity that is measured in terms of its current dollar value

An important use of the CPI is to adjust **nominal quantities**—quantities measured at their current dollar values—for the effects of inflation. To illustrate, suppose we know that the typical family in a certain metropolitan area had a total income of $20,000 in 1995 and $22,000 in the year 2000. Was this family economically better off in the year 2000 than in 1995?

Without any more information than this we might be tempted to say yes. After all, their income rose by 10 percent over the five-year period. But prices might also have been rising, as fast or faster than the family's income. Suppose the prices of the goods and services the family consumes rose 25 percent over the same period. Since the family's income rose only 10 percent, we would have to conclude that the family is worse off, in terms of the goods and services they can afford to buy, despite the increase in their *nominal,* or current-dollar, income.

***real quantity*** a quantity that is measured in physical terms—for example, in terms of quantities of goods and services

We can make a more precise comparison of the family's purchasing power in 1995 and 2000 by calculating their incomes in those years in *real* terms. In general, a **real quantity** is one that is measured in physical terms—for example, in terms of quantities of goods and services. To convert a nominal quantity into a real quantity, we must divide the nominal quantity by a price index for the period, as shown in Table 6.3. The calculations in the table show that in *real* or purchasing power terms, the family's income actually *decreased* by $2,400, or 12 percent of their initial real income of $20,000, between 1995 and 2000.

**TABLE 6.3**
**Comparing the Real Values of a Family's Income in 1995 and 2000**

| Year | Nominal family income | CPI | Real family income = Nominal family income/CPI |
|---|---|---|---|
| 1995 | $20,000 | 1.00 | $20,000/1.00 = $20,000 |
| 2000 | $22,000 | 1.25 | $22,000/1.25 = $17,600 |

***deflating (a nominal quantity)*** the process of dividing a nominal quantity by a price index (such as the CPI) to express the quantity in real terms

The problem for this family is that though their income has been rising in nominal (dollar) terms, it has not kept up with inflation. Dividing a nominal quantity by a price index to express the quantity in real terms is called **deflating** the nominal quantity. (Be careful not to confuse the idea of deflating a nominal quantity with deflation, or negative inflation. The two concepts are different.)

Dividing a nominal quantity by the current value of a price index to measure it in real or purchasing power terms is a very useful tool. It can be used to eliminate the effects of inflation from comparisons of any nominal quantity—workers' wages, health care expenditures, the components of the federal budget—over time. Why does this method work? In general, if you know both how many dollars you have spent on a given item and the item's price, you can figure out how many of the item you bought (by dividing your expenditures by the price). For example, if you spent $100 on hamburgers last month and hamburgers cost $2.50 each, you can determine that you purchased 40 hamburgers. Similarly, if you divide a family's dollar income or expenditures by a price index, which is a measure of the average price of the goods and services they buy, you will obtain a measure of the real quantity of goods and services they purchased. Such real quantities are sometimes referred to as *inflation-adjusted* quantities.

**EXAMPLE 6.3** **Home run hitters drive Cadillacs**

In 1930 the great baseball player Babe Ruth earned a salary of $80,000. When it was pointed out to him that he had earned more than President Hoover, Ruth replied, with some justification, "I had a better year than he did." In 1998 St. Louis Cardinals slugger Mark McGwire earned approximately $8.3 million in the process of breaking both Ruth's and Roger Maris's single-season home run records. Adjusting for inflation, who earned more, Ruth or McGwire?

To answer this question, we need to know that the CPI (using the average of 1982–1984 as the base year) was 0.167 in 1930 and 1.64 in 1998. Dividing Babe Ruth's salary by 0.167, we obtain approximately $479,000, which is Ruth's salary "in 1982–1984 dollars." In other words, to enjoy the same purchasing power during the 1982–1984 period as in 1930, the Babe would have needed a salary of $479,000. Dividing Mark McGwire's 1998 salary by the 1998 CPI, 1.64, yields a salary of $5.06 million in 1982–1984 dollars. We can now compare the salaries of the two power hitters. Although adjusting for inflation brings the two figures closer together (since part of McGwire's higher salary compensates for the increase in prices between 1930 and 1998), in real terms McGwire still earned more than 10 times Ruth's salary. Incidentally, McGwire also earned about 40 times what President Clinton earned.

Clearly, in comparing wages or earnings at two different points in time, we must adjust for changes in the price level. Doing so yields the **real wage**—the wage measured in terms of real purchasing power. The real wage for any given period is calculated by dividing the nominal (dollar) wage by the CPI for that period.

***real wage*** the wage paid to workers measured in terms of real purchasing power; the real wage for any given period is calculated by dividing the nominal (dollar) wage by the CPI for that period

### EXERCISE 6.4

**In 2001 Barry Bonds of the San Francisco Giants hit 73 home runs, breaking Mark McGwire's record. Bonds earned $10.3 million in 2001. In that year the CPI was 1.78. How did Bonds's real earnings compare to McGwire's?**

### EXAMPLE 6.4

#### Real wages of U.S. production workers

Production workers are nonsupervisory workers, such as those who work on factory assembly lines. According to the Bureau of Labor Statistics, http://stats.bls.gov, the average U.S. production worker earned $3.23 per hour in 1970 and $10.01 per hour in 1990. Compare the real wages for this group of workers in these years.

To find the real wage in 1970 and 1990, we need to know that the CPI was 0.388 in 1970 and 1.307 in 1990 (again using the 1982–1984 average as the base period). Dividing $3.23 by 0.388, we find that the real wage in 1970 was $8.32. Dividing $10.01 by 1.307, we find that the real wage in 1990 was only $7.66. In real or purchasing power terms, manufacturing workers' wages fell between 1970 and 1990, despite the fact that the nominal or dollar wage more than tripled.

Figure 6.1 shows nominal wages and real wages for U.S. production workers for the period 1960–2001. Notice the dramatic difference between the two trends.

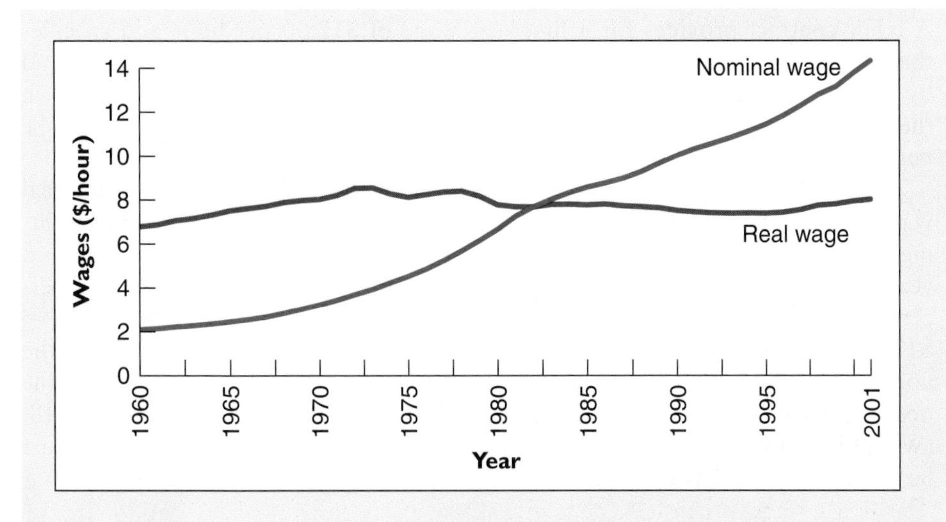


**FIGURE 6.1**
**Nominal and Real Wages for Production Workers, 1960–2001.**
Though nominal wages of production workers have risen dramatically since 1960, real wages have stagnated.

Looking only at nominal wages, one might conclude that production-line workers were much better paid in 2001 than in 1960. But once wages are adjusted for inflation, we see that in terms of buying power production-line workers' wages have stagnated since the early 1970s. This example illustrates the crucial importance of adjusting for inflation when comparing dollar values over time.

### EXERCISE 6.5

**In 1950 the minimum wage prescribed by federal law was $0.75 per hour. In 1997 the minimum wage was raised to $5.15 per hour. How does the real minimum wage in 1950 compare to that of 1997? The CPI was 0.241 in 1950 and 1.61 in 1997.**

## INDEXING TO MAINTAIN BUYING POWER

The consumer price index can also be used to convert real quantities to nominal quantities. Suppose, for example, that in the year 2000 the government paid certain Social Security recipients $1,000 per month in benefits. Let's assume that Congress would like the buying power of these benefits to remain constant over time so that the recipients' standard of living is unaffected by inflation. To achieve that goal, at what level should Congress set the monthly Social Security benefit in the year 2005?

The nominal, or dollar, benefit Congress should pay in the year 2005 to maintain the purchasing power of retired people depends on how much inflation has taken place between 2000 and 2005. Suppose that the CPI has risen 20 percent between 2000 and 2005. That is, on average the prices of the goods and services consumers buy have risen 20 percent over that period. For Social Security recipients to "keep up" with inflation, their benefit in the year 2005 must be $1,200 per month, or 20 percent more than it was in 2000. In general, to keep purchasing power constant, the dollar benefit must be increased each year by the percentage increase in the CPI.

***indexing*** the practice of increasing a nominal quantity each period by an amount equal to the percentage increase in a specified price index. Indexing prevents the purchasing power of the nominal quantity from being eroded by inflation

The practice of increasing a nominal quantity according to changes in a price index to prevent inflation from eroding purchasing power is called **indexing.** In the case of Social Security, federal law provides for the automatic indexing of benefits. Each year, without any action by Congress, benefits increase by an amount equal to the percentage increase in the CPI. Some labor contracts are indexed as well so that wages are adjusted fully or partially for changes in inflation (see Example 6.5).

### EXAMPLE 6.5 An indexed labor contract

A labor contract provides for a first-year wage of $12.00 per hour and specifies that the real wage will rise by 2 percent in the second year of the contract and by another 2 percent in the third year. The CPI is 1.00 in the first year, 1.05 in the second year, and 1.10 in the third year. Find the dollar wage that must be paid in the second and third years.

Because the CPI is 1.00 in the first year, both the nominal wage and the real wage are $12.00. Let $W_2$ stand for the nominal wage in the second year. Deflating by the CPI in the second year, we can express the real wage in the second year as $W_2/1.05$. The contract says that the second-year real wage must be 2 percent higher than the real wage in the first year, so $W_2/1.05 = \$12.00 \times 1.02 = \$12.24$. Multiplying through by 1.05 to solve for $W_2$, we get $W_2 = \$12.85$, the nominal wage required by the contract in the second year. In the third year the nominal wage $W_3$ must satisfy the equation $W_3/1.10 = \$12.24 \times 1.02 = \$12.48$. (Why?) Solving this equation for $W_3$ yields $13.73 as the nominal wage that must be paid in the third year.

**EXERCISE 6.6**

**The minimum wage is not indexed to inflation, but suppose it had been when it was introduced in 1950. What would the nominal minimum wage have been in 1997, the year that Congress raised the minimum wage to $5.15 by legislation? See Exercise 6.5 for the data necessary to answer the question.**

**Every few years there is a well-publicized battle in Congress over whether the minimum wage should be raised. Why do these heated legislative debates recur so regularly?**

Because the minimum wage is not indexed to inflation, its purchasing power falls as prices rise. Congress must therefore raise the nominal minimum wage periodically to keep the real value of the minimum wage from eroding. Ironically, despite the public's impression that Congress has raised the nominal minimum wage steeply over the years, the real minimum wage has fallen about one-third since 1970.

Why doesn't Congress index the minimum wage to the CPI and eliminate the need to reconsider it so often? Evidently, some members of Congress prefer to hold a highly publicized debate on the issue every few years—perhaps because it mobilizes both advocates and opponents of the minimum wage to make campaign donations to those members who represent their views.

**RECAP** **METHODS TO ADJUST FOR INFLATION**

*Deflating.* To correct a nominal quantity, such as a family's dollar income, for changes in the price level, divide it by a price index such as the CPI. This process, called *deflating* the nominal quantity, expresses the nominal quantity in terms of real purchasing power. If nominal quantities from two different years are deflated by a price index with the same base year, the purchasing power of the two deflated quantities can be compared.

*Indexing.* To ensure that a nominal payment, such as a Social Security benefit, represents a constant level of real purchasing power, increase the nominal quantity each year by a percentage equal to the rate of inflation for that year (a procedure known as *indexing*).

## DOES THE CPI MEASURE "TRUE" INFLATION?

You may have concluded that measuring inflation is straightforward, but as with GDP and the unemployment rate, the issue is not free from controversy. Indeed the question of whether U.S. inflation is properly measured has been the subject of serious debates in recent years. Because the CPI is one of the most important U.S. economic statistics, the issue is far from academic. Policymakers pay close attention to the latest inflation numbers when deciding what actions to take. Furthermore, because of the widespread use of indexing, changes in the CPI directly impact the government's budget. For example, if the CPI rises by 3 percent during a given year, by law Social Security benefits—which are a significant part of federal government spending—increase automatically by 3 percent. Many other government payments and private contracts, such as union labor contracts, are indexed to the CPI as well.

When a 1996 report concluded that changes in the CPI are a poor measure of "true" inflation, therefore, a major controversy ensued. The report, prepared by a commission headed by Michael Boskin, formerly the chief economic adviser to President George H. W. Bush, concluded that the official CPI inflation rate *overstates* the true inflation rate by as much as one to two percentage points a year. In other words, if the official CPI inflation rate is reported to be 3 percent, the "true" inflation rate might be 2 percent, or even 1 percent.

If this assessment is in fact correct, the indexing of Social Security and other government benefits to the CPI could be costing the federal government billions of dollars more than necessary every year. In addition, an overstated rate of inflation would lead to an underestimation of the true improvement in living standards over time. If the typical family's nominal income increases by 3 percent per year, and inflation is reported to be 3 percent per year, economists would conclude that American families are experiencing no increase in their real income. But if the "true" inflation rate is really 2 percent per year, then the family's real income is actually rising by 1 percent per year (the 3 percent increase in nominal income minus 2 percent inflation).

The Boskin Commission gave a number of reasons why the official inflation rate, based on the CPI, may overestimate the true rate of inflation. Two are particularly important. First, in practice government statisticians cannot always adjust adequately for changes in the *quality* of goods and services. Suppose a new personal computer has 20 percent more memory, computational speed, and data storage capacity than last year's model. Suppose too for the sake of illustration that its price is 20 percent higher. Has there been inflation in computer prices? Economists would say no; although consumers are paying 20 percent more for a computer, they are getting a 20 percent better machine. The situation is really no different from paying 20 percent more for a pizza that is 20 percent bigger. However, because quality change is difficult to measure precisely and because they have many thousands of goods and services to consider, government statisticians often miss or understate changes in quality. In general, whenever statisticians fail to adjust adequately for improvements in the quality of goods or services, they will tend to overstate inflation. This type of overstatement is called *quality adjustment bias.*

An extreme example of quality adjustment bias can occur whenever a totally new good becomes available. For example, the introduction of the first effective AIDS drugs significantly increased the quality of medical care received by AIDS patients. In practice, however, quality improvements that arise from totally new products are likely to be poorly captured by the CPI, if at all. The problem is that since the new good was not produced in the base year, there is no base-year price with which to compare the current price of the good. Government statisticians use various approaches to correct for this problem, such as comparing the cost of the new drug to the cost of the next-best therapies. But such methods are necessarily imprecise and open to criticism.

The second problem emphasized by the Boskin Commission arises from the fact that the CPI is calculated for a fixed basket of goods and services. This procedure does not allow for the possibility that consumers can switch from products whose prices are rising to those whose prices are stable or falling. Ignoring the fact that consumers can switch from more expensive to less expensive goods leads statisticians to overestimate the true increase in the cost of living.

Suppose, for instance, that people like coffee and tea equally well and in the base year consumed equal amounts of each. But then a frost hits a major coffee-producing nation, causing the price of coffee to double. The increase in coffee prices encourages consumers to forgo coffee and drink tea instead—a switch that doesn't make them much worse off, since they like coffee and tea equally well. However, the CPI, which measures the cost of buying the base-year basket of goods and services, will rise significantly when the price of coffee doubles. This rise in the CPI, which ignores the fact that people can substitute tea for coffee without being made significantly worse off, exaggerates the true increase in the cost of living. This type of overstatement of inflation is called *substitution bias.*

EXAMPLE 6.6

**Substitution bias**

Suppose the CPI basket for 1995, the base year, is as follows:

| Item | Expenditure |
|---|---|
| Coffee (50 cups at $1/cup) | $ 50.00 |
| Tea (50 cups at $1/cup) | 50.00 |
| Scones (100 at $1 each) | 100.00 |
| Total | $200.00 |

Assume that consumers are equally happy to drink coffee or tea with their scones. In 1995, coffee and tea cost the same, and the average person drinks equal amounts of coffee and tea.

In the year 2000, coffee has doubled in price to $2 per cup. Tea remains at $1 per cup, and scones are $1.50 each. What has happened to the cost of living as measured by the CPI? How does this result compare to the true cost of living?

To calculate the value of the CPI for the year 2000, we must first find the cost of consuming the 1995 basket of goods in that year. At year 2000 prices, 50 cups each of coffee and tea and 100 scones cost (50 × $2) + (50 × $1) + (100 × $1.50) = $300. Since consuming the same basket of goods cost $200 in 1995, the base year, the CPI in 2000 is $300/$200, or 1.50. This calculation leads us to conclude that the cost of living has increased 50 percent between 1995 and 2000.

However, we have overlooked the possibility that consumers can substitute a cheaper good (tea) for the more expensive one (coffee). Indeed, since consumers like coffee and tea equally well, when the price of coffee doubles they will shift entirely to tea. Their new consumption basket—100 cups of tea and 100 scones—is just as enjoyable to them as their original basket. If we allow for the substitution of less expensive goods, how much has the cost of living really increased? The cost of 100 cups of tea and 100 scones in the year 2000 is only $250, not $300. From the consumer's point of view, the true cost of living has risen by only $50, or 25 percent. The 50 percent increase in the CPI therefore overstates the increase in the cost of living as the result of substitution bias.

The Boskin Commission's findings have been controversial. While quality adjustment bias and substitution bias undoubtedly distort the measurement of inflation, estimating precisely how much of an overstatement they create is difficult. (If economists knew exactly how big these biases were, they could simply correct the data.) But the Bureau of Labor Statistics (the agency responsible for calculating the CPI) has recently made significant efforts to improve the quality of its data as a result of the Commission's report.

ECONOMIC NATURALIST 6.2

**Why is inflation in the health care sector apparently high?**

Government statisticians report inflation rates for different categories of goods and services, as well as for the overall consumer basket. According to the official measures, over recent decades the prices of medical services have tended to rise much more rapidly than the prices of other goods and services. Why is inflation in the health care sector apparently high?

Although inflation rates in the health care sector are high, some economists have argued that reported rates greatly overstate the true rate of inflation in that sector. The reason, claim critics, is the quality adjustment bias. Health care is a dynamic sector of the economy, in which ongoing technological change has significantly improved the quality of care. To the extent that official data fail to account for improvements in the quality of medical care, inflation in the health care sector will be overstated.

Economists Matthew Shapiro and James Wilcox[1] illustrated the problem with the example of changes in the treatment of cataracts, a cloudiness in the lens of the eye that impairs vision. The lens must still be removed surgically, but there have been important improvements in the procedure over the past 30 years. First, surgeons can now replace the defective lens with an artificial one, which improves the patient's vision considerably without contact lenses or thick glasses. Second, the techniques for making and closing the surgical incision have been substantially improved. Besides reducing complications and therefore follow-up visits, the new techniques can be performed in the physician's office, with no hospital stay (older techniques frequently required three nights in the hospital). Thus the new technologies have both improved patient outcomes and reduced the number of hours doctors and nurses spend on the procedure.

Shapiro and Wilcox point out that official measures of health care inflation are based primarily on data such as the doctor's hourly rate or the cost of a night in the hospital. They do not take into account either the reduction in a doctor's time or the shorter hospital stay now needed for procedures such as cataract surgery. Furthermore, Shapiro and Wilcox argue, official measures do not take adequate account of improvements in patient outcomes, such as the improved vision cataract patients now enjoy. Because of the failure to adjust for improvements in the quality of procedures, including increased productivity of medical personnel, official measures may significantly overstate inflation in the health care sector.

## THE COSTS OF INFLATION: NOT WHAT YOU THINK

In the late 1970s, when inflation was considerably higher than it is now, the public told poll takers that they viewed it as "public enemy number one"—that is, as the nation's most serious problem. Although U.S. inflation rates have not been very high in recent years, today many Americans remain concerned about inflation or the threat of inflation. Why do people worry so much about inflation? Detailed opinion surveys often find that many people are confused about the meaning of inflation and its economic effects. Before describing the true economic costs of inflation, which are real and serious, let's examine this confusion people experience about inflation and its costs.

***price level*** a measure of the overall level of prices at a particular point in time as measured by a price index such as the CPI

***relative price*** the price of a specific good or service *in comparison to* the prices of other goods and services

We need first to distinguish between the *price level* and the *relative price* of a good or service. The **price level** is a measure of the overall level of prices at a particular point in time as measured by a price index such as the CPI. Recall that the inflation rate is the percentage change in the price level from year to year. In contrast, a **relative price** is the price of a specific good or service *in comparison to* the prices of other goods and services. For example, if the price of oil were to rise by 10 percent while the prices of other goods and services were rising on average by 3 percent, the relative price of oil would increase. But if oil prices rise by 3 percent while other prices rise by 10 percent, the relative price of oil would decrease. That is, oil would become cheaper relative to other goods and services, even though it has not become cheaper in absolute terms.

Public opinion surveys suggest that many people are confused about the distinction between inflation, or an increase in the overall *price level,* and an increase in a specific *relative price.* Suppose that hostilities in the Middle East were to double the price of gas at the pump, leaving other prices unaffected. Appalled by the increase in gasoline prices, people might demand that the government do something about "this inflation." But while the increase in gas prices hurts consumers, is it an example of inflation? Gasoline is only one item in a consumer's budget, one

[1]"Mismeasurement in the Consumer Price Index: An Evaluation," in Ben Bernanke and Julio Rotemberg (eds.), NBER *Macroeconomics Annual,* 1996.

of the thousands of goods and services that people buy every day. Thus the increase in the price of gasoline might affect the overall price level, and hence the inflation rate, only slightly. In this example, inflation is not the real problem. What upsets consumers is the change in the *relative price* of oil, particularly compared to the price of labor (wages). By increasing the cost of using a car, the increase in the relative price of oil reduces the income people have left over to spend on other things.

Again, changes in relative prices do *not* necessarily imply a significant amount of inflation. For example, increases in the prices of some goods could well be counterbalanced by decreases in the prices of other goods, in which case the price level and the inflation rate would be largely unaffected. Conversely, inflation can be high without affecting relative prices. Imagine, for example, that all prices in the economy, including wages and salaries, go up exactly 10 percent each year. The inflation rate is 10 percent, but relative prices are not changing. Indeed, because wages (the price of labor) are increasing by 10 percent per year, people's ability to buy goods and services is unaffected by the inflation.

These examples show that changes in the average price level (inflation) and changes in the relative prices of specific goods are two quite different issues. The public's tendency to confuse the two is important, because the remedies for the two problems are different. To counteract changes in relative prices, the government would need to implement policies that affect the supply and demand for specific goods. In the case of an increase in oil prices, for example, the government could try to restore supplies by mediating the peace process in the Middle East. To counteract inflation, however, the government must resort (as we will see) to changes in macroeconomic policies, such as monetary or fiscal policies. If, in confusion, the public forces the government to adopt anti-inflationary policies when the real problem is a relative price change, the economy could actually be hurt by the effort. Here is an example of why economic literacy is important, both to policymakers and the general public.

**EXAMPLE 6.7**

**The price level, relative prices, and inflation**

Suppose the value of the CPI is 1.20 in the year 2000, 1.32 in 2001, and 1.40 in 2002. Assume also that the price of oil increases 8 percent between 2000 and 2001 and another 8 percent between 2001 and 2002. What is happening to the price level, the inflation rate, and the relative price of oil?

The price level can be measured by the CPI. Since the CPI is higher in 2001 than in 2000 and higher still in 2002 than in 2001, the price level is rising throughout the period. The inflation rate is the *percentage increase* in the CPI. Since the CPI increases by 10 percent between 2000 and 2001, the inflation rate between those years is 10 percent. However, the CPI increases only about 6 percent between 2001 and 2002 ($1.40/1.32 \approx 1.06$), so the inflation rate decreases to about 6 percent between those years. The decline in the inflation rate implies that although the price level is still rising, it is doing so at a slower pace than the year before.

The price of oil rises 8 percent between 2000 and 2001. But because the general inflation over that period is 10 percent, the relative price of oil—that is, its price *relative to all other goods and services*—falls by about 2 percent ($8\% - 10\% = -2\%$). Between 2001 and 2002 the price of oil rises by another 8 percent, while the general inflation rate is about 6 percent. Hence the relative price of oil rises between 2001 and 2002 by about 2 percent ($8\% - 6\%$).

## THE TRUE COSTS OF INFLATION

Having dispelled the common confusion between inflation and relative price changes, we are now free to address the true economic costs of inflation. There are a variety of such costs, each of which tends to reduce the efficiency of the economy. Five of the most important are discussed here.

## "SHOE-LEATHER" COSTS

As all shoppers know, cash is convenient. Unlike checks, which are not accepted everywhere, and credit cards, for which a minimum purchase is often required, cash can be used in almost any routine transaction. Businesses, too, find cash convenient to hold. Having plenty of cash on hand facilitates transactions with customers and reduces the need for frequent deposits and withdrawals from the bank.

Inflation raises the cost of holding cash to consumers and businesses. Consider a miser with $10,000 in $20 bills under his mattress. What happens to the buying power of his hoard over time? If inflation is zero so that on average the prices of goods and services are not changing, the buying power of the $10,000 does not change over time. At the end of a year the miser's purchasing power is the same as it was at the beginning of the year. But suppose the inflation rate is 10 percent. In that case, the purchasing power of the miser's hoard will fall by 10 percent each year. After a year, he will have only $9,000 in purchasing power. In general, the higher the rate of inflation, the less people will want to hold cash because of the loss of purchasing power that they will suffer.

Technically, currency is a debt owed by the government to the currency holder. So when currency loses value, the losses to holders of cash are offset by gains to the government, which now owes less in real terms to currency holders. Thus, from the point of view of society as a whole, the loss of purchasing power is not in itself a cost of inflation, because it does not involve wasted resources. (Indeed, no real goods or services were used up when the miser's currency hoard lost part of its value.) However, when faced with inflation, people are not likely to accept a loss in purchasing power but instead will take actions to try to "economize" on their cash holdings. For example, instead of drawing out enough cash for a month the next time they visit the bank, they will draw out only enough to last a week. The inconvenience of visiting the bank more often to minimize one's cash holdings is a real cost of inflation. Similarly, businesses will reduce their cash holdings by sending employees to the bank more frequently, or by installing computerized systems to monitor cash usage. To deal with the increase in bank transactions required by consumers and businesses trying to use less cash, banks will need to hire more employees and expand their operations.

The costs of more frequent trips to the bank, new cash management systems, and expanded employment in banks are real costs. They use up resources, including time and effort, that could be used for other purposes. Traditionally, the costs of economizing on cash have been called *shoe-leather costs*—the idea being that shoe leather is worn out during extra trips to the bank. Shoe-leather costs probably are not a significant problem in the United States today, where inflation is only 2 to 3 percent per year. But in economies with high rates of inflation, they can become quite significant.

**EXAMPLE 6.8** **Shoe-leather costs at Woodrow's Hardware**

Woodrow's Hardware needs $5,000 cash per day for customer transactions. Woodrow has a choice between going to the bank first thing on Monday morning to withdraw $25,000—enough cash for the whole week—or going to the bank first thing every morning for $5,000 each time. Woodrow puts the cost of going to the bank, in terms of inconvenience and lost time, at $4 per trip. Assume that funds left in the bank earn precisely enough interest to keep their purchasing power unaffected by inflation.

If inflation is zero, how often will Woodrow go to the bank? If it is 10 percent? In this example, what are the shoe-leather costs of a 10 percent inflation rate?

If inflation is zero, there is no cost to holding cash. Woodrow will go to the bank only once a week, incurring a shoe-leather cost of $4 per week. But if inflation is 10 percent, Woodrow may need to change his banking habits. If he continues to go to the bank only on Monday mornings, withdrawing $25,000 for

the week, what will be Woodrow's average cash holding over the week? At the beginning of each day, his cash holding will be as follows:

|  |  |
|---|---|
| Monday | $25,000 |
| Tuesday | 20,000 |
| Wednesday | 15,000 |
| Thursday | 10,000 |
| Friday | 5,000 |

Averaging the holdings on those five days, we can calculate that Woodrow's average cash holding at the beginning of each day is $75,000/5 = $15,000. If inflation is 10 percent a year, over the course of a year the cost to Woodrow of holding an average of $15,000 in cash equals 10 percent of $15,000, or $1,500.

On the other hand, if Woodrow goes to the bank every day, his average cash holding at the beginning of the day will be only $5,000. In that case, his losses from inflation will be $500 (10 percent of $5,000) a year. Will Woodrow start going to the bank every day when inflation reaches 10 percent? The *benefit* of changing his banking behavior is a loss of only $500 per year to inflation, rather than $1,500, or $1,000 saved. The *cost* of going to the bank every day is $4 per trip. Assuming Woodrow's store is open 50 weeks a year, going to the bank 5 days a week instead of 1 day a week adds 200 trips per year, at a total cost of $800. Since the $800 cost is less than the $1,000 benefit, Woodrow will begin going to the bank more often.

To repeat, the shoe-leather costs of a high inflation rate are the extra costs incurred to avoid holding cash. In this example they are the additional $800 per year associated with Woodrow's daily trips to the bank.

## "NOISE" IN THE PRICE SYSTEM

In Chapter 3 we described the remarkable economic coordination that is necessary to provide the right amount and the right kinds of food to New Yorkers every day. This feat is not orchestrated by some Food Distribution Ministry staffed by bureaucrats. It is done much better than a Ministry ever could by the workings of free markets, operating without central guidance.

How do free markets transmit the enormous amounts of information necessary to accomplish complex tasks like the provisioning of New York City? The answer, as we saw in Chapter 3, is through the price system. When the owners of French restaurants in Manhattan cannot find sufficient quantities of chanterelles, a particularly rare and desirable mushroom, they bid up its market price. Specialty food suppliers notice the higher price for chanterelles and realize that they can make a profit by supplying more chanterelles to the market. At the same time, price-conscious diners will shift to cheaper, more available mushrooms. The market for chanterelles will reach equilibrium only when there are no more unexploited opportunities for profit, and both suppliers and demanders are satisfied at the market price (the *equilibrium principle*). Multiply this example a million times, and you will gain a sense of how the price system achieves a truly remarkable degree of economic coordination.

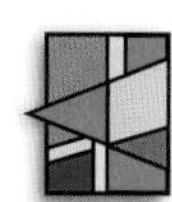

When inflation is high, however, the subtle signals that are transmitted through the price system become more difficult to interpret, much in the way that static, or "noise," makes a radio message harder to interpret. In an economy with little or no inflation, the supplier of specialty foodstuffs will immediately recognize the increase in chanterelle prices as a signal to bring more to market. If inflation is high, however, the supplier must ask whether a price increase represents a true increase in the demand for chanterelles or is just a result of the general inflation, which causes all food prices to rise. If the price rise reflects only inflation, the price of chanterelles *relative to other goods and services* has not really

Inflation adds static to the information conveyed by changes in prices.

changed. The supplier therefore should not change the quantity of mushrooms he brings to market.

In an inflationary environment, to discern whether the increase in chanterelle prices is a true signal of increased demand, the supplier needs to know not only the price of chanterelles but also what is happening to the prices of other goods and services. Since this information takes time and effort to collect, the supplier's response to the change in chanterelle prices is likely to be slower and more tentative.

In summary, price changes are the market's way of communicating information to suppliers and demanders. An increase in the price of a good or service, for example, tells demanders to economize on their use of the good or service and suppliers to bring more of it to market. But in the presence of inflation, prices are affected not only by changes in the supply and demand for a product but by changes in the general price level. Inflation creates static, or "noise," in the price system, obscuring the information transmitted by prices and reducing the efficiency of the market system. This reduction in efficiency imposes real economic costs.

## DISTORTIONS OF THE TAX SYSTEM

Just as some government expenditures, such as Social Security benefits, are indexed to inflation, many taxes are also indexed. In the United States, people with higher incomes pay a higher *percentage* of their income in taxes. Without indexing, an inflation that raises people's nominal incomes would force them to pay an increasing percentage of their income in taxes, even though their *real* incomes may not have increased. To avoid this phenomenon, which is known as *bracket creep,* Congress has indexed income tax brackets to the CPI. The effect of this indexation is that a family whose nominal income is rising at the same rate as inflation does not have to pay a higher percentage of income in taxes.

Although indexing has solved the problem of bracket creep, many provisions of the tax code have not been indexed, either because of lack of political support or because of the complexity of the task. As a result, inflation can produce unintended changes in the taxes people pay, which in turn may cause them to change their behavior in economically undesirable ways.

To illustrate, an important provision in the business tax code for which inflation poses problems is the *capital depreciation allowance,* which works as follows. Suppose a firm buys a machine for $1,000, expecting it to last for 10 years. Under U.S. tax law, the firm can take one-tenth of the purchase price, or $100, as a deduction from its taxable profits in each of the 10 years. By deducting a fraction of the purchase price from its taxable profits, the firm reduces its taxes. The exact amount of the yearly tax reduction is the tax rate on corporate profits times $100.

The idea behind this provision of the tax code is that the wearing out of the machine is a cost of doing business that should be deducted from the firm's profit. Also, in giving firms a tax break for investing in new machinery, Congress intended to encourage firms to modernize their plants. Yet capital depreciation allowances are not indexed to inflation. Suppose that, at a time when the inflation rate is high, a firm is considering purchasing a $1,000 machine. The managers know that the purchase will allow them to deduct $100 per year from taxable profits for the next 10 years. But that $100 is a fixed amount that is not indexed to inflation. Looking forward, managers will recognize that 5, 6, or 10 years into the future, the real value of the $100 tax deduction will be much lower than at present because of inflation. They will have less incentive to buy the machine and may decide not to make the investment at all. Indeed, many studies have found that a high rate of inflation can significantly reduce the rate at which firms invest in new factories and equipment.

Because the complex U.S. tax code contains hundreds of provisions and tax rates that are not indexed, inflation can seriously distort the incentives provided by the tax system for people to work, save, and invest. The resulting adverse effects on economic efficiency and economic growth represent a real cost of inflation.

## UNEXPECTED REDISTRIBUTION OF WEALTH

Yet another concern about inflation is that it may arbitrarily redistribute wealth from one group to another. Consider a group of union workers who signed a contract setting their wages for the next three years. If those wages are not indexed to inflation, then the workers will be vulnerable to upsurges in the price level. Suppose, for example, that inflation is much higher than expected over the three years of the contract. In that case the buying power of the workers' wages—their real wages—will be less than anticipated when they signed the contract.

From society's point of view, is the buying power that workers lose to inflation really "lost"? The answer is no; the loss in their buying power is exactly matched by an unanticipated gain in the employer's buying power, because the real cost of paying the workers is less than anticipated. In other words, the effect of the inflation is not to *destroy* purchasing power but to *redistribute* it, in this case from the workers to the employer. If inflation had been *lower* than expected, the workers would have enjoyed greater purchasing power than they anticipated and the employer would have been the loser.

Another example of the redistribution caused by inflation takes place between borrowers (debtors) and lenders (creditors). Suppose one of the authors of this book wants to buy a house on a lake and borrows $150,000 from the bank to pay for it. Shortly after signing the mortgage agreement, he learns that inflation is likely to be much higher than expected. How should he react to the news? Perhaps as a public-spirited macroeconomist the author should be saddened to hear that inflation is rising, but as a consumer he should be pleased. In real terms, the dollars with which he will repay his loan in the future will be worth much less than expected. The loan officer should be distraught, because the dollars the bank will receive from the author will be worth less, in purchasing power terms, than expected at contract signing. Once again, no real wealth is "lost" to the inflation; rather, the borrower's gain is just offset by the lender's loss. *In general, unexpectedly high inflation rates help borrowers at the expense of lenders,* because borrowers are able to repay their loans in less valuable dollars. Unexpectedly low inflation rates, in contrast, help lenders and hurt borrowers by forcing borrowers to repay in dollars that are worth more than expected when the loan was made.

Although redistributions caused by inflation do not directly destroy wealth, but only transfer it from one group to another, they are still bad for the economy. Our economic system is based on incentives. For it to work well, people must know that if they work hard, save some of their income, and make wise financial investments, they will be rewarded in the long run with greater real wealth and a better standard of living. Some observers have compared a high-inflation economy to a casino, in which wealth is distributed largely by luck—that is, by random fluctuations in the inflation rate. In the long run, a "casino economy" is likely to perform poorly, as its unpredictability discourages people from working and saving. (Why bother if inflation can take away your savings overnight?) Rather, a high-inflation economy encourages people to use up resources in trying to anticipate inflation and protect themselves against it.

### INTERFERENCE WITH LONG-RUN PLANNING

The fifth and final cost of inflation we will examine is its tendency to interfere with the long-run planning of households and firms. Many economic decisions take place within a long time horizon. Planning for retirement, for example, may begin when workers are in their twenties or thirties. And firms develop long-run investment and business strategies that look decades into the future.

Clearly, high and erratic inflation can make long-term planning difficult. Suppose, for example, that you want to enjoy a certain standard of living when you retire. How much of your income do you need to save to make your dreams a reality? That depends on what the goods and services you plan to buy will cost 30 or 40 years from now. With high and erratic inflation, even guessing what your chosen lifestyle will cost by the time you retire is extremely difficult. You may end up saving too little and having to compromise on your retirement plans; or you may save too much, sacrificing more than you need to during your working years. Either way, inflation will have proved costly.

In summary, inflation damages the economy in a variety of ways. Some of its effects are difficult to quantify and are therefore controversial. But most economists agree that a low and stable inflation rate is instrumental in maintaining a healthy economy.

### THE TRUE COSTS OF INFLATION

The public sometimes confuses changes in relative prices (such as the price of oil) with inflation, which is a change in the overall level of prices. This confusion can cause problems, because the remedies for undesired changes in relative prices and for inflation are different.

There are a number of true costs of inflation, which together tend to reduce economic growth and efficiency. These include:

- Shoe-leather costs, or the costs of economizing on cash (for example, by making more frequent trips to the bank or installing a computerized cash management system).
- "Noise" in the price system, which occurs when general inflation makes it difficult for market participants to interpret the information conveyed by prices.
- Distortions of the tax system, for example, when provisions of the tax code are not indexed.
- Unexpected redistributions of wealth, as when higher-than-expected inflation hurts wage earners to the benefit of employers or hurts creditors to the benefit of debtors.
- Interference with long-term planning, arising because people find it difficult to forecast prices over long periods.

## HYPERINFLATION

***hyperinflation*** a situation in which the inflation rate is extremely high

Although there is some disagreement about whether an inflation rate of, say, 5 percent per year imposes important costs on an economy, few economists would question the fact that an inflation rate of 500 percent or 1,000 percent per year disrupts economic performance. A situation in which the inflation rate is extremely high is called **hyperinflation.** Although there is no official threshold above which inflation becomes hyperinflation, inflation rates in the range of 500 to 1,000 percent per year would surely qualify. In the past few decades episodes of hyperinflation have occurred in several Latin American countries (including Bolivia, Argentina, and Brazil), in Israel, and in several countries attempting to make the transition from communism to capitalism, including Russia. The United States has never experienced hyperinflation, although the short-lived Confederate States of America suffered severe inflation during the Civil War. Between 1861 and 1865 prices in the Confederacy rose to 92 times their prewar levels.

Hyperinflation greatly magnifies the costs of inflation. For example, shoe-leather costs—a relatively minor consideration in times of low inflation—become quite important during hyperinflation, when people may visit the bank two or three times per day to hold money for as short a time as possible. With prices changing daily or even hourly, markets work quite poorly, slowing economic growth. Massive redistributions of wealth take place, impoverishing many. Not surprisingly, episodes of hyperinflation rarely last more than a few years; they are so disruptive that they quickly lead to public outcry for relief.

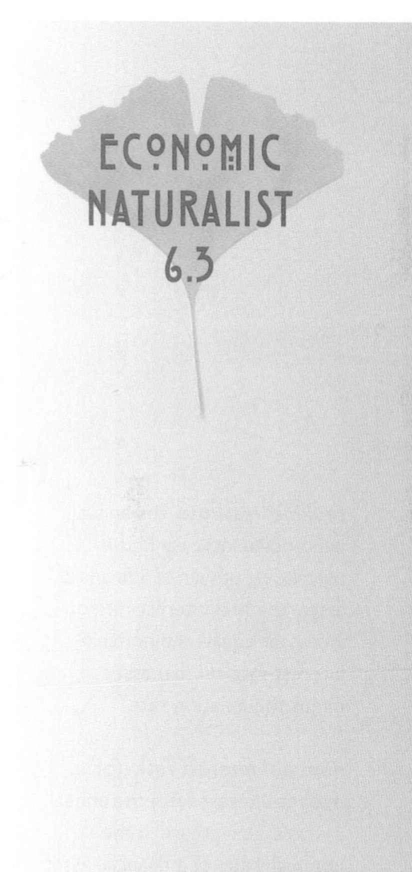


**How costly is high inflation?**

Economic theory suggests that high inflation rates, especially those associated with hyperinflation, reduce economic efficiency and growth. Most economists believe that the economic costs associated with high inflation outweigh the perceived benefits, yet we continue to see episodes of high inflation throughout the world. In reality, how costly are high inflation rates?

Economists Stanley Fischer, Ratna Sahay, and Carlos A. Végh[2] examined the economic performance of 133 market economies over the period 1960–96 and uncovered 45 episodes of high inflation (12-month inflation rates greater than 100%) among 25 different countries. They found that, while uncommon, episodes of high inflation impose significant economic costs on the countries experiencing them. During periods of high inflation, these countries saw real GDP per person fall by an average of 1.6% per year, real consumption per person fall by an average of 1.3% per year, and real investment per person fall by an average of 3.3% per year. During low inflation years these same countries experienced positive growth in each of these variables. In addition, during periods of high inflation, these countries' trade and government budget deficits were larger than during low inflation years.

Falling output and consumption levels caused by high inflation reduce the economic well-being of households and firms, and have a disproportionate effect on poor workers, who are least likely to have their wages indexed to the inflation rate and thus avoid a real loss in purchasing power. As pointed out in the last section, high inflation rates also distort relative prices in the marketplace, leading to a misallocation of resources that can have long-term economic consequences. Falling investment in new capital caused by high inflation, for example, leads not only to a slowdown in current economic activity but also to reduced growth rates of future output. Because of these adverse economic effects, policymakers have an incentive to keep inflation rates low.

## INFLATION AND INTEREST RATES

So far we have focused on the measurement and economic costs of inflation. Another important aspect of inflation is its close relationship to other key macroeconomic variables. For example, economists have long realized that during periods of high inflation interest rates tend to be high as well. We will close this chapter with a look at the relationship between inflation and interest rates, which will provide a useful background in the chapters to come.

[2]"Modern Hyper- and High Inflations," *Journal of Economic Literature,* Vol. 11 (September 2002), pp. 837–880.

## INFLATION AND THE REAL INTEREST RATE

Earlier in our discussion of the ways in which inflation redistributes wealth, we saw that inflation tends to hurt creditors and help debtors by reducing the value of the dollars with which debts are repaid. The effect of inflation on debtors and creditors can be explained more precisely using an economic concept called the *real interest rate.* An example will illustrate.

Suppose that there are two neighboring countries, Alpha and Beta. In Alpha, whose currency is called the alphan, the inflation rate is zero and is expected to remain at zero. In Beta, where the currency is the betan, the inflation rate is 10 percent and is expected to remain at that level. Bank deposits pay 2 percent annual interest in Alpha and 10 percent annual interest in Beta. In which countries are bank depositors getting a better deal?

You may answer "Beta," since interest rates on deposits are higher in that country. But if you think about the effects of inflation, you will recognize that Alpha, not Beta, offers the better deal to depositors. To see why, think about the change over a year in the real purchasing power of deposits in the two countries. In Alpha, someone who deposits 100 alphans in the bank on January 1 will have 102 alphans on December 31. Because there is no inflation in Alpha, on average prices are the same at the end of the year as they were at the beginning. Thus the 102 alphans the depositor can withdraw represent a 2 percent increase in buying power.

In Beta, the depositor who deposits 100 betans on January 1 will have 110 betans by the end of the year—10 percent more than she started with. But the prices of goods and services in Beta, we have assumed, will also rise by 10 percent. Thus the Beta depositor can afford to buy precisely the same amount of goods and services at the end of the year as she could at the beginning; she gets no increase in buying power. So the Alpha depositor has the better deal, after all.

***real interest rate*** the annual percentage increase in the purchasing power of a financial asset; the real interest rate on any asset equals the nominal interest rate on that asset minus the inflation rate

***nominal interest rate*** (or market interest rate) the annual percentage increase in the nominal value of a financial asset

Economists refer to the annual percentage increase in the *real* purchasing power of a financial asset as the **real interest rate,** or the *real rate of return,* on that asset. In our example, the real purchasing power of deposits rises by 2 percent per year in Alpha and by 0 percent per year in Beta. So the real interest rate on deposits is 2 percent in Alpha and 0 percent in Beta. The real interest rate should be distinguished from the more familiar market interest rate, also called the *nominal interest rate.* The **nominal interest rate** is the annual percentage increase in the nominal, or dollar, value of an asset.

As the example of Alpha and Beta illustrates, we can calculate the real interest rate for any financial asset, from a checking account to a government bond, by subtracting the rate of inflation from the market or nominal interest rate on that asset. So in Alpha, the real interest rate on deposits equals the nominal interest rate (2 percent) minus the inflation rate (0 percent), or 2 percent. Likewise in Beta, the real interest rate equals the nominal interest rate (10 percent) minus the inflation rate (10 percent), or 0 percent.

We can write this definition of the real interest rate in mathematical terms:

$$r = i - \pi,$$

where $r$ = the real interest rate,
$i$ = the nominal, or market, interest rate,
$\pi$ = the inflation rate.

**EXAMPLE 6.9**

### Real interest rates in the 1970s, 1980s, and 1990s

Following are interest rates on government bonds for selected years in the 1970s, 1980s, and 1990s. In which of these years did the financial investors who bought government bonds get the best deal? The worst deal?

| Year | Interest rate (%) | Inflation rate (%) |
|---|---|---|
| 1970 | 6.5 | 5.7 |
| 1975 | 5.8 | 9.1 |
| 1980 | 11.5 | 13.5 |
| 1985 | 7.5 | 3.6 |
| 1990 | 7.5 | 5.4 |
| 1995 | 5.5 | 2.8 |
| 1999 | 4.7 | 2.5 |

Financial investors and lenders do best when the real (not the nominal) interest rate is high, since the real interest rate measures the increase in their purchasing power. We can calculate the real interest rate for each year by subtracting the inflation rate from the nominal interest rate. The results are 0.8 percent for 1970, −3.3 percent for 1975, −2.0 percent for 1980, 3.9 percent for 1985, 2.1 percent for 1990, 2.7 percent for 1995, and 2.2 percent for 1999. For purchasers of government bonds, the best of these years was 1985, when they enjoyed a real return of 3.9 percent. The worst year was 1975, when their real return was actually negative. In other words, despite receiving 5.8 percent nominal interest, financial investors ended up losing buying power in 1975, as the inflation rate exceeded the interest rate earned by their investments.

Figure 6.2 shows the real interest rate in the United States since 1960 as measured by the nominal interest rate paid on the federal government's debt minus the inflation rate. Note that the real interest rate was negative in the 1970s but reached historically high levels in the mid-1980s.

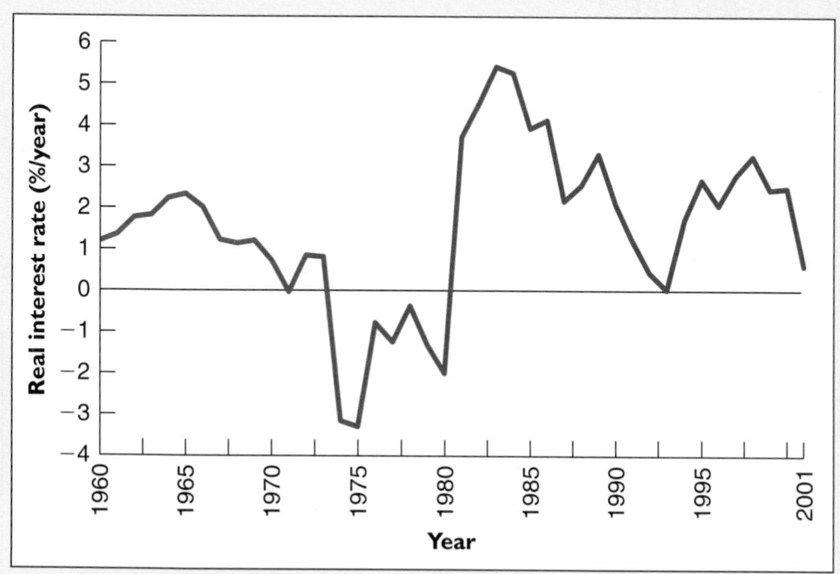


SOURCE: *Economic Report of the President* (http://w3.access.gpo.gov/eop/) and authors' calculations.

**FIGURE 6.2**
**The Real Interest Rate in the United States, 1960–2001.**
The real interest rate is the nominal interest rate—here the interest rate on funds borrowed by the federal government for a term of 3 months—minus the rate of inflation. In the United States, the real interest rate was negative in the 1970s but reached historically high levels in the mid-1980s.

## EXERCISE 6.7

**You have some funds to invest but are unimpressed with the low interest rates your bank offers. You consult a broker, who suggests a bond issued by the government of a small island nation. The broker points out that these bonds pay 25 percent interest—much more than your bank—and**

**that the island's government has never failed to repay its debts. What should be your next question?**

The concept of the real interest rate helps to explain more precisely why an unexpected surge in inflation is bad for lenders and good for borrowers. For any given nominal interest rate that the lender charges the borrower, the higher the inflation rate, the lower the real interest rate the lender actually receives. So unexpectedly high inflation leaves the lender worse off. Borrowers, on the other hand, are better off when inflation is unexpectedly high, because their real interest rate is lower than anticipated.

Although unexpectedly high inflation hurts lenders and helps borrowers, a high rate of inflation that is *expected* may not redistribute wealth at all, because expected inflation can be built into the nominal interest rate. Suppose, for example, that the lender requires a real interest rate of 2 percent on new loans. If the inflation rate is confidently expected to be zero, the lender can get a 2 percent real interest rate by charging a nominal interest rate of 2 percent. But if the inflation rate is expected to be 10 percent, the lender can still ensure a real interest rate of 2 percent by charging a nominal interest rate of 12 percent. Thus high inflation, if it is *expected,* need not hurt lenders—as long as the lenders can adjust the nominal interest they charge to reflect the expected inflation rate.

**EXERCISE 6.8**

**What is the real rate of return to holding cash? (*Hint:* Does cash pay interest?) Does this real rate of return depend on whether the rate of inflation is correctly anticipated? How does your answer relate to the idea of shoe-leather costs?**

## THE FISHER EFFECT

Earlier we mentioned the observation that interest rates tend to be high when inflation is high and low when inflation is low. This relationship can be seen in Figure 6.3, which shows both the U.S. inflation rate and a nominal interest rate (the rate at which the government borrows for short periods) from 1960 to the present. Notice that nominal interest rates have tended to be high in periods of high inflation, such as the late 1970s, and relatively low in periods of low inflation, such as the early 1960s and the late 1990s.

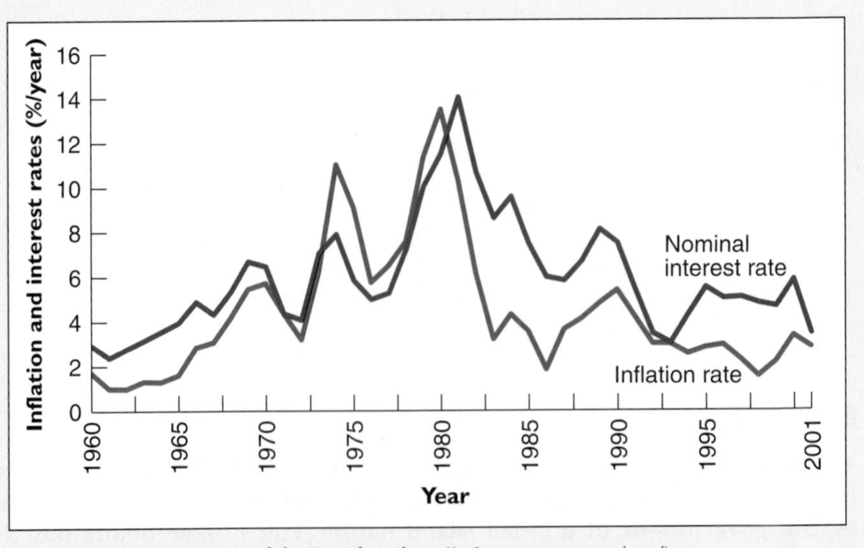


**FIGURE 6.3**
**Inflation and Interest Rates in the United States, 1960–2001.**
Nominal interest rates tend to be high when inflation is high and low when inflation is low, a phenomenon called the Fisher effect.

SOURCE: *Economic Report of the President* (http://w3.access.gpo.gov/eop/).

Why do interest rates tend to be high when inflation is high? Our discussion of real interest rates provides the answer. Suppose inflation has recently been high, so borrowers and lenders anticipate that it will be high in the near future. We would expect lenders to raise their nominal interest rate so that their real rate of return will be unaffected. For their part, borrowers are willing to pay higher nominal interest rates when inflation is high, because they understand that the higher nominal interest rate only serves to compensate the lender for the fact that the loan will be repaid in dollars of reduced real value—in real terms, their cost of borrowing is unaffected by an equal increase in the nominal interest rate and the inflation rate. Conversely, when inflation is low, lenders do not need to charge so high a nominal interest rate to ensure a given real return. Thus nominal interest rates will be high when inflation is high and low when inflation is low. This tendency for nominal interest rates to follow inflation rates is called the **Fisher effect,** after the early twentieth-century American economist Irving Fisher, who first pointed out the relationship.

***Fisher effect*** the tendency for nominal interest rates to be high when inflation is high and low when inflation is low

## SUMMARY

- The basic tool for measuring inflation is the *consumer price index,* or CPI. The CPI measures the cost of purchasing a fixed basket of goods and services in any period relative to the cost of the same basket of goods and services in a base year. The *inflation rate* is the annual percentage rate of change in the price level as measured by a *price index* such as the CPI.

- The official U.S. inflation rate, based on the CPI, may overstate the true inflation rate for two reasons: First, it may not adequately reflect improvements in the quality of goods and services. Second, the method of calculating the CPI ignores the fact that consumers can substitute cheaper goods and services for more expensive ones.

- A *nominal quantity* is a quantity that is measured in terms of its current dollar value. Dividing a nominal quantity, such as a family's income or a worker's wage in dollars, by a price index, such as the CPI, expresses that quantity in terms of real purchasing power. This procedure is called *deflating* the nominal quantity. If nominal quantities from two different years are deflated by a common price index, the purchasing power of the two quantities can be compared. To ensure that a nominal payment, such as a Social Security benefit, represents a constant level of real purchasing power, the nominal payment should be increased each year by a percentage equal to the inflation rate. This method of adjusting nominal payments to maintain their purchasing power is called *indexing.*

- The public sometimes confuses increases in the *relative prices* for specific goods or services with inflation, which is an increase in the general price level. Since the remedies for a change in relative prices are different from the remedies for inflation, this confusion can cause problems.

- Inflation imposes a number of true costs on the economy, including "shoe-leather" costs, which are the real resources that are wasted as people try to economize on cash holdings; "noise" in the price system; distortions of the tax system; unexpected redistributions of wealth; and interference with long-run planning. Because of these costs, most economists agree that sustained economic growth is more likely if inflation is low and stable. *Hyperinflation,* a situation in which the inflation rate is extremely high, greatly magnifies the costs of inflation and is highly disruptive to the economy.

- The *real interest rate* is the annual percentage increase in the purchasing power of a financial asset. It is equal to the *nominal,* or *market, interest rate* minus the inflation rate. When inflation is unexpectedly high, the real interest rate is lower than anticipated, which hurts lenders but benefits borrowers. When inflation is unexpectedly low, lenders benefit and borrowers are hurt. To obtain a given real rate of return, lenders must charge a high nominal interest rate when inflation is high and a low nominal interest rate when inflation is low. The tendency for nominal interest rates to be high when inflation is high and low when inflation is low is called the *Fisher effect.*

## KEY TERMS

## REVIEW QUESTIONS

1. Explain why changes in the cost of living for any particular individual or family may differ from changes in the official cost-of-living index, the CPI.
2. What is the difference between the *price level* and the *rate of inflation* in an economy?
3. Why is it important to adjust for inflation when comparing nominal quantities (for example, workers' average wages) at different points in time? What is the basic method for adjusting for inflation?
4. Describe how indexation might be used to guarantee that the purchasing power of the wage agreed to in a multi-year labor contract will not be eroded by inflation.
5. Give two reasons why the official inflation rate may understate the "true" rate of inflation. Illustrate by examples.
6. "It's true that unexpected inflation redistributes wealth, from creditors to debtors, for example. But what one side of the bargain loses, the other side gains. So from the perspective of the society as a whole, there is no real cost." Do you agree? Discuss.
7. How does inflation affect the real return on holding cash?
8. True or false: If both the potential lender and the potential borrower correctly anticipate the rate of inflation, inflation will not redistribute wealth from the creditor to the debtor. Explain.

## PROBLEMS

1. Government survey takers determine that typical family expenditures each month in the year designated as the base year are as follows:

   20 pizzas at $10 each
   Rent of apartment, $600 per month
   Gasoline and car maintenance, $100
   Phone service (basic service plus 10 long-distance calls), $50

   In the year following the base year, the survey takers determine that pizzas have risen to $11 each, apartment rent is $640, gasoline and maintenance has risen to $120, and phone service has dropped in price to $40.
   a. Find the CPI in the subsequent year and the rate of inflation between the base year and the subsequent year.
   b. The family's nominal income rose by 5 percent between the base year and the subsequent year. Are they worse off or better off in terms of what their income is able to buy?

2. Here are values of the CPI (multiplied by 100) for each year from 1990 to 2001. For each year beginning with 1991, calculate the rate of inflation from the previous year. What happened to inflation rates over the 1990s?

| Year | CPI |
|---|---|
| 1990 | 130.7 |
| 1991 | 136.2 |
| 1992 | 140.3 |
| 1993 | 144.5 |
| 1994 | 148.2 |
| 1995 | 152.4 |
| 1996 | 156.9 |
| 1997 | 160.5 |
| 1998 | 163.0 |
| 1999 | 166.6 |
| 2000 | 172.2 |
| 2001 | 177.1 |

3. According to the U.S. Census Bureau (http://www.census.gov/), nominal income for the typical family of four in the United States (median income) was $24,332 in 1980, $32,777 in 1985, $41,451 in 1990, and $62,228 in 2000. In purchasing power terms, how did family income compare in each of those four years? You will need to know that the CPI (multiplied by 100, 1982–1984 = 100) was 82.4 in 1980, 107.6 in 1985, 130.7 in 1990, and 172.2 in 2000. In general terms, how would your answer be affected if the Boskin Commission's conclusions about the CPI were confirmed?

4. A recent report found that the real entry-level wage for college graduates declined by 8 percent between 1990 and 1997. The nominal entry-level wage in 1997 was $13.65 per hour. Assuming that the findings are correct, what was the nominal entry-level wage in 1990? You will need to use data from Problem 2.

5. Here is a hypothetical income tax schedule, expressed in nominal terms, for the year 2002:

| Family income | Taxes due (percent of income) |
|---|---|
| ≤$20,000 | 10 |
| $20,001–$30,000 | 12 |
| $30,001–$50,000 | 15 |
| $50,001–$80,000 | 20 |
| >$80,000 | 25 |

The legislature wants to ensure that families with a given real income are not pushed up into higher tax brackets by inflation. The CPI (times 100) is 175 in 2002 and 185 in 2004. How should the income tax schedule above be adjusted for the year 2004 to meet the legislature's goal?

6. The typical consumer's food basket in the base year 2000 is as follows:

30 chickens at $3.00 each
10 hams at $6.00 each
10 steaks at $8.00 each

A chicken feed shortage causes the price of chickens to rise to $5.00 each in the year 2001. Hams rise to $7.00 each, and the price of steaks is unchanged.

a. Calculate the change in the "cost-of-eating" index between 2000 and 2001.
b. Suppose that consumers are completely indifferent between two chickens and one ham. For this example, how large is the substitution bias in the official "cost-of-eating" index?

7. Here are the actual per-gallon prices for unleaded regular gasoline for June of each year between 1978 and 1986, together with the values of the CPIs for those years. For each year from 1979 to 1986, find the CPI inflation rate and the change in the relative price of gasoline, both from the previous year. Would it be fair to say that most of the changes in gas prices during this period were due to general inflation, or were factors specific to the oil market playing a role as well?

| Year | Gasoline price ($/gallon) | CPI (1982–1984 = 1.00) |
|---|---|---|
| 1978 | 0.663 | 0.652 |
| 1979 | 0.901 | 0.726 |
| 1980 | 1.269 | 0.824 |
| 1981 | 1.391 | 0.909 |
| 1982 | 1.309 | 0.965 |
| 1983 | 1.277 | 0.996 |
| 1984 | 1.229 | 1.039 |
| 1985 | 1.241 | 1.076 |
| 1986 | 0.955 | 1.136 |

8. Repeat Example 6.8 from the text (shoe-leather costs at Woodrow's Hardware). Calculate shoe-leather costs (relative to the original situation, in which Woodrow goes to the bank once a week) assuming that
   a. Inflation is 5 percent rather than 10 percent.
   b. Inflation is 5 percent and Woodrow's trips to the banks cost $2 each.
   c. Inflation remains at 10 percent and a trip to the bank costs $4, but Woodrow needs $10,000 per day to transact with customers.

9. On January 1, 2000, Albert invested $1,000 at 6 percent interest per year for three years. The CPI on January 1, 2000, stood at 100. On January 1, 2001, the CPI (times 100) was 105, on January 1, 2002, it was 110, and on January 1, 2003, the day Albert's investment matured, the CPI was 118. Find the real rate of interest earned by Albert in each of the three years and his total real return over the three-year period. Assume that interest earnings are reinvested each year and themselves earn interest.

10. Frank is lending $1,000 to Sarah for two years. Frank and Sarah agree that Frank should earn a 2 percent real return per year.
    a. The CPI (times 100) is 100 at the time that Frank makes the loan. It is expected to be 110 in one year and 121 in two years. What nominal rate of interest should Frank charge Sarah?
    b. Suppose Frank and Sarah are unsure about what the CPI will be in two years. Show how Frank and Sarah could index Sarah's annual repayments to ensure that Frank gets an annual 2 percent real rate of return.

11.* The Bureau of Labor Statistics has found that the base-year expenditures of the typical consumer break down as follows:

| | |
|---|---|
| Food and beverages | 17.8% |
| Housing | 42.8% |
| Apparel and upkeep | 6.3% |
| Transportation | 17.2% |
| Medical care | 5.7% |
| Entertainment | 4.4% |
| Other goods, services | 5.8% |
| Total | 100.0% |

Suppose that since the base year the prices of food and beverages have increased by 10 percent, the price of housing has increased by 5 percent, and the price of medical care has increased by 10 percent. Other prices are unchanged. Find the CPI for the current year.

## ANSWERS TO IN-CHAPTER EXERCISES

6.1 The cost of the family's basket in 1995 remains at $680, as in Table 6.1. If the rent on their apartment falls to $400 in 2000, the cost of reproducing the 1995 basket of goods and services in 2000 is $620 ($400 for rent + $150 for hamburgers + $70 for movie tickets). The CPI for 2000 is accordingly $620/$680, or 0.912. So in this example, the cost of living fell nearly 9 percent between 1995 and 2000.

6.2 To construct your own personal price index, you would need to determine the basket of goods and services that you personally purchased in the base year. Your personal price index in each period would then be defined as the cost of your personal basket in that period relative to its cost in the base year. To the extent that your mix of purchases differs from that of the typical American consumer, your cost-of-living index will differ from the official CPI. For example, if in the base year you spent a higher share of your budget than the typical American on goods and services that have risen relatively rapidly in price, your personal inflation rate will be higher than the CPI inflation rate.

*Problem marked with an asterisk (*) is more difficult.

6.3 The percentage changes in the CPI in each year from the previous year are as follows:

| | |
|---|---|
| 1930 | −2.3% = (0.167 − 0.171)/0.171 |
| 1931 | −9.0% |
| 1932 | −9.9% |
| 1933 | −5.1% |

Negative inflation is called deflation. The experience of the 1930s, when prices were falling, contrasts sharply with the 1970s, during which prices rose rapidly.

6.4 Bonds's real earnings, in 1982–1987 dollars, were \$10.3 million/1.78, or \$5.79 million. Bonds earned about 14 percent more than McGwire in real terms.

6.5 The real minimum wage in 1950 is \$0.75/0.241, or \$3.11 in 1982–1984 dollars. The real minimum wage in 1997 is \$5.15/1.61, or \$3.20 in 1982–1984 dollars. So the real minimum wage in 1997 is only slightly higher than what it was in 1950.

6.6 The increase in the cost of living between 1950 and 1997 is reflected in the ratio of the 1997 CPI to the 1950 CPI, or 1.61/0.241 = 6.68. That is, the cost of living in 1997 was 6.68 times what it was in 1950. If the minimum wage were indexed to preserve its purchasing power, it would have been 6.68 times higher in 1997 than in 1950, or 6.68 × \$0.75 = \$5.01. Since the actual minimum wage in 1997 was \$5.15, the effect of formal indexation in this case would have been negligible.

6.7 You should be concerned about the real return on your investment, not your nominal return. To calculate your likely real return, you need to know not only the nominal interest paid on the bonds of the island nation but also the prevailing inflation rate in that country. So your next question should be, "What is the rate of inflation in this country likely to be over the period that I am holding these bonds?"

6.8 The real rate of return to cash, as with any asset, is the nominal interest rate less the inflation rate. But cash pays no interest; that is, the nominal interest rate on cash is zero. Therefore, the real rate of return on cash is just minus the inflation rate. In other words, cash loses buying power at a rate equal to the rate of inflation. This rate of return depends on the actual rate of inflation and does not depend on whether the rate of inflation is correctly anticipated.

If inflation is high so that the real rate of return on cash is very negative, people will take actions to try to reduce their holdings of cash, such as going to the bank more often. The costs associated with trying to reduce holdings of cash are what economists call shoe-leather costs.

# PART 3

# THE ECONOMY IN THE LONG RUN

For millennia the great majority of the world's inhabitants eked out a spare existence by tilling the soil. Only a small proportion of the population lived above the level of subsistence, learned to read and write, or traveled more than a few miles from their birthplaces. Large cities grew up, serving as imperial capitals and centers of trade, but the great majority of urban populations lived in dire poverty, subject to malnutrition and disease.

Then, about three centuries ago, a fundamental change occurred. Spurred by technological advances and entrepreneurial innovations, a process of economic growth began. Sustained over many years, this growth in the economy's productive capacity has transformed almost every aspect of how we live—from what we eat and wear to how we work and play. What caused this economic growth? And why have some countries enjoyed substantially greater rates of growth than others? As Nobelist Robert E. Lucas Jr. put it in a classic article on economic development, "The consequences for human welfare involved in questions like these are simply staggering: Once one starts to think about them, it is hard to think about anything else." Although most people would attach less significance to these questions than Lucas did, they are undoubtedly of very great importance.

The subject of Part 3 is the behavior of the economy in the long run, including the factors that cause the economy to grow and develop. Chapter 7 begins by tackling directly the causes and consequences of economic growth. A key conclusion of the chapter is that improvements in average labor productivity are the primary source of rising living standards; hence policies to improve living standards should focus on stimulating productivity. Chapter 8 studies long-term trends in the labor market, analyzing the long-run effects of economic growth on real wages and employment opportunities. As the creation of new capital goods is an important factor underlying rising productivity, Chapter 9 examines the processes of saving and capital formation. Chapter 10 introduces the concept of money and examines its relationship to inflation in the long run. In addition, this chapter introduces the Federal Reserve System and describes how the "Fed" controls the money supply to promote economic stability. Chapter 11 discusses the role of banks, bond markets, and stock markets in allocating saving to productive uses, as well as the role of international capital flows, which facilitate the allocation of saving across countries.

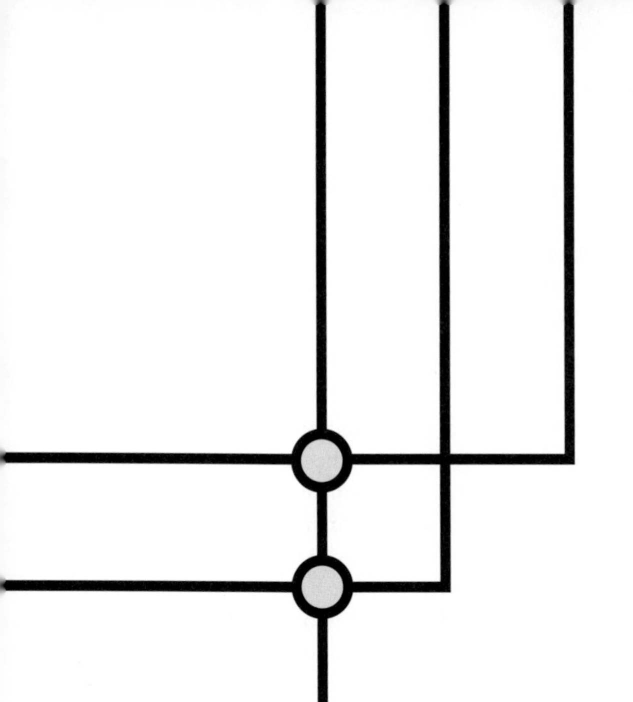

CHAPTER

7

# ECONOMIC GROWTH, PRODUCTIVITY, AND LIVING STANDARDS

One of us once attended a conference on the effects of economic growth and development on society. A speaker at the conference posed the following question: "Which would you rather be? An ordinary, middle-class American living today, or the richest person in America at the time of George Washington?"

A member of the audience spoke out immediately: "I can answer that question in one word. Dentistry."

The answer drew a laugh, perhaps because it reminded people of George Washington's famous wooden teeth. But it was a good answer. Dentistry in early America—whether the patient was rich or poor—was a primitive affair. Most dentists simply pulled a patient's rotten teeth, with a shot of whiskey for anesthetic.

Other types of medical care were not much better than dentistry. Eighteenth-century doctors had no effective weapons against tuberculosis, typhoid fever, diphtheria, influenza, pneumonia, and other communicable diseases. Such illnesses, now quite treatable, were major killers in Washington's time. Infants and children were particularly susceptible to deadly infectious diseases, especially whooping cough and measles. Even a well-to-do family often lost two or three children to these illnesses. Washington, an unusually large and vigorous

Would you rather be a rich person living in the eighteenth century or a middle-class person living in the twenty-first century?

man, lived to the age of 67, but the average life expectancy during his era was probably not much more than 40 years.

Medical care is not the only aspect of ordinary life that has changed drastically over the past two centuries. Author Stephen Ambrose, in his account of the Lewis and Clark expedition, described the limitations of transportation and communication in early America:

> A critical fact in the world of 1801 was that nothing moved faster than the speed of a horse. No human being, no manufactured item, no bushel of wheat, no side of beef (or any beef on the hoof for that matter), no letter, no information, no idea, order, or instruction of any kind moved faster, and, as far as Jefferson's contemporaries were able to tell, nothing ever would.
>
> And except on a racetrack, no horse moved very fast. Road conditions in the United States ranged from bad to abominable, and there weren't very many of them. The best highway in the country ran from Boston to New York; it took a light stagecoach . . . three full days to make the 175-mile journey. The hundred miles from New York to Philadelphia took two full days.[1]

Today New Yorkers can go to Philadelphia by train in an hour and a half. What would George Washington have thought of that? And how would nineteenth-century pioneers, who crossed the continent by wagon train, have reacted to the idea that their great-grandchildren would be able to have breakfast in New York and lunch the same day in San Francisco?

No doubt you can think of other enormous changes in the way average people live, even over the past few decades. Computer technologies and the Internet have changed the ways people work and study in just a few years, for example. Though these changes are due in large part to scientific advances, such discoveries *by themselves* usually have little effect on most people's lives. New scientific knowledge leads to widespread improvements in living standards only when it is commercially applied. Better understanding of the human immune system, for example, has little impact unless it leads to new therapies or drugs. And a new drug will do little to help unless it is affordable to those who need it.

A tragic illustration of this point is the AIDS epidemic in Africa. Although some new drugs will moderate the effects of the virus that causes AIDS, they are so expensive that they are of little practical value in poverty-stricken African nations grappling with the disease. But even if the drugs were affordable, they would have limited benefit without modern hospitals, trained health professionals, and adequate nutrition and sanitation. In short, most improvements in a nation's living standard are the result not just of scientific and technological advances but of an economic system that makes the benefits of those advances available to the average person.

In this chapter we will explore the sources of economic growth and rising living standards in the modern world. We will begin by reviewing the remarkable economic growth in the industrialized countries, as measured by real GDP per person. Since the mid-nineteenth century (and earlier in some countries), a radical transformation in living standards has occurred in these countries. What explains this transformation? The key to rising living standards is a *continuing increase in average labor productivity,* which depends on several factors, from the skills and motivation workers bring to their jobs to the legal and social environment in which they work. We will analyze each of these factors and discuss its implications for government policies to promote growth. We will also discuss the costs of rapid economic growth and consider whether there may be limits to the amount of economic growth a society can achieve.

[1]Stephen E. Ambrose, *Undaunted Courage: Meriwether Lewis, Thomas Jefferson, and the Opening of the American West,* New York: Touchstone (Simon & Schuster), 1996, p. 52.

## THE REMARKABLE RISE IN LIVING STANDARDS: THE RECORD

The advances in health care and transportation mentioned in the beginning of this chapter illustrate only a few of the impressive changes that have taken place in people's material well-being over the past two centuries, particularly in industrialized countries like the United States. To study the factors that affect living standards systematically, however, we must go beyond anecdotes and adopt a specific measure of economic well-being in a particular country and time.

In Chapter 5 we introduced the concept of real GDP as a basic measure of the level of economic activity in a country. Recall that, in essence, real GDP measures the physical volume of goods and services produced within a country's borders during a specific period, such as a quarter or a year. Consequently, real GDP *per person* provides a measure of the quantity of goods and services available to the typical resident of a country at a particular time. Although real GDP per person is certainly not a perfect indicator of economic well-being, as we saw in Chapter 5, it is positively related to a number of pertinent variables, such as life expectancy, infant health, and literacy. Lacking a better alternative, economists have focused on real GDP per person as a key measure of a country's living standard and stage of economic development.

Figure 4.2 showed the remarkable growth in real GDP per person that occurred in the United States between 1900 and 2001. For comparison, Table 7.1 shows real GDP per person in eight major countries in selected years from 1870 to 2000. Figure 7.1 displays the same data graphically for five of the eight countries.

The data in Table 7.1 and Figure 7.1 tell a dramatic story. For example, in the United States (which was already a relatively wealthy industrialized country in 1870), real GDP per person grew more than 11-fold between 1870 and 2000. In Japan, real GDP per person grew more than 25 times over the same period. Underlying these statistics is an amazingly rapid process of economic growth and transformation, through which in just a few generations relatively poor agrarian societies became highly industrialized economies—with average standards of living that could scarcely have been imagined in 1870. As Figure 7.1 shows, a significant part of this growth has occurred since 1950, particularly in Japan.

**TABLE 7.1**
**Real GDP per Person in Selected Countries, 1870–2000 (in 1995 U.S. Dollars)**

| Country | 1870 | 1913 | 1950 | 1979 | 2000 | Annual % change 1870–2000 | Annual % change 1950–2000 |
|---|---|---|---|---|---|---|---|
| Australia | 5,626 | 7,385 | 9,561 | 18,033 | 24,708 | 1.1 | 1.9 |
| Canada | 2,447 | 5,791 | 9,362 | 20,899 | 26,604 | 1.8 | 2.1 |
| France | 2,249 | 4,401 | 6,049 | 17,801 | 22,447 | 1.8 | 2.6 |
| Germany | 1,205 | 2,320 | 5,005 | 18,014 | 23,247 | 2.3 | 3.1 |
| Italy | 2,248 | 3,167 | 4,042 | 13,331 | 21,930 | 1.8 | 3.4 |
| Japan | 963 | 1,825 | 2,216 | 16,899 | 24,772 | 2.5 | 4.8 |
| United Kingdom | 3,500 | 5,374 | 7,832 | 14,889 | 21,142 | 1.4 | 2.0 |
| United States | 2,843 | 6,745 | 11,921 | 22,480 | 32,629 | 1.9 | 2.0 |

SOURCES: Derived from Angus Maddison, *Phases of Capitalist Development,* Oxford: Oxford University Press, reprinted 1988, Tables A2, B2–B4. Rebased to 1995 and updated to 2000 by the authors using OECD *Quarterly National Accounts.* "Germany" refers to West Germany in 1950 and 1979.

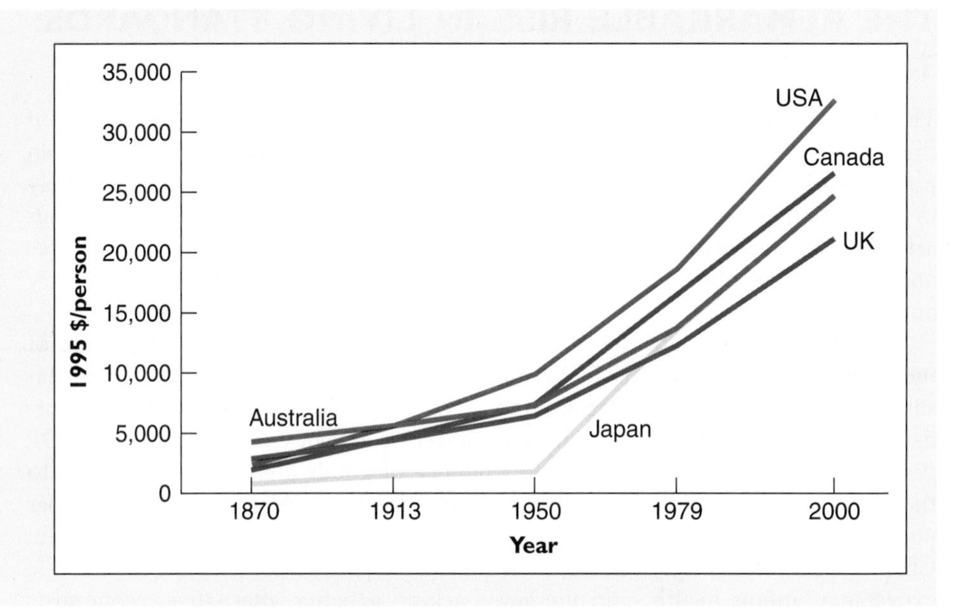


**FIGURE 7.1**
**Real GDP per Person in Five Industrialized Countries, 1870–2000.**
Economic growth has been especially rapid since the 1950s, particularly in Japan.

A note of caution is in order. The farther back in time we go, the less precise are historical estimates of real GDP. Most governments did not keep official GDP statistics until after World War II; production records from earlier periods are often incomplete or of questionable accuracy. Comparing economic output over a century or more is also problematic because many goods and services that are produced today were unavailable—indeed, inconceivable—in 1870. How many nineteenth-century horse-drawn wagons, for example, would be the economic equivalent of a BMW 328i automobile or a Boeing 757 jet? Despite the difficulty of making precise comparisons, however, we can say with certainty that the variety, quality, and quantity of available goods and services increased enormously in industrialized countries during the nineteenth and twentieth centuries, a fact reflected in the data on real GDP per capita.

## WHY "SMALL" DIFFERENCES IN GROWTH RATES MATTER

The last two columns of Table 7.1 show the annual growth rates of real GDP per person, both for the entire 1870–2000 period and the more recent years, 1950–2000. At first glance these growth rates don't seem to differ much from country to country. For example, for the period 1870–2000, the highest growth rate is 2.5 percent (Japan) and the lowest is 1.1 percent (Australia). But consider the long-run effect of this seemingly "small" difference in annual growth rates. In 1870, in terms of output per person, Australia was by far the richest of the eight countries listed in Table 7.1, with a real GDP per person nearly six times that of Japan. Yet by 2000 Japan had not just caught up to but exceeded Australia. This remarkable change in economic fortunes is the result of the apparently small difference between a 1.1 percent growth rate and a 2.5 percent growth rate, maintained over 130 years.

The fact that what seem to be small differences in growth rates can have large long-run effects results from what is called the *power of compound interest.*

**EXAMPLE 7.1**

### Compound interest (1)

In 1800 your great-great-grandfather deposited $10.00 in a checking account at 4 percent interest. Interest is compounded annually (so that interest paid at the end of each year receives interest itself in later years). Great-Grandpa's will specified that the account be turned over to his most direct descendant (you) in the year 2000. When you withdrew the funds in that year, how much was the account worth?

The account was worth \$10.00 in 1800; $\$10.00 \times 1.04 = \$10.40$ in 1801; $\$10.00 \times 1.04 \times 1.04 = \$10.00 \times (1.04)^2 = \$10.82$ in 1802; and so on. Since 200 years elapsed between 1800, when the deposit was made, and the year 2000, when the account was closed, the value of the account in the year 2000 was $\$10.00 \times (1.04)^{200}$, or $\$10.00 \times 1.04$ to the 200th power. Using a calculator, you will find that \$10.00 times 1.04 to the 200th power is \$25,507.50—a good return for a \$10.00 deposit!

**Compound interest**—an arrangement in which interest is paid not only on the original deposit but on all previously accumulated interest—is distinguished from *simple interest,* in which interest is paid only on the original deposit. If your great-grandfather's account had been deposited at 4 percent simple interest, it would have accumulated only 40 cents each year (4 percent of the original \$10.00 deposit), for a total value of $\$10.00 + 200 \times \$0.40 = \$90.00$ after 200 years. The tremendous growth in the value of his account came from the compounding of the interest—hence the phrase "the power of compound interest."

***compound interest*** the payment of interest not only on the original deposit but on all previously accumulated interest

**EXAMPLE 7.2**

### Compound interest (2)

Refer to Example 7.1. What would your great-grandfather's \$10.00 deposit have been worth after 200 years if the annual interest rate had been 2 percent? 6 percent?

At 2 percent interest the account would be worth \$10.00 in 1800; $\$10.00 \times 1.02 = \$10.20$ in 1801; $\$10.00 \times (1.02)^2 = \$10.40$ in 1802; and so on. In the year 2000 the value of the account would be $\$10.00 \times (1.02)^{200}$, or \$524.85. If the interest rate were 6 percent, after 200 years the account would be worth $\$10.00 \times (1.06)^{200}$, or \$1,151,259.04. Let's summarize the results of Examples 7.1 and 7.2:

| Interest rate (%) | Value of \$10 after 200 years |
|---|---|
| 2 | \$524.85 |
| 4 | \$25,507.50 |
| 6 | \$1,151,259.04 |

The power of compound interest is that even at relatively low rates of interest, a small sum, compounded over a long enough period, can greatly increase in value. A more subtle point, illustrated by this example, is that small differences in interest rates matter a lot. The difference between a 2 percent and a 4 percent interest rate doesn't seem tremendous, but over a long period of time it implies large differences in the amount of interest accumulated on an account. Likewise, the effect of switching from a 4 percent to a 6 percent interest rate is enormous, as our calculations show.

Economic growth rates are similar to compound interest rates. Just as the value of a bank deposit grows each year at a rate equal to the interest rate, so the size of a nation's economy expands each year at the rate of economic growth. This analogy suggests that even a relatively modest rate of growth in output per person—say, 1 to 2 percent per year—will produce tremendous increases in average living standard over a long period. And relatively small *differences* in growth rates, as in the case of Australia versus Japan, will ultimately produce very different living standards. Over the long run, then, the rate of economic growth is an extremely important variable. Hence, government policy changes or other factors that affect the long-term growth rate even by a small amount will have a major economic impact.

**EXERCISE 7.1**

**Suppose that real GDP per capita in the United States had grown at 2.5 percent per year, as Japan's did, instead of the actual 1.9 percent per year, from 1870 to 2000. How much larger would real GDP per person have been in the United States in 2000?**

## WHY NATIONS BECOME RICH: THE CRUCIAL ROLE OF AVERAGE LABOR PRODUCTIVITY

What determines a nation's economic growth rate? To get some insight into this vital question, we will find it useful to express real GDP per person as the product of two terms: average labor productivity and the share of the population that is working.

To do this, let $Y$ equal total real output (as measured by real GDP, for example), $N$ equal the number of employed workers, and $POP$ equal the total population. Then real GDP per person can be written as $Y/POP$; average labor productivity, or output per employed worker, equals $Y/N$; and the share of the population that is working is $N/POP$. The relationship between these three variables is

$$\frac{Y}{POP} = \frac{Y}{N} \times \frac{N}{POP},$$

which, as you can see by canceling out $N$ on the right-hand side of the equation, always holds exactly. In words, this basic relationship is

Real GDP per person = Average labor productivity ×
Share of population employed.

This expression for real GDP per person tells us something very basic and intuitive: The quantity of goods and services that each person can consume depends on (1) how much each worker can produce and (2) how many people (as a fraction of the total population) are working. Furthermore, because real GDP per person equals average labor productivity times the share of the population that is employed, real GDP per person can *grow* only to the extent that there is *growth* in worker productivity and/or the fraction of the population that is employed.

Figures 7.2 and 7.3 show the U.S. figures for the three key variables in the relationship above for the period 1960–2000. Figure 7.2 shows both real GDP per person and real GDP per worker (average labor productivity). Figure 7.3 shows the portion of the entire U.S. population (not just the working-age population) that was employed during that period. Once again, we see that the expansion in output per person in the United States has been impressive. Between 1960 and 2000, real GDP per person in the U.S. grew by 160 percent. Thus in 2000, the average American enjoyed about 2 1/2 times as many goods and services as in 1960. Figures 7.2 and 7.3 show that increases in both labor productivity and the share of the population holding a job contributed to this rise in living standard.

Let's look a bit more closely at these two contributing factors, beginning with the share of the population that is employed. As Figure 7.3 shows, between 1960 and 2000 the number of people employed in the United States rose from 36 to 49 percent of the entire population, a remarkable increase. The growing tendency of women to work outside the home (see Economic Naturalist 5.1) was the most important reason for this rise in employment. Another factor leading to higher rates of employment was an increase in the share of the general population that

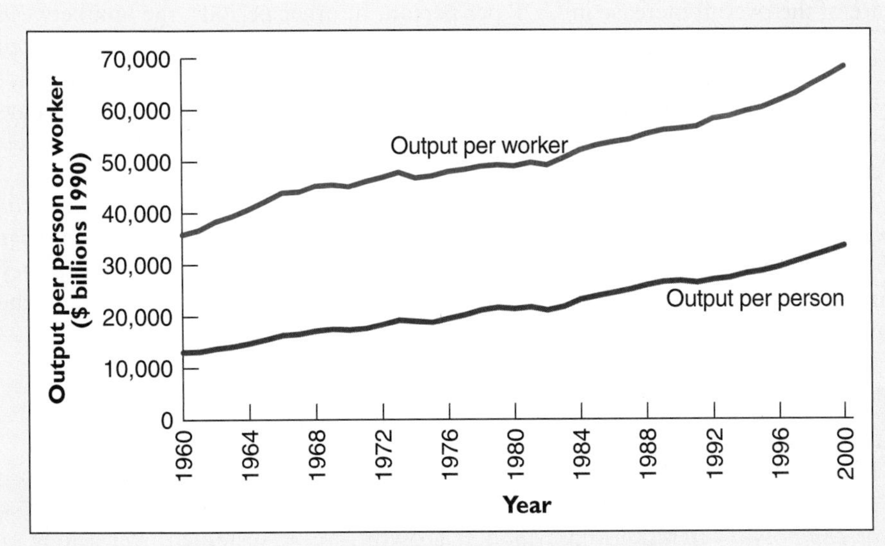


SOURCE: *Economic Report of the President* (http://w3.access.gpo.gov/eop/).

**FIGURE 7.2**
**Real GDP per Person and Average Labor Productivity in the United States, 1960–2000.**
Real output per person in the United States grew 160 percent between 1960 and 2000, and real output per worker (average labor productivity) grew by 92 percent.

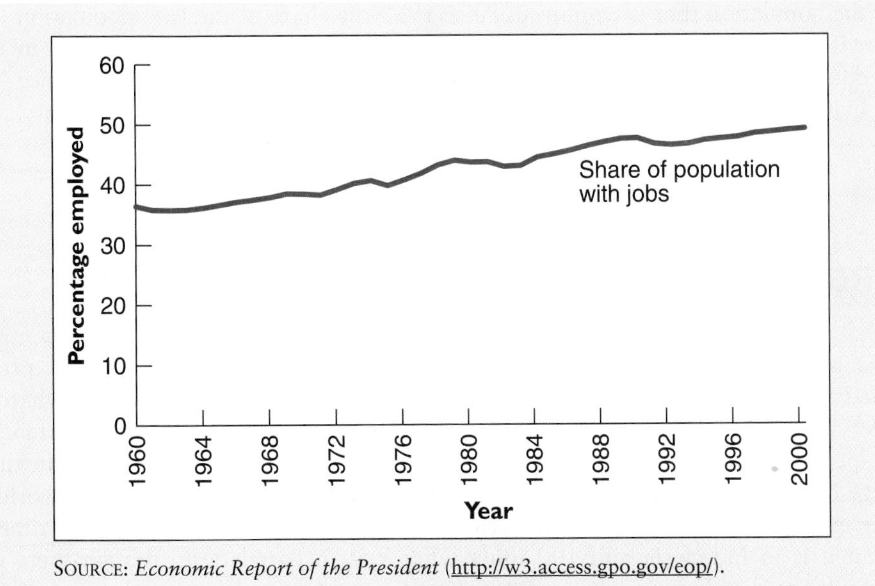


SOURCE: *Economic Report of the President* (http://w3.access.gpo.gov/eop/).

**FIGURE 7.3**
**Share of the U.S. Population Employed, 1960–2000.**
The share of the U.S. population holding a job increased from 36 percent in 1960 to 49 percent in 2000.

is of working age (ages 16 to 65). The coming of age of the "baby boom" generation, born in the years after World War II, and to a lesser extent the immigration of young workers from other countries, helped cause this growth in the workforce.

Although the rising share of the U.S. population with jobs contributed significantly to the increase in real GDP per person during the past four decades, this trend almost certainly will not continue in the future. Women's participation in the labor force seems unlikely to continue rising at the same rate as in the past four decades. More important, the baby boom generation, now in its prime years of employment, will begin to reach retirement age around the year 2010. As more and more baby boomers retire, the fraction of the population that is employed will begin to drop, probably significantly. In the long run, then, the improvement in living standards brought about by the rising share of Americans with jobs will likely prove transitory.

What about the other factor that determines output per person, average labor productivity? As Figure 7.2 shows, between 1960 and 2000, average labor productivity in the United States increased by 92 percent, accounting for a sizable share of the overall increase in GDP per person. In other periods, the link between average labor productivity and output per person in the United States has often been even stronger, since in most earlier periods the share of the population holding jobs was more stable than it has been recently. (See Figure 4.2 for the behavior of real GDP per person and average labor productivity in the United States over the period 1900–2000.)

This quick look at recent data supports a more general conclusion. *In the long run, increases in output per person arise primarily from increases in average labor productivity.* In simple terms, the more people can produce, the more they can consume. To understand why economies grow, then, we must understand the reasons for increased labor productivity.

**ECONOMIC GROWTH AND PRODUCTIVITY**

Real GDP per person, a basic indicator of living standards, has grown dramatically in the industrialized countries. This growth reflects the *power of compound interest:* Even a modest growth rate, if sustained over a long period of time, can lead to large increases in the size of the economy.

Output per person equals average labor productivity times the share of the population that is employed. Since 1960 the share of the U.S. population with jobs has risen significantly, but this variable is likely to decline in coming decades. In the long run, increases in output per person and hence living standards arise primarily from increases in average labor productivity.

## THE DETERMINANTS OF AVERAGE LABOR PRODUCTIVITY

What determines the productivity of the average worker in a particular country at a particular time? Popular discussions of this issue often equate worker productivity with the willingness of workers of a given nationality to work hard. Everything else being equal, a culture that promotes hard work certainly tends to increase worker productivity. But intensity of effort alone cannot explain the huge differences in average labor productivity that we observe around the world. For example, average labor productivity in the United States is about 24 times what it is in Indonesia and 100 times what it is in Bangladesh, though there is little doubt that Indonesians and Bangladeshis work very hard.

In this section we will examine six factors that appear to account for the major differences in average labor productivity, both between countries and between generations. Later in the chapter we will discuss how economic policies can influence these factors to spur productivity and growth.

### HUMAN CAPITAL

To illustrate the factors that determine average labor productivity, we introduce two prototypical assembly line workers, Lucy and Ethel.

**EXAMPLE 7.3**

**Lucy and Ethel on the assembly line**

Lucy and Ethel have jobs wrapping chocolate candies and placing them into boxes. Lucy, a novice wrapper, can wrap only 100 candies per hour. Ethel, who has had on-the-job training, can wrap 300 candies per hour. Lucy and Ethel each work 40 hours per week. Find average labor productivity, in terms of candies wrapped per week and candies wrapped per hour, (a) for Lucy, (b) for Ethel, and (c) for Lucy and Ethel as a team.

We have defined average labor productivity in general terms as output per worker. Note, though, that the measurement of average labor productivity depends on the time period that is specified. For example, the data presented in Figure 7.2 tell us how much the average worker produces *in a year*. In this example we are concerned with how much Lucy and Ethel can produce *per hour* of work or *per week* of work. Any one of these ways of measuring labor productivity is equally valid, as long as we are clear about the time unit we are using.

Lucy and Ethel's hourly productivities are given in the problem: Lucy can wrap 100 candies per hour and Ethel can wrap 300. Lucy's weekly productivity is (40 hours/week) × (100 candies wrapped/hour) = 4,000 wrapped candies per week. Ethel's weekly productivity is (40 hours/week) × (300 candies wrapped/hour), or 12,000 candies per week.

Together Lucy and Ethel can wrap 16,000 candies per week. As a team, their average weekly productivity is (16,000 candies wrapped)/(2 weeks of work), or 8,000 candies per week. Their average hourly productivity as a team is (16,000 candies wrapped)/(80 hours of work) = 200 candies per hour. Notice that, taken as a team, the two women's productivity lies midway between their individual productivities.

The Everett Collection

How productive are these workers?

Ethel is more productive than Lucy because she has had on-the-job training, which has allowed her to develop her candy-wrapping skills to a higher level than Lucy's. Because of her training, Ethel can produce more than Lucy can in a given number of hours.

## EXERCISE 7.2

**Suppose Ethel attends additional classes in candy wrapping and learns how to wrap 500 candies per hour. Find the output per week and output per hour for Lucy and Ethel, both individually and as a team.**

Economists would explain the difference in the two women's performance by saying that Ethel has more human capital than Lucy. *Human capital* comprises the talents, education, training, and skills of workers. Workers with a large stock of human capital are more productive than workers with less training. For example, a secretary who knows how to use a word processing program will be able to type more letters than one who doesn't; an auto mechanic who is familiar with computerized diagnostic equipment will be able to fix engine problems that less well-trained mechanics could not.

ECONOMIC NATURALIST 7.1

### Why did West Germany and Japan recover so successfully from the devastation of World War II?

Germany and Japan sustained extensive destruction of their cities and industries during World War II and entered the postwar period impoverished. Yet within 30 years both countries had not only been rebuilt but had become worldwide industrial and economic leaders. What accounts for these "economic miracles"?

Many factors contributed to the economic recovery of West Germany and Japan from World War II, including the substantial aid provided by the United States to Europe under the Marshall Plan and to Japan during the U.S. occupation. Most economists agree, however, that high levels of *human capital* played a crucial role in both countries.

At the end of the war Germany's population was exceptionally well educated, with a large number of highly qualified scientists and engineers. The country also had (and still does today) an extensive apprentice system that provided on-the-job training to young workers. As a result, Germany had a skilled industrial workforce. In addition, the area that became West Germany benefited substantially from an influx of skilled workers from East Germany and the rest of Soviet-controlled

Europe, including 20,000 trained engineers and technicians. Beginning as early as 1949, this concentration of human capital contributed to a major expansion of Germany's technologically sophisticated, highly productive manufacturing sector. By 1960 West Germany was a leading exporter of high-quality manufactured goods, and its citizens enjoyed one of the highest standards of living in Europe.

Japan, which probably sustained greater physical destruction in the war than Germany, also began the postwar period with a skilled and educated labor force. In addition, occupying American forces restructured the Japanese school system and encouraged all Japanese to obtain a good education. Even more so than the Germans, however, the Japanese emphasized on-the-job training. As part of a lifetime employment system, under which workers were expected to stay with the same company their entire career, Japanese firms invested extensively in worker training. The payoff to these investments in human capital was a steady increase in average labor productivity, particularly in manufacturing. By the 1980s Japanese manufactured goods were among the most advanced in the world and Japan's workers among the most skilled.

Although high levels of human capital were instrumental in the rapid economic growth of West Germany and Japan, human capital alone cannot create a high living standard. A case in point is Soviet-dominated East Germany, which had a level of human capital similar to West Germany's after the war but did not enjoy the same economic growth. For reasons we will discuss later in the chapter (see Economic Naturalist 7.3), the communist system imposed by the Soviets utilized East Germany's human capital far less effectively than the economic systems of Japan and West Germany.

Human capital is analogous to physical capital (such as machines and factories) in that it is acquired primarily through the investment of time, energy, and money. For example, to learn how to use a word processing program, a secretary might need to attend a technical school at night. The cost of going to school includes not only the tuition paid but also the *opportunity cost* of the secretary's time spent attending class and studying. The benefit of the schooling is the increase in wages the secretary will earn when the course has been completed.

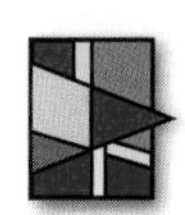

We know by the *cost-benefit principle* that the secretary should learn word processing only if the benefits exceed the costs, including the opportunity costs. In general, then, we would expect to see people acquire additional education and skills when the difference in the wages paid to skilled and unskilled workers is significant.

## PHYSICAL CAPITAL

Workers' productivity depends not only on their skills and effort but on the tools they have to work with. Even the most skilled surgeon cannot perform open-heart surgery without sophisticated equipment, and an expert computer programmer is of limited value without a computer. These examples illustrate the importance of *physical capital,* such as factories and machines. More and better capital allows workers to produce more efficiently, as Example 7.4 shows.

**EXAMPLE 7.4**

### Lucy and Ethel get automated

Refer to Example 7.3. Lucy and Ethel's boss acquired an electric candy-wrapping machine, which is designed to be operated by one worker. Using this machine, an untrained worker can wrap 500 candies per hour. What are Lucy and Ethel's hourly and weekly outputs now? Will the answer change if the boss gets a second machine? A third?

Suppose for the sake of simplicity that a candy-wrapping machine must be assigned to one worker only. (This assumption rules out sharing arrangements, in which one worker uses the machine on the day shift and another on the night shift.) If the boss buys just one machine, she will assign it to Lucy. (Why? See Exercise 7.3.) Now Lucy will be able to wrap 500 candies per hour, while Ethel can wrap only 300 per hour. Lucy's weekly output will be 20,000 wrapped candies (40 hours × 500 candies wrapped per hour). Ethel's weekly output is still 12,000 wrapped candies (40 hours × 300 candies wrapped per hour). Together they can now wrap 32,000 candies per week, or 16,000 candies per week each. On an hourly basis, average labor productivity for the two women taken together is 32,000 candies wrapped per 80 hours of work, or 400 candies wrapped per hour—twice their average labor productivity before the boss bought the machine.

With two candy-wrapping machines available, both Lucy and Ethel could use a machine. Each could wrap 500 candies per hour, for a total of 40,000 wrapped candies per week. Average labor productivity for both women taken together would be 20,000 wrapped candies per week, or 500 wrapped candies per hour.

What would happen if the boss purchased a third machine? With only two workers, a third machine would be useless: it would add nothing to either total output or average labor productivity.

## EXERCISE 7.3

**Using the assumptions made in Examples 7.3 and 7.4, explain why the boss should give the single available candy-wrapping machine to Lucy rather than Ethel. (*Hint:* Use *the principle of increasing opportunity cost,* introduced in Chapter 3.)**

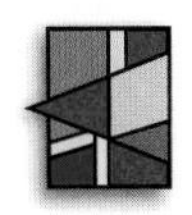

The candy-wrapping machine is an example of a *capital good,* which was defined in Chapter 5 as a long-lived good, which is itself produced and used to produce other goods and services. Capital goods include machines and equipment (such as computers, earthmovers, or assembly lines) as well as buildings (such as factories or office buildings).

Capital goods like the candy-wrapping machine enhance workers' productivity. Table 7.2 summarizes the results from Examples 7.3 and 7.4. For each number of machines the boss might acquire (column 1), Table 7.2 gives the total weekly output of Lucy and Ethel taken together (column 2), the total number of hours worked by the two women (column 3), and average output per hour (column 4), equal to total weekly output divided by total weekly hours.

Table 7.2 demonstrates two important points about the effect of additional capital on output. First, for a given number of workers, adding more capital generally increases both total output and average labor productivity. For example, adding the first candy-wrapping machine increases weekly output (column 2) by 16,000 candies and average labor productivity (column 4) by 200 candies wrapped per hour.

The second point illustrated by Table 7.2 is that the more capital is already in place, the smaller the benefits of adding extra capital. Notice that the first

**TABLE 7.2**
**Capital, Output, and Productivity in the Candy-Wrapping Factory**

| (1) Number of machines (capital) | (2) Total number of candies wrapped each week (output) | (3) Total hours worked per week | (4) Candies wrapped per hour worked (productivity) |
|---|---|---|---|
| 0 | 16,000 | 80 | 200 |
| 1 | 32,000 | 80 | 400 |
| 2 | 40,000 | 80 | 500 |
| 3 | 40,000 | 80 | 500 |

machine adds 16,000 candies to total output, but the second machine adds only 8,000. The third machine, which cannot be used since there are only two workers, does not increase output or productivity at all. This result illustrates a general principle of economics, called *diminishing returns to capital.* According to the principle of **diminishing returns to capital,** if the amount of labor and other inputs employed is held constant, then the greater the amount of capital already in use, the less an additional unit of capital adds to production. In the case of the candy-wrapping factory, diminishing returns to capital implies that the first candy-wrapping machine acquired adds more output than the second, which in turn adds more output than the third.

***diminishing returns to capital*** if the amount of labor and other inputs employed is held constant, then the greater the amount of capital already in use, the less an additional unit of capital adds to production

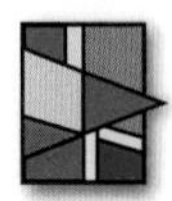

Diminishing returns to capital are a natural consequence of firms' incentive to use each piece of capital as productively as possible. To maximize output, managers will assign the first machine that a firm acquires to the most productive use available, the next machine to the next most productive use, and so on—an illustration of the *principle of increasing opportunity cost,* or *low-hanging-fruit principle* (Chapter 3). When many machines are available, all the highly productive ways of using them already have been exploited. Thus adding yet another machine will not raise output or productivity by very much. If Lucy and Ethel are already operating two candy-wrapping machines, there is little point to buying a third machine, except perhaps as a replacement or spare.

The implications of Table 7.2 can be applied to the question of how to stimulate economic growth. First, increasing the amount of capital available to the workforce will tend to increase output and average labor productivity. The more adequately equipped workers are, the more productive they will be. Second, the degree to which productivity can be increased by an expanding stock of capital is limited. Because of diminishing returns to capital, an economy in which the quantity of capital available to each worker is already very high will not benefit much from further expansion of the capital stock.

Is there empirical evidence that giving workers more capital makes them more productive? Figure 7.4 shows the relationship between average labor productivity (real GDP per worker) and the amount of capital per worker in 15 major countries, including the 8 industrialized countries listed in Table 7.1. The figure shows a strong relationship between the amounts of capital per worker and productivity, consistent with the theory. Note, though, that the relationship between capital and productivity is somewhat weaker for the richest countries. For example, Germany has more capital per worker than the United States, but German workers are less productive than American workers on average. Diminishing returns to capital

**FIGURE 7.4**
**Average Labor Productivity and Capital per Worker in 15 Countries, 1990.**
Countries with large amounts of capital per worker also tend to have high average labor productivity, as measured by real GDP per worker.

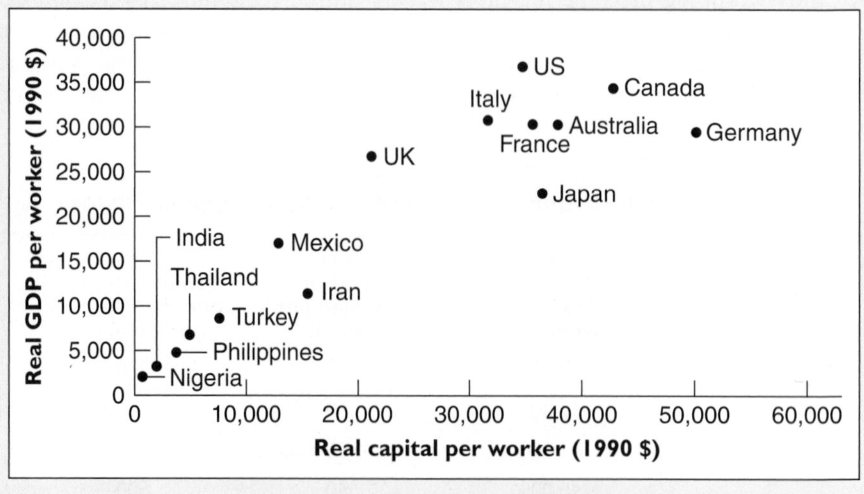


SOURCE: Penn World Tables (www.nber.org). Countries included are those listed in Table 7.1, plus all countries with populations of 40 million or more for which data are available.

may help to explain the weakening of the relationship between capital and productivity at high levels of capital. In addition, Figure 7.4 does not account for many other differences among countries, such as differences in economic systems or government policies. Thus we should not expect to see a perfect relationship between the two variables.

## LAND AND OTHER NATURAL RESOURCES

Besides capital goods, other inputs to production help to make workers more productive, among them land, energy, and raw materials. Fertile land is essential to agriculture, and modern manufacturing processes make intensive use of energy and raw materials.

In general, an abundance of natural resources increases the productivity of the workers who use them. For example, a farmer can produce a much larger crop in a land-rich country like the United States or Australia than in a country where the soil is poor or arable land is limited in supply. With the aid of modern farm machinery and great expanses of land, today's American farmers are so productive that even though they constitute less than 3 percent of the population, they provide enough food not only to feed the country but to export to the rest of the world.

Although there are limits to a country's supply of arable land, many other natural resources, such as petroleum and metals, can be obtained through international markets. Because resources can be obtained through trade, countries need not possess large quantities of natural resources within their own borders to achieve economic growth. Indeed, a number of countries have become rich without substantial natural resources of their own, including Japan, Hong Kong, Singapore, and Switzerland. Just as important as possessing natural resources is the ability to use them productively—for example, by means of advanced technologies.

## TECHNOLOGY

Besides human capital, physical capital, and natural resources, a country's ability to develop and apply new, more productive technologies will help to determine its productivity. Consider just one industry, transportation. Two centuries ago, as suggested by the quote from Stephen Ambrose in the beginning of the chapter, the horse and wagon were the primary means of transportation—a slow and costly method indeed. But in the nineteenth century, technological advances such as the steam engine supported the expansion of riverborne transportation and the development of a national rail network. In the twentieth century, the invention of the internal combustion engine and the development of aviation, supported by the construction of an extensive infrastructure of roads and airports, have produced increasingly rapid, cheap, and reliable transport. Technological change has clearly been a driving force in the transportation revolution.

New technologies can improve productivity in industries other than the one in which they are introduced. Once farmers could sell their produce only in their local communities, for example. Now the availability of rapid shipping and refrigerated transport allows farmers to sell their products virtually anywhere in the world. With a broader market in which to sell, farmers can specialize in those products best suited to local land and weather conditions. Similarly, factories can obtain their raw materials wherever they are cheapest and most abundant, produce the goods they are most efficient at manufacturing, and sell their products wherever they will fetch the best price. Both these examples illustrate *the principle of comparative advantage,* that overall productivity increases when producers concentrate on those activities at which they are relatively most efficient (see Chapter 3).

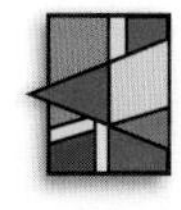

Numerous other technological developments led to increased productivity, including advances in communication and medicine and the introduction of computer technology. All indications are that the Internet will have a major impact on the U.S. economy, not just in retailing but in many other sectors. In fact, *most economists would probably agree that new technologies are the single most important source of productivity improvement,* and hence of economic growth in general.

However, economic growth does not automatically follow from breakthroughs in basic science. To make the best use of new knowledge, an economy needs entrepreneurs who can exploit scientific advances commercially, as well as a legal and political environment that encourages the practical application of new knowledge.

**EXERCISE 7.4**

**A new kind of wrapping paper has been invented that makes candy-wrapping quicker and easier. The use of this paper *increases* the number of candies a person can wrap by hand by 200 per hour, and the number of candies a person can wrap by machine by 300 per hour. Using the data from Examples 7.3 and 7.4, construct a table like Table 7.2 that shows how this technological advance affects average labor productivity. Do diminishing returns to capital still hold?**

## ENTREPRENEURSHIP AND MANAGEMENT

***entrepreneurs*** people who create new economic enterprises

The productivity of workers depends in part on the people who help to decide what to produce and how to produce it: entrepreneurs and managers. **Entrepreneurs** are people who create new economic enterprises. Because of the new products, services, technological processes, and production methods they introduce, entrepreneurs are critical to a dynamic, healthy economy. In the late nineteenth and early twentieth centuries, individuals like Henry Ford and Alfred Sloan (automobiles), Andrew Carnegie (steel), John D. Rockefeller (oil), and J. P. Morgan (finance) played central roles in the development of American industry—and, not incidentally, amassed huge personal fortunes in the process. These people and others like them (including contemporary entrepreneurs like Bill Gates) have been criticized for some of their business practices, in some cases with justification. Clearly, though, they and dozens of other prominent business leaders of the past century have contributed significantly to the growth of the U.S. economy. Henry Ford, for example, developed the idea of mass production, which lowered costs sufficiently to bring automobiles within reach of the average American family. Ford began his business in his garage, a tradition that has been maintained by thousands of innovators ever since.

Entrepreneurship, like any form of creativity, is difficult to teach, although some of the supporting skills, like financial analysis and marketing, can be learned in college or business school. How, then, does a society encourage entrepreneurship? History suggests that the entrepreneurial spirit will always exist; the challenge to society is to channel entrepreneurial energies in economically productive ways. For example, economic policymakers need to ensure that taxation is not so heavy, and regulation not so inflexible, that small businesses—some of which will eventually become big businesses—cannot get off the ground. Sociological factors may play a role as well. Societies in which business and commerce are considered to be beneath the dignity of refined, educated people are less likely to produce successful entrepreneurs (see Economic Naturalist 7.2). In the United States, for the most part, business has been viewed as a respectable activity. Overall, a social and economic milieu that allows entrepreneurship to flourish appears to promote economic growth and rising productivity, perhaps especially so in high-technology eras like our own.

EXAMPLE 7.5

**Lucy and Ethel get discovered**

A television producer "discovers" Lucy and Ethel working on the production line at the candy factory and makes them stars of their own TV series. The show is a huge success and is seen by millions. How does the producer's discovery of Lucy and Ethel affect their average labor productivity?

Because Lucy and Ethel are so popular, and because television advertising fees are based on the number of people who watch a show, Lucy and Ethel are likely to increase the TV network's revenues by many millions of dollars. This increase in revenues is a measure of Lucy and Ethel's contribution to GDP. Their average labor productivity (in per-hour terms) is their contribution to GDP divided by the number of hours they work on the show. Clearly, Lucy's and Ethel's average labor productivity as TV stars will be many times what it was when they were candy wrappers.

This example illustrates the importance of creative entrepreneurship: If not for the TV producer's entrepreneurial skills, Lucy and Ethel would still be working on the assembly line, where their average labor productivity would be much lower.

ECONOMIC NATURALIST 7.2

**Why did medieval China stagnate economically?**

The Sung period in China (A.D. 960–1270) was one of considerable technological sophistication; its inventions included paper, waterwheels, water clocks, gunpowder, and possibly the compass. Yet no significant industrialization occurred, and in subsequent centuries Europe saw more economic growth and technological innovation than China. Why did medieval China stagnate economically?

According to research by economist William Baumol,[2] the main impediment to industrialization during the Sung period was a social system that inhibited entrepreneurship. Commerce and industry were considered low-status activities, not fit for an educated person. In addition, the emperor had the right to seize his subjects' property and to take control of their business enterprises—a right that greatly reduced his subjects' incentives to undertake business ventures. The most direct path to status and riches in medieval China was to go through a system of demanding civil service examinations given by the government every three years. The highest scorers on these national examinations were granted lifetime positions in the imperial bureaucracy, where they wielded much power and often became wealthy, in part through corruption. Not surprisingly, medieval China did not develop a dynamic entrepreneurial class, and consequently its scientific and technological advantages did not translate into sustained economic growth. China's experience shows why scientific advances alone cannot guarantee economic growth; to have economic benefits, scientific knowledge must be commercially applied through new products and new, more efficient means of producing goods and services.

Although entrepreneurship may be more glamorous, managers—the people who run businesses on a daily basis—also play an important role in determining average labor productivity. Managerial jobs span a wide range of positions, from the supervisor of the loading dock to the CEO (chief executive officer) at the helm of a *Fortune* 500 company. Managers work to satisfy customers, deal with

[2]"Entrepreneurship: Productive, Unproductive, and Destructive," *Journal of Political Economy*, October 1990, pp. 893–921.

suppliers, organize production, obtain financing, assign workers to jobs, and motivate them to work hard and effectively. Such activities enhance labor productivity. For example, in the 1970s and 1980s, Japanese managers introduced new production methods that greatly increased the efficiency of Japanese manufacturing plants. Among them was the *just-in-time* inventory system, in which suppliers deliver production components to the factory just when they are needed, eliminating the need for factories to stockpile components. Japanese managers also pioneered the idea of organizing workers into semi-independent production teams, which allowed workers more flexibility and responsibility than the traditional assembly line. Managers in the United States and other countries studied the Japanese managerial techniques closely and adopted many of them.

## THE POLITICAL AND LEGAL ENVIRONMENT

So far we have emphasized the role of the private sector in increasing average labor productivity. But government too has a role to play in fostering improved productivity. One of the key contributions government can make is to provide a *political and legal environment* that encourages people to behave in economically productive ways—to work hard, save and invest wisely, acquire useful information and skills, and provide the goods and services that the public demands.

One specific function of government that appears to be crucial to economic success is the establishment of *well-defined property rights*. Property rights are well defined when the law provides clear rules for determining who owns what resources (through a system of deeds and titles, for example) and how those resources can be used. Imagine living in a society in which a dictator, backed by the military and the police, could take whatever he wanted, and regularly did so. In such a country, what incentive would you have to raise a large crop or to produce other valuable goods and services? Very little, since much of what you produced would likely be taken away from you. Unfortunately, in many countries of the world today, this situation is far from hypothetical.

Political and legal conditions affect the growth of productivity in other ways, as well. Political scientists and economists have documented the fact that *political instability* can be detrimental to economic growth. This finding is reasonable, since entrepreneurs and savers are unlikely to invest their resources in a country whose government is unstable, particularly if the struggle for power involves civil unrest, terrorism, or guerrilla warfare. On the other hand, a political system that promotes the *free and open exchange of ideas* will speed the development of new technologies and products. For example, some economic historians have suggested that the decline of Spain as an economic power was due in part to the advent of the Spanish Inquisition, which permitted no dissent from religious orthodoxy. Because of the Inquisition's persecution of those whose theories about the natural world contradicted Church doctrine, Spanish science and technology languished, and Spain fell behind more tolerant nations like the Netherlands.

### EXERCISE 7.5

**A Bangladeshi worker who immigrates to America is likely to find that his average labor productivity is much higher in the United States than it was at home. The worker is, of course, the same person he was when he lived in Bangladesh. How can the simple act of moving to the United States increase the worker's productivity? What does your answer say about the incentive to immigrate?**

**Why did communism fail?**

ECONOMIC NATURALIST 7.3

For more than 70 years, from the Russian revolution in 1917 until the collapse of the Soviet Union in 1991, communism was believed by many to pose a major challenge to market-based economic systems. Yet, by the time of the Soviet Union's breakup, the poor economic record of communism had become apparent. Indeed, low living standards in communist countries, compared to those achieved in the West, were a major reason for the popular discontent that brought down the communist system in Europe. Economically speaking, why did communism fail?

The poor growth records of the Soviet Union and other communist countries did not reflect a lack of resources or economic potential. The Soviet Union had a highly educated workforce; a large capital stock; a vast quantity of natural resources, including land and energy; and access to sophisticated technologies. Yet, at the time of its collapse, output per person in the Soviet Union was probably less than one-seventh what it was in the United States.

Most observers would agree that the political and legal environment that established the structure of the communist economic system was a major cause of its ultimate failure. The economic system of the Soviet Union and other communist countries had two main elements: First, the capital stock and other resources were owned by the government rather than by individuals or private corporations. Second, most decisions regarding production and distribution were made and implemented by a government planning agency rather than by individuals and firms interacting through markets. This system performed poorly, we now understand, for several reasons.

One major problem was *the absence of private property rights.* With no ability to acquire a significant amount of private property, Soviet citizens had little incentive to behave in economically productive ways. The owner of an American or Japanese firm is strongly motivated to cut costs and to produce goods that are highly valued by the public, because the owner's income is determined by the firm's profitability. In contrast, the performance of a Soviet firm manager was judged on whether the manager produced the quantities of goods specified by the government's plan—irrespective of the quality of the goods produced or whether consumers wanted them. Soviet managers had little incentive to reduce costs or produce better, more highly valued products, as any extra profits would accrue to the government and not to the manager; nor was there any scope for entrepreneurs to start new businesses. Likewise, workers had little reason to work hard or effectively under the communist system, as pay rates were determined by the government planning agency rather than by the economic value of what the workers produced.

A second major weakness of the communist system was the *absence of free markets.* In centrally planned economies, markets are replaced by detailed government plans that specify what should be produced and how. But, as we saw in the example of New York City's food supply (Chapter 3), the coordination of even relatively basic economic activities can be extremely complex and require a great deal of information, much of which is dispersed among many people. In a market system, changes in prices both convey information about the goods and services people want and provide suppliers the incentives to bring these goods and services to market. Indeed, as we know from the *equilibrium principle,* a market in equilibrium leaves individuals with no unexploited opportunities. Central planners in communist countries proved far less able to deal with this complexity than decentralized markets. As a result, under communism consumers suffered constant shortages and shoddy goods.

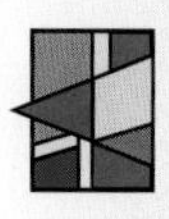

After the collapse of communism, many formerly communist countries began the difficult transition to a market-oriented economic system. Changing an entire economic system (the most extreme example of a *structural policy*) is a slow and difficult task, and many countries saw economic conditions worsen at first rather than improve. *Political instability* and the absence of a modern *legal framework,*

particularly laws applying to commercial transactions, have often hampered the progress of reforms. However, a number of formerly communist countries, including Poland, the Czech Republic, and the former East Germany, have succeeded in implementing Western-style market systems and have begun to achieve significant economic growth.

**DETERMINANTS OF AVERAGE LABOR PRODUCTIVITY**

Key factors determining average labor productivity in a country include:

- The skills and training of workers, called *human capital*
- The quantity and quality of *physical capital*—machines, equipment, and buildings
- The availability of land and other *natural resources*
- The sophistication of the *technologies* applied in production
- The effectiveness of *management* and *entrepreneurship*
- The broad *social and legal environment*

## THE WORLDWIDE PRODUCTIVITY SLOWDOWN—AND RECOVERY?

During the 1950s and 1960s most of the major industrialized countries saw rapid growth in real GDP and average labor productivity. In the 1970s, however, productivity growth began to slow around the world. Slower growth in real GDP and in average living standards followed.

The slowdown in the growth of labor productivity is documented in Table 7.3, which gives data for five major industrialized countries. Note the sharp decline in productivity growth in all eight countries during 1973–1979 compared to 1960–1973. Japan's case was particularly striking: Its productivity growth rate fell from 7.6 percent per year in 1960–1973 to 2.7 percent in 1973–1979. In the United States, annual productivity growth fell from 2.3 percent before 1973 to just 0.6 percent per year during 1973–1979. During the period 1979–2000, productivity growth improved somewhat in the United States and the United Kingdom, but in all five countries, the rate of productivity improvement since 1979 has been much slower than it was prior to 1973.

**TABLE 7.3**
**Average Labor Productivity Growth Rates in Selected Countries, 1960–2000**

| | Percentage growth, annual rates | | |
|---|---|---|---|
| **Country** | **1960–1973** | **1973–1979** | **1979–2000** |
| France | 4.6 | 2.3 | 1.8 |
| Germany | 4.0 | 2.6 | 2.0 |
| Japan | 7.6 | 2.7 | 2.0 |
| United Kingdom | 2.8 | 1.3 | 1.7 |
| United States | 2.3 | 0.6 | 1.7 |

SOURCE: Growth rates of real GDP per employed person, Bureau of Labor Statistics and OECD.

The sudden decline in worldwide productivity growth around 1973 is puzzling to economists and policymakers alike. What might have caused it? Some analysts have suggested answers that are specific to the United States, such as the perceived decline in the quality of public education. However, such explanations are not very convincing, since they do not explain why so many other countries had the same problem. In the 1970s and 1980s many economists thought that the fourfold increase in oil prices that followed the Arab–Israeli war (1973) might have caused the slowdown. However, oil prices (relative to the prices of other goods) have long since returned to pre-1973 levels, yet productivity growth has not. Thus oil prices are no longer thought to have played a critical role in the slowdown.

One view of the slowdown in productivity since 1973 is that (at least in part) it is not a real phenomenon but the result of *poor measurement of productivity growth*. According to this argument, many of the productivity improvements that occur in modern services-oriented economies are difficult to capture in economic statistics. For example, the computerization of inventories allows supermarkets to offer customers a wider variety of products, with less chance that a particular product will be out of stock. ATM machines allow people to make banking transactions 24 hours a day, not just when the bank is open. And many medical procedures can be done far more quickly, safely, and painlessly today than just a few years ago (see Economic Naturalist 6.2). In theory, all these improvements in the quality of services should be captured in real GDP, and hence in productivity measures. In reality, accurate measurement of improvements in quality is difficult, as we saw when discussing biases in the CPI in Chapter 6, and some improvements may be missed. If the productivity slowdown is not real but reflects only poor measurement, then economists need not worry about it.

Another explanation, which is not mutually exclusive with the first, has been called the *technological depletion hypothesis*.[3] According to this hypothesis, the high rates of productivity in the 1950s and 1960s reflected an unusual period of "catch-up" after the Depression and the Second World War. Although scientific and technical advances continued to be made during the 1930s and 1940s (many of which arose from military research), depression and war prevented them from being adapted to civilian use. During the 1950s and 1960s, the backlog of technological breakthroughs was applied commercially, producing high rates of productivity growth at first and then a sharp decline in new technological opportunities. Once the catch-up period was over, productivity growth slowed. According to this hypothesis, then, the slowdown in productivity growth since the 1970s reflects a dearth of technological opportunities relative to the immediate postwar period. From this perspective, the 1950s and 1960s were the exception, and the period since the 1970s represents a return to more normal rates of productivity growth.

Although the rate of productivity growth over the past two decades has generally been low, in recent years there have been some hints of a possible recovery in productivity, particularly in the United States. Between 1991 and 2000, U.S. productivity growth averaged nearly 2.3 percent per year, close to the rate achieved prior to 1973. Will this increase in productivity growth be sustained? Optimists argue that the United States is currently leading a new technological revolution, sparked by advances in computers, communications, genetics, and other fields, which will allow productivity to continue to grow indefinitely. Others are more cautious, arguing that the long-run commercial value of the new technologies has yet to be proven. A great deal is riding on which view will turn out to be correct.

[3]See, for example, William Nordhaus, "Economic Policy in the Face of Declining Productivity Growth," *European Economic Review,* May/June 1982, pp. 131–158.

## THE COSTS OF ECONOMIC GROWTH

Both this chapter and Chapter 5 emphasized the positive effects of economic growth on the average person's living standard. But should societies always strive for the highest possible rate of economic growth? The answer is no. Even if we accept for the moment the idea that increased output per person is always desirable, attaining a higher rate of economic growth does impose costs on society.

What are the costs of increasing economic growth? The most straightforward is the cost of creating new capital. We know that by expanding the capital stock we can increase future productivity and output. But, to increase the capital stock, we must divert resources that could otherwise be used to increase the supply of consumer goods. For example, to add more robot-operated assembly lines, a society must employ more of its skilled technicians in building industrial robots and fewer in designing video games. To build new factories, more carpenters and lumber must be assigned to factory construction and less to finishing basements or renovating family rooms. In short, high rates of investment in new capital require people to tighten their belts, consume less, and save more—a real economic cost.

Should a country undertake a high rate of investment in capital goods at the sacrifice of consumer goods? The answer depends on the extent that people are willing and able to sacrifice consumption today to have a bigger economic pie tomorrow. In a country that is very poor, or is experiencing an economic crisis, people may prefer to keep consumption relatively high and savings and investment relatively low. The midst of a thunderstorm is not the time to be putting something aside for a rainy day! But in a society that is relatively well off, people may be more willing to make sacrifices to achieve higher economic growth in the future.

Consumption sacrificed to capital formation is not the only cost of achieving higher growth. In the United States in the nineteenth and early twentieth centuries, periods of rapid economic growth were often times in which many people worked extremely long hours at dangerous and unpleasant jobs. While those workers helped to build the economy that Americans enjoy today, the costs were great in terms of reduced leisure time and, in some cases, workers' health and safety.

Other costs of growth include the cost of the research and development that is required to improve technology and the costs of acquiring training and skill (human capital). The fact that a higher living standard tomorrow must be purchased at the cost of current sacrifices is an example of the *scarcity principle,* that having more of one good thing usually means having less of another. Because achieving higher economic growth imposes real economic costs, we know from the *cost-benefit principle* that higher growth should be pursued only if the benefits outweigh the costs.

## PROMOTING ECONOMIC GROWTH

If a society decides to try to raise its rate of economic growth, what are some of the measures that policymakers might take to achieve this objective? Here is a short list of suggestions, based on our discussion of the factors that contribute to growth in average labor productivity and, hence, output per person.

### POLICIES TO INCREASE HUMAN CAPITAL

Because skilled and well-educated workers are more productive than unskilled labor, governments in most countries try to increase the human capital of their citizens by supporting education and training programs. In the United States, government provides public education through high school and grants extensive support to post-secondary schools, including technical schools, colleges, and universities. Publicly funded early intervention programs like Head Start also attempt to build human capital by helping disadvantaged children prepare for school. To a lesser degree than some other countries, the U.S. government also funds job training for unskilled youths and retraining for workers whose skills have become obsolete.

**Why do almost all countries provide free public education?**

Why do almost all countries provide free public education?

All industrial countries provide their citizens free public education through high school, and most subsidize college and other post-secondary schools. Why?

Americans are so used to the idea of free public education that this question may seem odd. But why should the government provide free education when it does not provide even more essential goods and services, such as food or medical care, for free, except to the most needy? Furthermore, educational services can be, and indeed commonly are, supplied and demanded on the private market, without the aid of the government.

An important argument for free or at least subsidized education is that the private demand curve for educational services does not include all the social benefits of education. (Recall the *equilibrium principle* of Chapter 3, which states in part that a market in equilibrium may not exploit all gains achievable from collective action.) For example, the democratic political system relies on an educated citizenry to operate effectively—a factor that an individual demander of educational services has little reason to consider. From a narrower economic perspective, we might argue that individuals do not capture the full economic returns from their schooling. For example, people with high human capital, and thus high earnings, pay more taxes—funds that can be used to finance government services and aid the less fortunate. Because of income taxation, the private benefit to acquiring human capital is less than the social benefit, and the demand for education on the private market may be less than optimal from society's viewpoint. Similarly, educated people are more likely than others to contribute to technological development, and hence to general productivity growth, which may benefit many other people besides themselves. Finally, another argument for public support of education is that poor people who would like to invest in human capital may not be able to do so because of insufficient income.

The Nobel laureate Milton Friedman, among many economists, suggested that these arguments may justify government grants, called educational *vouchers*, to help citizens purchase educational services in the private sector, but they do *not* justify the government providing education directly, as through the public school system. Defenders of public education, on the other hand, argue that the government should have some direct control over education in order to set standards and monitor quality. What do you think?

## POLICIES THAT PROMOTE SAVING AND INVESTMENT

Average labor productivity increases when workers can utilize a sizable and modern capital stock. To support the creation of new capital, government can encourage high rates of saving and investment in the private sector. Many provisions in the U.S. tax code are designed expressly to stimulate households to save and firms to invest. For example, a household that opens an Individual Retirement Account (IRA) is able to save for retirement without paying taxes on either the funds deposited in the IRA or the interest earned on the account. (However, taxes are due when the funds are withdrawn at retirement.) The intent of IRA legislation is to make saving more financially attractive to American households. Similarly, at various times Congress has instituted an investment tax credit, which reduces the tax bills of firms that invest in new capital. Private-sector saving and investment are discussed in greater detail in Chapter 9.

Government can contribute directly to capital formation through *public investment*, or the creation of government-owned capital. Public investment includes the building of roads, bridges, airports, dams, and, in some countries, energy and communications networks. The construction of the U.S. interstate highway system, begun during the administration of President Eisenhower, is often cited as an example of successful public investment. The interstate system substantially reduced long-haul transportation costs in the United States, improving productivity throughout the economy. Today, the web of computers and communications links we call the Internet is having a similar effect. This project, too, received crucial government funding in its early stages. Many research studies have confirmed that government investment in the *infrastructure,* the public capital that supports private-sector economic activities, can be a significant source of growth.

## POLICIES THAT SUPPORT RESEARCH AND DEVELOPMENT

Productivity is enhanced by technological progress, which in turn requires investment in research and development (R&D). In many industries private firms have adequate incentive to conduct research and development activities. There is no need, for example, for the government to finance research for developing a better underarm deodorant. But some types of knowledge, particularly basic scientific knowledge, may have widespread economic benefits that cannot be captured by a single private firm. The developers of the silicon computer chip, for example, were instrumental in creating huge new industries, yet they received only a small portion of the profits flowing from their inventions. Because society in general, rather than the individual inventors, may receive much of the benefit from basic research, government may need to support basic research, as it does through agencies such as the National Science Foundation. The federal government also sponsors a great deal of applied research, particularly in military and space applications. To the extent that national security allows, the government can increase growth by sharing the fruits of such research with the private sector. For example, the Global Positioning System (GPS), which was developed originally for military purposes, is now available in private passenger vehicles, helping drivers find their way.

## THE LEGAL AND POLITICAL FRAMEWORK

Although economic growth comes primarily from activities in the private sector, the government plays an essential role in providing the framework within which the private sector can operate productively. We have discussed the importance of secure property rights and a well-functioning legal system, of an economic environment that encourages entrepreneurship, and of political stability and the free and open exchange of ideas. Government policymakers should also consider the potential effects of tax and regulatory policies on activities that increase productivity, such as investment, innovation, and risk taking. Policies that affect the legal and political framework are examples of *structural macroeconomic policies* (see Chapter 4).

## THE POOREST COUNTRIES: A SPECIAL CASE?

Radical disparities in living standards exist between the richest and poorest countries of the world (see Table 5.4 for some data). Achieving economic growth in the poorest countries is thus particularly urgent. Are the policy prescriptions of this section relevant to those countries, or are very different types of measures necessary to spur growth in the poorest nations?

To a significant extent, the same factors and policies that promote growth in richer countries apply to the poorest countries as well. Increasing human capital by supporting education and training, increasing rates of saving and investment, investing in public capital and infrastructure, supporting research and development, and encouraging entrepreneurship are all measures that will enhance economic growth in poor countries.

However, to a much greater degree than in richer countries, most poor countries need to improve the legal and political environment that underpins their economies. For example, many developing countries have poorly developed or corrupt legal systems, which discourage entrepreneurship and investment by creating uncertainty about property rights. Taxation and regulation in developing countries are often heavy-handed and administered by inefficient bureaucracies, to the extent that it may take months or years to obtain the approvals needed to start a small business or expand a factory. Regulation is also used to suppress market forces in poor countries; for example, the government, rather than the market, may determine the allocation of bank credit or the prices for agricultural products. Structural policies that aim to ameliorate these problems are important preconditions for generating growth in the poorest countries. But probably most important—and most difficult, for some countries—is establishing political stability and the rule of law. Without political stability, domestic and foreign savers will be reluctant to invest in the country, and economic growth will be difficult if not impossible to achieve.

Can rich countries help poor countries to develop? Historically, richer nations have tried to help by providing financial aid through loans or grants from individual countries (foreign aid) or by loans made by international agencies, such as the World Bank. Experience has shown, however, that financial aid to countries that do not undertake structural reforms, such as reducing excessive regulation or improving the legal system, is of limited value. To make their foreign aid most effective, rich countries should help poor countries achieve political stability and undertake the necessary reforms to the structure of their economies.

## ARE THERE LIMITS TO GROWTH?

Earlier in this chapter we saw that even relatively low rates of economic growth, if sustained for a long period, will produce huge increases in the size of the economy. This fact raises the question of whether economic growth can continue indefinitely without depleting natural resources and causing massive damage to the global environment. Does the basic truth that we live in a finite world of finite resources imply that, ultimately, economic growth must come to an end?

The concern that economic growth may not be sustainable is not a new one. An influential 1972 book, *The Limits to Growth*,[4] reported the results of computer simulations that suggested that unless population growth and economic expansion were halted, the world would soon be running out of natural resources, drinkable water, and breathable air. This book, and later works in the same vein, raise some fundamental questions that cannot be done full justice here. However, in some ways its conclusions are misleading.

One problem with the "limits to growth" thesis lies in its underlying concept of economic growth. Those who emphasize the environmental limits on growth assume implicitly that economic growth will always take the form of more of what we have now—more smoky factories, more polluting cars, more fast-food restaurants. If that were indeed the case, then surely there would be limits to the growth the planet can sustain. But growth in real GDP does not necessarily take such a form. Increases in real GDP can also arise from new or higher-quality products. For example, not too long ago tennis rackets were relatively simple items made primarily of wood. Today they are made of newly invented synthetic materials and designed for optimum performance using sophisticated computer simulations. Because these new high-tech tennis rackets are more valued by consumers than the old wooden ones, they increase the real GDP. Likewise, the introduction of new pharmaceuticals has contributed to economic growth, as have the expanded

[4]Donella H. Meadows, Dennis L. Meadows, Jørgen Randers, and William W. Behrens, III, *The Limits to Growth*, New York: New American Library, 1972.

number of TV channels, digital sound, and Internet-based sales. Thus, economic growth need not take the form of more and more of the same old stuff; it can mean newer, better, and perhaps cleaner and more efficient goods and services.

A second problem with the "limits to growth" conclusion is that it overlooks the fact that increased wealth and productivity expand society's capacity to take measures to safeguard the environment. In fact, the most polluted countries in the world are not the richest but those that are in a relatively early stage of industrialization (see Economic Naturalist 7.5). At this stage countries must devote the bulk of their resources to basic needs—food, shelter, health care—and continued industrial expansion. In these countries, clean air and water may be viewed as a luxury rather than a basic need. In more economically developed countries, where the most basic needs are more easily met, extra resources are available to keep the environment clean. Thus continuing economic growth may lead to less, not more, pollution.

A third problem with the pessimistic view of economic growth is that it ignores the power of the market and other social mechanisms to deal with scarcity. During the oil-supply disruptions of the 1970s, newspapers were filled with headlines about the energy crisis and the imminent depletion of world oil supplies. Yet 30 years later, the world's known oil reserves are actually *greater* than they were in the 1970s.

Today's energy situation is so much better than was expected 30 years ago because the market went to work. Reduced oil supplies led to an increase in prices that changed the behavior of both demanders and suppliers. Consumers insulated their homes, purchased more energy-efficient cars and appliances, and switched to alternative sources of energy. Suppliers engaged in a massive hunt for new reserves, opening up major new sources in Latin America, China, and the North Sea. In short, market forces solved the energy crisis.

In general, shortages in any resource will trigger price changes that induce suppliers and demanders to deal with the problem. Simply extrapolating current economic trends into the future ignores the power of the market system to recognize shortages and make the necessary corrections. Government actions spurred by political pressures, such as the allocation of public funds to preserve open space or reduce air pollution, can be expected to supplement market adjustments.

Despite the shortcomings of the "limits to growth" perspective, most economists would agree that not all the problems created by economic growth can be dealt with effectively through the market or the political process. Probably most important, global environmental problems, such as the possibility of global warming or the ongoing destruction of rain forests, are a particular challenge for existing economic and political institutions. Environmental quality is not bought and sold in markets and thus will not automatically reach its optimal level through market processes (recall the *equilibrium principle*). Nor can local or national governments effectively address problems that are global in scope. Unless international mechanisms are established for dealing with global environmental problems, these problems may become worse as economic growth continues.

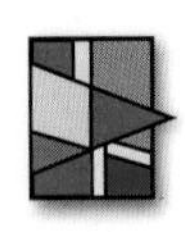

ECONOMIC NATURALIST 7.5

**Why is the air quality so poor in Mexico City?**

Developing countries like Mexico, which are neither fully industrialized nor desperately poor, often have severe environmental problems. Why?

One concern about economic growth is that it will cause ever-increasing levels of environmental pollution. Empirical studies show, however, that the relationship between pollution and real GDP per person is more like an inverted U (see Figure 7.5). In other words, as countries move from very low levels of real GDP per person to "middle-income" levels, most measures of pollution tend to worsen, but environmental quality improves as real GDP per person rises even further. One study of the relationship between air quality and real GDP per person found that the level of real GDP per person at which air quality is the worst—indicated by

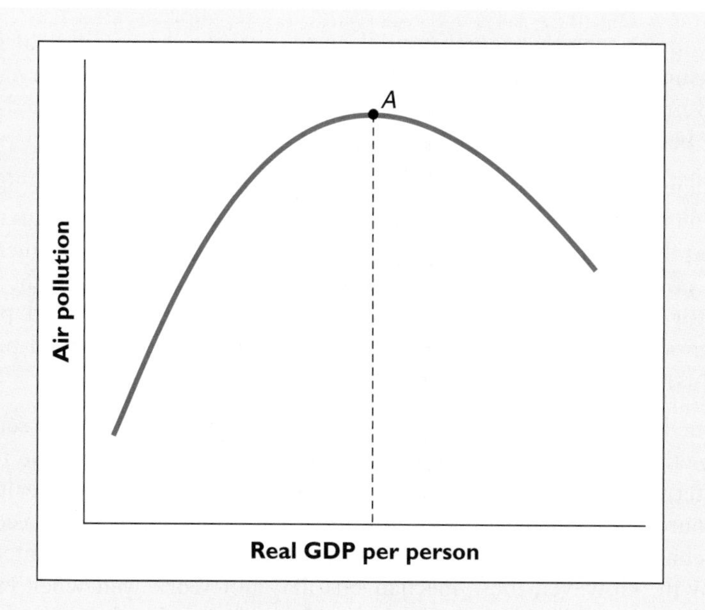


**FIGURE 7.5**
**The Relationship between Air Pollution and Real GDP per Person.**
Empirically, air pollution increases with real GDP per person up to a point and then begins to decline. Maximum air pollution (point *A*) occurs at a level of real GDP per person roughly equal to that of Mexico.

point *A* in Figure 7.5—is roughly equal to the average income level in Mexico.[5] And indeed, the air quality in Mexico City is exceptionally poor, as any visitor to that sprawling metropolis can attest.

That pollution may worsen as a country industrializes is understandable, but why does environmental quality improve when real GDP per person climbs to very high levels? There are a variety of explanations for this phenomenon. Compared to middle-income economies, the richer economies are relatively more concentrated in "clean," high-value services like finance and software production as opposed to pollution-intensive industries like heavy manufacturing. Rich economies are also more likely to have the expertise to develop sophisticated and cost-effective antipollution technologies. But the main reason the richer economies tend to be cleaner is the same reason that the homes of rich people are generally cleaner and in better condition than the homes of the poor. As income rises above the level necessary to fulfill basic needs, more resources remain to dedicate to "luxuries" like a clean environment (the *scarcity principle*). For the rich family, the extra resources will pay for a cleaning service; for the rich country, they will pay for pollution control devices in factories and on automobiles. Indeed, antipollution laws are generally tougher and more strictly enforced in rich countries than in middle-income and poor countries.

## RECAP ECONOMIC GROWTH: DEVELOPMENTS AND ISSUES

- Labor productivity growth slowed throughout the industrialized world in the early 1970s. One possible explanation is that productivity growth has become harder to measure; another is the technological depletion hypothesis, that technological opportunities occur less frequently today than in the immediate postwar period. Some evidence suggests a recent resurgence in productivity growth in the United States.

[5]Gene M. Grossman and Alan B. Krueger, "Environmental Impacts of a North American Free Trade Agreement," in Peter Garber, ed., *The Mexico–U.S. Free Trade Agreement,* Cambridge, MA: MIT Press, 1993. See also Grossman and Krueger, "Economic Growth and the Environment," *Quarterly Journal of Economics,* May 1995, pp. 353–78; and World Bank, *World Development Report: Development and the Environment,* 1992.

- Economic growth has substantial costs, notably the sacrifice of current consumption that is required to free resources for creating new capital and new technologies. Higher rates of growth should be pursued only if the benefits outweigh the costs.
- Policies for promoting economic growth include policies to increase human capital (education and training); policies that promote saving and capital formation; policies that support research and development; and the provision of a legal and political framework within which the private sector can operate productively. Deficiencies in the legal and political framework (for example, official corruption or poorly defined property rights) are a special problem for many developing countries.
- Some have argued that finite resources imply ultimate limits to economic growth. This view overlooks the facts that growth can take the form of better, rather than more, goods and services; that increased wealth frees resources to safeguard the environment; and that political and economic mechanisms exist to address many of the problems associated with growth. However, these mechanisms may not work well when environmental or other problems arising from economic growth are global in scope.

## SUMMARY


- Over the past two centuries the industrialized nations saw enormous improvements in living standards, as reflected in large increases in real GDP per person. Because of the power of *compound interest,* relatively small differences in growth rates, if continued over long periods, can produce large differences in real GDP per person and average living standards. Thus, the rate of long-term economic growth is an economic variable of critical importance.

- Real GDP per person is the product of average labor productivity (real GDP per employed worker) and the share of the population that is employed. Growth in real GDP per person can occur only through growth in average labor productivity, in the share of the population that is working, or both. In the period since 1960, increases in the share of the U.S. population holding a job contributed significantly to rising real GDP per person. But in the past four decades, as in most periods, the main source of the increase in real GDP per person was rising average labor productivity.

- Among the factors that determine labor productivity are the talents, education, training, and skills of workers, or human capital; the quantity and quality of the physical capital that workers use; the availability of land and other natural resources; the application of technology to the production and distribution of goods and services; the effectiveness of *entrepreneurs* and managers; and the broad social and legal environment. Because of *diminishing returns to capital,* beyond a certain point expansion of the capital stock is not the most effective way to increase average labor productivity. Economists generally agree that new technologies are the most important single source of improvements in productivity.

- Since the 1970s the industrial world has experienced a slowdown in productivity growth. Some economists have suggested that the "slowdown" is more the result of an inability to measure increases in the quality of output than of any real economic change. Others have suggested that the exploitation of a backlog of technological opportunities following the Great Depression and World War II led to unusually high growth rates in the 1950s and 1960s, a view called the technological depletion hypothesis. In this view, the slower growth in U.S. productivity since about 1970 in fact reflects a return to a more normal rate of growth. U.S. productivity growth has picked up since about 1991, however, possibly as the result of new technologies.

- Economic growth has costs as well as benefits. Prominent among them is the need to sacrifice current consumption to achieve a high rate of investment in new capital goods; other costs of growing more quickly include extra work effort and the costs of research and development. Thus more economic growth is not necessarily better; whether increased economic growth is desirable depends on whether the benefits of growth outweigh the costs.

- Among the ways in which government can stimulate economic growth are by adopting policies that encourage the creation of human capital; that promote saving and investment, including public investment in infrastructure; that support research and development, particularly in the basic sciences; and that provide a legal and political framework that supports private-sector activities. The poorest countries, with poorly developed legal, tax, and regulatory systems, are often in the greatest need of an improved legal and political framework and increased political stability.

- Are there limits to growth? Arguments that economic growth must be constrained by environmental problems and the limits of natural resources ignore the fact that economic growth can take the form of increasing quality as well as increasing quantity. Indeed, increases in output can provide additional resources for cleaning up the environment. Finally, the market system, together with political processes, can solve many of the problems associated with economic growth. On the other hand, global environmental problems, which can be handled neither by the market nor by individual national governments, have the potential to constrain economic growth.

## KEY TERMS

compound interest (173)

diminishing returns to capital (180)

entrepreneurs (182)

## REVIEW QUESTIONS

1. What has happened to real GDP per person in the industrialized countries over the past century? What implications does this have for the average person?
2. Why do economists consider growth in average labor productivity to be the key factor in determining long-run living standards?
3. What is *human capital*? Why is it economically important? How is new human capital created?
4. You have employed five workers of varying physical strength to dig a ditch. Workers without shovels have zero productivity in ditchdigging. How should you assign shovels to workers if you don't have enough shovels to go around? How should you assign any additional shovels that you obtain? Using this example, discuss (a) the relationship between the availability of physical capital and average labor productivity and (b) the concept of diminishing returns to capital.
5. Discuss how talented entrepreneurs and effective managers can enhance average labor productivity.
6. What major contributions can the government make to the goal of increasing average labor productivity?
7. What explanations have been offered for the slowdown in productivity growth observed in industrial countries since the early 1970s?
8. Discuss the following statement: "Because the environment is fragile and natural resources are finite, ultimately economic growth must come to an end."

## PROBLEMS

1. Richland's real GDP per person is $10,000, and Poorland's real GDP per person is $5,000. However, Richland's real GDP per person is growing at 1 percent per year and Poorland's is growing at 3 percent per year. Compare real GDP per person in the two countries after 10 years and after 20 years. Approximately how many years will it take Poorland to catch up to Richland?
2. Refer to Table 7.3 for growth rates of average labor productivity over the periods 1960–1973, 1973–1979, and 1979–2000. Suppose that growth of average labor productivity in the United States had continued at its 1960–1973 rate until 2000. Proportionally, how much higher would U.S. average labor productivity in 2000 have been, compared to its actual value? (*Note:* You do not need to know the actual values of average labor productivity in any year to solve this problem.) Does your answer shed light on why economists consider the post-1973 productivity slowdown to be an important issue?
3. The "graying of America" will substantially increase the fraction of the population that is retired in the decades to come. To illustrate the implications for U.S. living standards, suppose that over the 40 years following 2000 the share of the population that is working returns to its 1960 level, while average labor productivity increases by as much as it did during 1960–2000. Under this scenario, what would be the net change in real GDP per person between 2000 and 2040? The following data will be useful:

| | Average labor productivity | Share of population employed |
|---|---|---|
| 1960 | $35,836 | 36.4% |
| 2000 | $66,588 | 48.9% |

4. Here are data for Canada, Germany, and Japan on the ratio of employment to population in 1979 and 2000:

| | 1979 | 2000 |
|---|---|---|
| Canada | 0.45 | 0.46 |
| Germany | 0.34 | 0.33 |
| Japan | 0.47 | 0.51 |

Using data from Table 7.1, find average labor productivity for each country in 1979 and in 2000. How much of the increase in output per person in each country over the 1979–2000 period is due to increased labor productivity? To increased employment relative to population?

5. Joanne has just completed high school and is trying to determine whether to go to junior college for two years or go directly to work. Her objective is to maximize the savings she will have in the bank five years from now. If she goes directly to work she will earn $20,000 per year for each of the next five years. If she goes to junior college, for each of the next two years she will earn nothing—indeed, she will have to borrow $6,000 each year to cover tuition and books. This loan must be repaid in full three years after graduation. If she graduates from junior college, in each of the subsequent three years her wages will be $38,000 per year. Joanne's total living expenses and taxes, excluding tuition and books, equal $15,000 per year.
   a. Suppose for simplicity that Joanne can borrow and lend at 0 percent interest. On purely economic grounds, should she go to junior college or work?
   b. Does your answer to part a change if she can earn $23,000 per year with only a high school degree?
   c. Does your answer to part a change if Joanne's tuition and books cost $8,000 per year?
   d.* Suppose that the interest rate at which Joanne can borrow and lend is 10 percent per year, but other data are as in part a. Savings are deposited at the end of the year they are earned and receive (compound) interest at the end of each subsequent year. Similarly, the loans are taken out at the end of the year in which they are needed, and interest does not accrue until the end of the subsequent year. Now that the interest rate has risen, should Joanne go to college or go to work?

6. The Good'n'Fresh Grocery Store has two checkout lanes and four employees. Employees are equally skilled, and all are able either to operate a register (checkers) or bag groceries (baggers). The store owner assigns one checker and one bagger to each lane. A lane with a checker and a bagger can check out 40 customers per hour. A lane with a checker only can check out 25 customers per hour.
   a. In terms of customers checked out per hour, what is total output and average labor productivity for the Good'n'Fresh Grocery Store?
   b. The owner adds a third checkout lane and register. Assuming that no employees are added, what is the best way to reallocate the workers to tasks? What is total output and average labor productivity (in terms of customers checked out per hour) now?
   c. Repeat part b for the addition of a fourth checkout lane, and a fifth. Do you observe diminishing returns to capital in this example?

7. Harrison, Carla, and Fred are housepainters. Harrison and Carla can paint 100 square feet per hour using a standard paintbrush, and Fred can paint 80 square feet per hour. Any of the three can paint 200 square feet per hour using a roller.
   a. Assume Harrison, Carla, and Fred have only paintbrushes at their disposal. What is the average labor productivity, in terms of square feet per painter-hour, for the three painters taken as a team? Assume that the three painters always work the same number of hours.
   b. Repeat part a for the cases in which the team has one, two, three, or four rollers available. Are there diminishing returns to capital?
   c. An improvement in paint quality increases the area that can be covered per hour (by either brushes or rollers) by 20 percent. How does this technological improvement

*Problem marked by an asterisk (*) is more difficult.

affect your answers to part b? Are there diminishing returns to capital? Does the technological improvement increase or reduce the economic value of an additional roller?

8. Hester's Hatchery raises fish. At the end of the current season she has 1,000 fish in the hatchery. She can harvest any number of fish that she wishes, selling them to restaurants for $5 apiece. Because big fish make little fish, for every fish that she leaves in the hatchery this year she will have two fish at the end of next year. The price of fish is expected to be $5 each next year as well. Hester relies entirely on income from current fish sales to support herself.
   a. How many fish should Hester harvest if she wants to maximize the growth of her stock of fish from this season to next season?
   b. Do you think maximizing the growth of her fish stock is an economically sound strategy for Hester? Why or why not? Relate to the text discussion on the costs of economic growth.
   c. How many fish should Hester harvest if she wants to maximize her current income? Do you think this is a good strategy?
   d. Explain why Hester is unlikely to harvest either all or none of her fish, but instead will harvest some and leave the rest to reproduce.
9. True or False: For advances in basic science to translate into improvements in standards of living, they must be supported by favorable economic conditions. Discuss, using concrete examples where possible to illustrate your arguments.
10. Write a short essay evaluating the U.S. economy in terms of each of the six determinants of average labor productivity discussed in the text. Are there any areas in which the U.S. is exceptionally strong, relative to other countries? Areas where the U.S. is less strong than some other countries? Illustrate your arguments with numbers from the *Statistical Abstract of the United States* (available online at www.census.gov/statab/www/) and other sources, as appropriate.

## ANSWERS TO IN-CHAPTER EXERCISES

7.1 If the United States had grown at the Japanese rate for the period 1870–2000, real GDP per person in 2000 would have been ($2,843) × ($1.025^{130}$) = $70,450. Actual GDP per person in the United States in 2000 was $32,629, so at the higher rate of growth output per person would have been $70,450/$32,629 = 2.16 times higher.

7.2 As before Lucy can wrap 4,000 candies per week, or 100 candies per hour. Ethel can wrap 500 candies per hour, and working 40 hours weekly she can wrap 20,000 candies per week. Together Lucy and Ethel can wrap 24,000 candies per week. Since they work a total of 80 hours between them, their output per hour as a team is 24,000 candies wrapped per 80 hours = 300 candies wrapped per hour, midway between their hourly productivities as individuals.

7.3 Because Ethel can wrap 300 candies per hour by hand, the benefit of giving Ethel the machine is 500 − 300 = 200 additional candies wrapped per hour. Because Lucy wraps only 100 candies per hour by hand, the benefit of giving Lucy the machine is 400 additional candies wrapped per hour. So the benefit of giving the machine to Lucy is greater than of giving it to Ethel. Equivalently, if the machine goes to Ethel, then Lucy and Ethel between them can wrap 500 + 100 = 600 candies per hour, but if Lucy uses the machine the team can wrap 300 + 500 = 800 candies per hour. So output is increased by letting Lucy use the machine.

7.4 Now, working by hand, Lucy can wrap 300 candies per hour and Ethel can wrap 500 candies per hour. With a machine, either Lucy or Ethel can wrap 800 candies per hour. As in Exercise 7.3, the benefit of giving a machine to Lucy (500 candies per hour) exceeds the benefit of giving a machine to Ethel (300 candies per hour), so if only one machine is available Lucy should use it.

The table analogous to Table 7.2 now looks like this:

**Relationship of Capital, Output, and Productivity in the Candy-Wrapping Factory**

| Number of machines (*K*) | Candies wrapped per week (*Y*) | Total hours worked (*N*) | Average hourly labor productivity (*Y/N*) |
|---|---|---|---|
| 0 | 32,000 | 80 | 400 |
| 1 | 52,000 | 80 | 650 |
| 2 | 64,000 | 80 | 800 |
| 3 | 64,000 | 80 | 800 |

Comparing this table with Table 7.2, you can see that technological advance has increased labor productivity for any value of *K*, the number of machines available.

Adding one machine increases output by 20,000 candies wrapped per week, adding the second machine increases output by 12,000 candies wrapped per week, and adding the third machine does not increase output at all (because there is no worker available to use it). So diminishing returns to capital still hold after the technological improvement.

7.5 Although the individual worker is the same person he was in Bangladesh, by coming to the United States he gains the benefit of factors that enhance average labor productivity in this country, relative to his homeland. These include more and better capital to work with, more natural resources per person, more advanced technologies, sophisticated entrepreneurs and managers, and a political-legal environment that is conducive to high productivity. It is not guaranteed that the value of the immigrant's human capital will rise (it may not, for example, if he speaks no English and has no skills applicable to the U.S. economy), but normally it will.

Since increased productivity leads to higher wages and living standards, on economic grounds the Bangladeshi worker has a strong incentive to immigrate to the United States if he is able to do so.

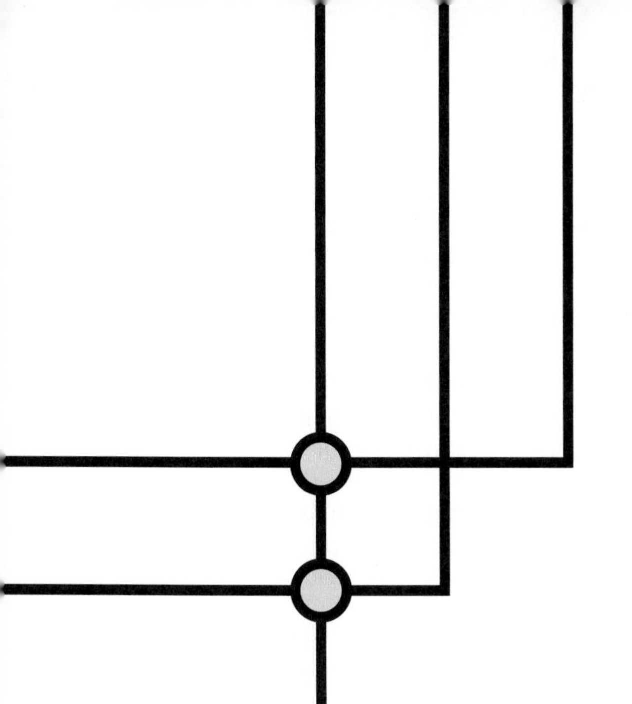

CHAPTER 10

# MONEY, PRICES, AND THE FEDERAL RESERVE

The Aztec people, who dominated central Mexico until the coming of the Spanish in the sixteenth century, are perhaps best remembered today for their elaborate ceremonies that culminated in human sacrifice. But the Aztec Empire also had a sophisticated economic system, which included active trade not only in a wide variety of agricultural goods—such as corn, tomatoes, and peanuts—but also in manufactured items, including jewelry, sandals, cloaks, baskets, alcohol, and weapons. Slaves and captives—some designated for sacrifice—were also bought and sold. Trade was carried out in local markets, notably in the Aztec capital city, located where Mexico City now stands, and over long distances, as far north as present-day Arizona, for example.[1]

Despite ample supplies of gold, the Aztecs did not use metallic coins to carry out their market transactions, as most European peoples did. Instead, they often used chocolate or, more specifically, cacao "beans" (the seeds of the cacao plant). Prices of goods or services were quoted in cacao beans, in much the same way that today we quote prices in terms of dollars. Cacao beans could also be used to purchase goods, to pay for a service, or to make change—for example, to balance a transaction when a good of greater value was traded for one of lesser value. For larger items, prices were quoted in terms of bags of approximately 24,000 cacao beans, although large purchases might actually be paid for with something easier to carry than bags of cacao beans, such as woven cloaks.

[1]See Jack Weatherford, *The History of Money: From Sandstone to Cyberspace,* New York: Crown Publishers, 1997, for a discussion of the use of money in many economies from the ancients to the present day.

Cacao beans were, for the Aztecs, a form of money. In everyday language, people use the word *money* to mean "income" or "wealth," as in "That job pays good money" or "I wish I had as much money as she does." Economists, however, give a much more specific meaning to the term. To the economist, **money** is any asset that can be used in making purchases.

***money*** any asset that can be used in making purchases

Common examples of money in the modern world are currency and coin. A checking account balance represents another asset that can be used in making payments (as when you write a check to pay for your weekly groceries) and so is also counted as money. In contrast, shares of stock, for example, cannot be used directly in most transactions. Stock must first be sold—that is, converted into cash or a checking account deposit—before further transactions, such as buying your groceries, can be made.

Historically, a wide variety of objects have been used as money, including not only cacao beans but also gold and silver coins, shells, beads, feathers, and, on the Island of Yap, large, immovable boulders. Prior to the use of metallic coins, by far the most common form of money was the cowrie, a type of shell found in the South Pacific. Cowries were used as money in some parts of Africa until very recently, being officially accepted for payment of taxes in Uganda until the beginning of the twentieth century. Today money can be virtually intangible, as in the case of your checking account balance, which exists only in the form of an entry in your bank's computer.

In this chapter we discuss the role of money in modern economies: why it is important, how it is measured, how it is created. Money plays a major role in everyday economic transactions but, as we will see, it is also quite important at the macro level. For example, as we mentioned in Chapter 4, one of the three main types of macroeconomic policy, monetary policy, relates primarily to decisions about how much money should be allowed to circulate in the economy. In the United States, monetary policy is made by the Federal Reserve, the nation's central bank. Because the Federal Reserve, or Fed, determines the nation's money supply, this chapter also introduces the Fed and discusses some of the policy tools at its disposal. Finally, the chapter discusses the important relationship between the amount of money in circulation and the rate of inflation in an economy.

## MONEY AND ITS USES

***medium of exchange*** an asset used in purchasing goods and services

***barter*** the direct trade of goods or services for other goods or services

Why do people use money? Money has three principal uses: a *medium of exchange*, a *unit of account*, and a *store of value*.

Money serves as a **medium of exchange** when it is used to purchase goods and services, as when you pay cash for a newspaper or write a check to cover your utilities bill. This is perhaps money's most crucial function. Think about how complicated daily life would become if there were no money. Without money, all economic transactions would have to be in the form of **barter**, which is the direct trade of goods or services for other goods or services.

© Catherine Karnow/Corbis

In a world without money, she could eat only by finding someone willing to trade food for a musical performance.

Barter is highly inefficient because it requires that each party to a trade has something that the other party wants, a so-called double coincidence of wants. For example, under a barter system, a musician could get her dinner only by finding someone willing to trade food for a musical performance. Finding such a match of needs, where each party happens to want exactly what the other person has to offer, would be difficult to do on a regular basis. In a world with money, the musician's problem is considerably simpler. First, she must find someone who is willing to pay money for her musical performance. Then, with the money received, she can purchase the food and other goods and services that she needs. In a society that uses money, it is not necessary that the person who wants to hear music and the person willing to provide food to the musician be one and the same. In other words, there need not be a double coincidence of wants for trades of goods and services to take place.

By eliminating the problem of having to find a double coincidence of wants in order to trade, the use of money in a society permits individuals to specialize in producing particular goods or services, as opposed to having every family or village produce most of what it needs. Specialization greatly increases economic efficiency and material standards of living, as we discussed in Chapter 2 (*the principle of comparative advantage*). This usefulness of money in making transactions explains why savers hold money, even though money generally pays a low rate of return. Cash, for example, pays no interest at all, and the balances in checking accounts usually pay a lower rate of interest than could be obtained in alternative financial investments.

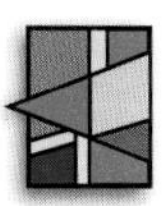

Money's second function is as a *unit of account*. As a **unit of account,** money is the basic yardstick for measuring economic value. In the United States virtually all prices—including the price of labor (wages) and the prices of financial assets, such as shares of General Motors stock—are expressed in dollars. Expressing economic values in a common unit of account allows for easy comparisons. For example, grain can be measured in bushels and coal in tons, but to judge whether 20 bushels of grain is economically more or less valuable than a ton of coal, we express both values in dollar terms. The use of money as a unit of account is closely related to its use as a medium of exchange; because money is used to buy and sell things, it makes sense to express prices of all kinds in money terms.

***unit of account*** a basic measure of economic value

As a **store of value,** its third function, money is a way of holding wealth. For example, the miser who stuffs cash in his mattress or buries gold coins under the old oak tree at midnight is holding wealth in money form. Likewise, if you regularly keep a balance in your checking account, you are holding part of your wealth in the form of money. Although money is usually the primary medium of exchange or unit of account in an economy, it is not the only store of value. There are numerous other ways of holding wealth, such as owning stocks, bonds, or real estate.

***store of value*** an asset that serves as a means of holding wealth

For most people, money is not a particularly good way to hold wealth, apart from its usefulness as a medium of exchange. Unlike government bonds and other types of financial assets, most forms of money pay no interest, and there is always the risk of cash being lost or stolen. However, cash has the advantage of being anonymous and difficult to trace, making it an attractive store of value for smugglers, drug dealers, and others who want their assets to stay out of the view of the Internal Revenue Service.

### Private money: Ithaca Hours and LETS

Since money is such a useful tool, why is money usually issued only by governments? Are there examples of privately issued money?

Money is usually issued by the government, not private individuals, but in part this reflects legal restrictions on private money issuance. Where the law allows, private moneys do sometimes emerge.[2] For example, privately issued currencies circulate in more than 30 U.S. communities. In Ithaca, New York, a private currency known as "Ithaca Hours" has circulated since 1991. Instituted by town resident Paul Glover, each Ithaca Hour is equivalent to $10, the average hourly wage of workers in the county. The bills, printed with specially developed inks to prevent counterfeiting, honor local people and the environment. An estimated 1,600 individuals and businesses have earned and spent Hours. Founder Paul Glover argues that the use of Hours, which can't be spent elsewhere, induces people to do more of their shopping in the local economy.

A more high-tech form of private money is associated with computerized trading systems called LETS, for local electronic trading system. These are quite popular in Australia, New Zealand, and Great Britain (the United States has

[2]Barbara A. Good, "Private Money: Everything Old Is New Again," Federal Reserve Bank of Cleveland, *Economic Commentary,* April 1, 1998.

about 10 of them). Participants in a LETS post a list of goods and services they would like to buy or sell. When transactions are made, the appropriate number of "computer credits" is subtracted from the buyer's account and added to the seller's account. People are allowed to have negative balances in their accounts, so participants have to trust other members not to abuse the system by buying many goods and services and then quitting. LETS credits exist in the computer only and are never in the form of paper or metal. In this respect LETS may foreshadow the electronic monetary systems of the future.

What do Ithaca Hours and LETS credits have in common? By functioning as a medium of exchange, each facilitates trade within a community.

## MEASURING MONEY

How much money, defined as financial assets usable for making purchases, is there in the U.S. economy at any given time? This question is not simple to answer because in practice it is not easy to draw a clear distinction between those assets that should be counted as money and those that should not. Dollar bills are certainly a form of money, and a van Gogh painting certainly is not. However, brokerage firms now offer accounts that allow their owners to combine financial investments in stocks and bonds with check-writing and credit card privileges. Should the balances in these accounts, or some part of them, be counted as money? It is difficult to tell.

Economists skirt the problem of deciding what is and isn't money by using several alternative definitions of money, which vary in how broadly the concept of money is defined. A relatively "narrow" definition of the amount of money in the U.S. economy is called M1. **M1** is the sum of currency outstanding and balances held in checking accounts. A broader measure of money, called **M2**, includes all the assets in M1 plus some additional assets that are usable in making payments, but at greater cost or inconvenience than currency or checks. Table 10.1 lists the components of M1 and M2 and also gives the amount of

**M1** sum of currency outstanding and balances held in checking accounts

**M2** All the assets in M1 plus some additional assets that are usable in making payments but at greater cost or inconvenience than currency or checks

**TABLE 10.1**
**Components of M1 and M2, July 2002**

| | | |
|---|---|---|
| **M1** | | **1,197.8** |
| Currency | 615.1 | |
| Demand deposits | 303.8 | |
| Other checkable deposits | 270.3 | |
| Travelers' checks | 8.6 | |
| **M2** | | **5,641.2** |
| M1 | 1,197.8 | |
| Savings deposits | 2,552.8 | |
| Small-denomination time deposits | 920.8 | |
| Money market mutual funds | 969.8 | |

NOTES: Billions of dollars, adjusted for seasonal variations. In M1, currency refers to cash and coin. Demand deposits are non-interest-bearing checking accounts, and "other checkable deposits" includes checking accounts that bear interest. M2 includes all the components of M1, balances in savings accounts, "small-denomination" (under $100,000) deposits held at banks for a fixed term, and money market mutual funds (MMMFs). MMMFs are organizations that sell shares, use the proceeds to buy safe assets (like government bonds), and often allow their shareholders some check-writing privileges.

SOURCE: Federal Reserve Bank of St. Louis, FRED database, www.stls.frb.org/research/aggreg.html, or Federal Reserve release H.6.

each type of asset outstanding as of July 2002. For most purposes, however, it is sufficient to think of money as the sum of currency outstanding and balances in checking accounts, or M1.

**MONEY AND ITS USES**

*Money* is any asset that can be used in making purchases, such as currency or a checking account. Money serves as a *medium of exchange* when it is used to purchase goods and services. The use of money as a medium of exchange eliminates the need for *barter* and the difficulties of finding a "double coincidence of wants." Money also serves as a *unit of account* and a *store of value.*

In practice, two basic measures of money are M1 and M2. M1, a more narrow measure, is made up primarily of currency and balances held in checking accounts. The broader measure, M2, includes all the assets in M1 plus some additional assets usable in making payments.

## COMMERCIAL BANKS AND THE CREATION OF MONEY

What determines the amount of money in the economy? If the economy's supply of money consisted entirely of currency, the answer would be simple: The supply of money would just be equal to the value of the currency created and circulated by the government. However, as we have seen, in modern economies the money supply consists not only of currency but also of deposit balances held by the public in commercial, that is, private, banks. The determination of the money supply in a modern economy thus depends in part on the behavior of commercial banks and their depositors.

To see how the existence of commercial banks affects the money supply, we will use the example of a fictional country, the Republic of Gorgonzola. Initially, we assume, Gorgonzola has no commercial banking system. To make trading easier and eliminate the need for barter, the government directs the central bank of Gorgonzola to put into circulation a million identical paper notes, called guilders. The central bank prints the guilders and distributes them to the populace. At this point the Gorgonzolan money supply is a million guilders.

However, the citizens of Gorgonzola are unhappy with a money supply made up entirely of paper guilders, since the notes may be lost or stolen. In response to the demand for safekeeping of money, some Gorgonzolan entrepreneurs set up a system of commercial banks. At first, these banks are only storage vaults where people can deposit their guilders. When people need to make a payment they can either physically withdraw their guilders or, more conveniently, write a check on their account. Checks give the banks permission to transfer guilders from the account of the person paying by check to the account of the person to whom the check is made out. With a system of payments based on checks the paper guilders need never leave the banking system, although they flow from one bank to another as a depositor of one bank makes a payment to a depositor in another bank. Deposits do not pay interest in this economy; indeed, the banks can make a profit only by charging depositors fees in exchange for safeguarding their cash.

Let's suppose for now that people prefer bank deposits to cash and so deposit all of their guilders with the commercial banks. With all guilders in the vaults of banks, the balance sheet of all of Gorgonzola's commercial banks taken together is as shown in Table 10.2.

The *assets* of the commercial banking system in Gorgonzola are the paper guilders sitting in the vaults of all the individual banks. The banking system's *liabilities* are the deposits of the banks' customers, since checking account balances represent money owed by the banks to the depositors.

**TABLE 10.2**
**Consolidated Balance Sheet of Gorgonzolan Commercial Banks (Initial)**

| Assets | | Liabilities | |
|---|---|---|---|
| Currency | 1,000,000 guilders | Deposits | 1,000,000 guilders |

***bank reserves*** cash or similar assets held by commercial banks for the purpose of meeting depositor withdrawals and payments

***100 percent reserve banking*** a situation in which banks' reserves equal 100 percent of their deposits

Cash or similar assets held by banks are called **bank reserves.** In this example, bank reserves, for all the banks taken together, equal 1,000,000 guilders—the currency listed on the asset side of the consolidated balance sheet. Banks hold reserves to meet depositors' demands for cash withdrawals or to pay checks drawn on their depositors' accounts. In this example, the bank reserves of 1,000,000 guilders equal 100 percent of banks' deposits, which are also 1,000,000 guilders. A situation in which bank reserves equal 100 percent of bank deposits is called **100 percent reserve banking.**

Bank reserves are held by banks in their vaults, rather than circulated among the public, and thus are *not* counted as part of the money supply. However, bank deposit balances, which can be used in making transactions, *are* counted as money. So, after the introduction of "safekeeper" banks in Gorgonzola, the money supply, equal to the value of bank deposits, is 1,000,000 guilders, which is the same as it was prior to the introduction of banks.

After a while, to continue the story, the commercial bankers of Gorgonzola begin to realize that keeping 100 percent reserves against deposits is not necessary. True, a few guilders flow in and out of the typical bank as depositors receive payments or write checks, but for the most part the stacks of paper guilders just sit there in the vaults, untouched and unused. It occurs to the bankers that they can meet the random inflow and outflow of guilders to their banks with reserves that are less than 100 percent of their deposits. After some observation, the bankers conclude that keeping reserves equal to only 10 percent of deposits is enough to meet the random ebb and flow of withdrawals and payments from their individual banks. The remaining 90 percent of deposits, the bankers realize, can be lent out to borrowers to earn interest.

So the bankers decide to keep reserves equal to 100,000 guilders, or 10 percent of their deposits. The other 900,000 guilders they lend out at interest to Gorgonzolan cheese producers who want to use the money to make improvements to their farms. After the loans are made, the balance sheet of all of Gorgonzola's commercial banks taken together has changed, as shown in Table 10.3.

**TABLE 10.3**
**Consolidated Balance Sheet of Gorgonzolan Commercial Banks after One Round of Loans**

| Assets | | Liabilities | |
|---|---|---|---|
| Currency (= reserves) | 100,000 guilders | Deposits | 1,000,000 guilders |
| Loans to farmers | 900,000 guilders | | |

***reserve-deposit ratio*** bank reserves divided by deposits

***fractional-reserve banking system*** a banking system in which bank reserves are less than deposits so that the reserve-deposit ratio is less than 100 percent

After the loans are made, the banks' reserves of 100,000 guilders no longer equal 100 percent of the banks' deposits of 1,000,000 guilders. Instead, the **reserve-deposit ratio,** which is bank reserves divided by deposits, is now equal to 100,000/1,000,000, or 10 percent. A banking system in which banks hold fewer reserves than deposits so that the reserve-deposit ratio is less than 100 percent is called a **fractional-reserve banking system.**

Notice that 900,000 guilders have flowed out of the banking system (as loans to farmers) and are now in the hands of the public. But we have assumed that private citizens prefer bank deposits to cash for making transactions. So ultimately people will redeposit the 900,000 guilders in the banking system. After these deposits are made, the consolidated balance sheet of the commercial banks is as in Table 10.4.

**TABLE 10.4**
**Consolidated Balance Sheet of Gorgonzolan Commercial Banks after Guilders Are Redeposited**

| Assets | | Liabilities | |
|---|---|---|---|
| Currency (= reserves) | 1,000,000 guilders | Deposits | 1,900,000 guilders |
| Loans to farmers | 900,000 guilders | | |

Notice that bank deposits, and hence the economy's money supply, now equal 1,900,000 guilders. In effect, the existence of the commercial banking system has permitted the creation of new money. These deposits, which are liabilities of the banks, are balanced by assets of 1,000,000 guilders in reserves and 900,000 guilders in loans owed to the banks.

The story does not end here. On examining their balance sheets, the bankers are surprised to see that they once again have "too many" reserves. With deposits of 1,900,000 guilders and a 10 percent reserve-deposit ratio, they need only 190,000 guilders in reserves. But they have 1,000,000 guilders in reserves—810,000 too many. Since lending out their excess guilders is always more profitable than leaving them in the vault, the bankers proceed to make another 810,000 guilders in loans. Eventually these loaned-out guilders are redeposited in the banking system, after which the consolidated balance sheet of the banks is as shown in Table 10.5.

**TABLE 10.5**
**Consolidated Balance Sheet of Gorgonzolan Commercial Banks after Two Rounds of Loans and Redeposits**

| Assets | | Liabilities | |
|---|---|---|---|
| Currency (= reserves) | 1,000,000 guilders | Deposits | 2,710,000 guilders |
| Loans to farmers | 1,710,000 guilders | | |

Now the money supply has increased to 2,710,000 guilders, equal to the value of bank deposits. Despite the expansion of loans and deposits, however, the bankers find that their reserves of 1,000,000 guilders *still* exceed the desired level of 10 percent of deposits, which are 2,710,000 guilders. And so yet another round of lending will take place.

### EXERCISE 10.1

**Determine what the balance sheet of the banking system of Gorgonzola will look like after a third round of lending to farmers and redeposits of guilders into the commercial banking system. What is the money supply at that point?**

The process of expansion of loans and deposits will only end when reserves equal 10 percent of bank deposits, because as long as reserves exceed 10 percent of deposits the banks will find it profitable to lend out the extra reserves. Since reserves at the end of every round equal 1,000,000 guilders, for the reserve-deposit ratio to equal 10 percent, total deposits must equal 10,000,000 guilders. Further, since the balance sheet must balance, with assets equal to liabilities, we know as well that at the end of the process loans to cheese producers must equal 9,000,000 guilders. If loans equal 9,000,000 guilders, then bank assets, the sum of loans and reserves (1,000,000 guilders), will equal 10,000,000 guilders, which is the same as bank liabilities (bank deposits). The final consolidated balance sheet is as shown in Table 10.6.

**TABLE 10.6**
**Final Consolidated Balance Sheet of Gorgonzolan Commercial Banks**

| Assets | | Liabilities | |
|---|---|---|---|
| Currency (= reserves) | 1,000,000 guilders | Deposits | 10,000,000 guilders |
| Loans to farmers | 9,000,000 guilders | | |

The money supply, which is equal to total deposits, is 10,000,000 guilders at the end of the process. We see that the existence of a fractional-reserve banking system has multiplied the money supply by a factor of 10, relative to the economy with no banks or the economy with 100 percent reserve banking. Put another way, with a 10 percent reserve-deposit ratio, each guilder deposited in the banking system can "support" 10 guilders worth of deposits.

To find the money supply in this example more directly, we observe that deposits will expand through additional rounds of lending as long as the ratio of bank reserves to bank deposits exceeds the reserve-deposit ratio desired by banks. When the actual ratio of bank reserves to deposits equals the desired reserve-deposit ratio, the expansion stops. So ultimately, deposits in the banking system satisfy the following relationship:

$$\frac{\text{Bank reserves}}{\text{Bank deposits}} = \text{Desired reserve-deposit ratio.}$$

This equation can be rewritten to solve for bank deposits:

$$\text{Bank deposits} = \frac{\text{Bank reserves}}{\text{Desired reserve-deposit ratio}}. \tag{10.1}$$

In Gorgonzola, since all the currency in the economy flows into the banking system, bank reserves equal 1,000,000 guilders. The reserve-deposit ratio desired by banks is 0.10. Therefore, using Equation 10.1, we find that bank deposits equal (1,000,000 guilders)/0.10, or 10 million guilders, the same answer we found in the consolidated balance sheet of the banks, Table 10.6.

**EXERCISE 10.2**

**Find deposits and the money supply in Gorgonzola if the banks' desired reserve-deposit ratio is 5 percent rather than 10 percent. What if the total amount of currency circulated by the central bank is 2,000,000 guilders and the desired reserve-deposit ratio remains at 10 percent?**

## THE MONEY SUPPLY WITH BOTH CURRENCY AND DEPOSITS

In the example of Gorgonzola we assumed that all money is held in the form of deposits in banks. In reality, of course, people keep only part of their money holdings in the form of bank accounts and hold the rest in the form of currency. Fortunately, allowing for the fact that people hold both currency and bank deposits does not greatly complicate the determination of the money supply, as Example 10.1 shows.

EXAMPLE 10.1

### The money supply with both currency and deposits

Suppose that the citizens of Gorgonzola choose to hold a total of 500,000 guilders in the form of currency and to deposit the rest of their money in banks. Banks keep reserves equal to 10 percent of deposits. What is the money supply in Gorgonzola?

The money supply is the sum of currency in the hands of the public and bank deposits. Currency in hands of the public is given as 500,000 guilders. What is the quantity of bank deposits? Since 500,000 of the 1,000,000 guilders issued by the central bank are being used by the public in the form of currency, only the remaining 500,000 guilders is available to serve as bank reserves. We know that deposits equals bank reserves divided by the reserve-deposit ratio, so deposits are 500,000 guilders/0.10 = 5,000,000 guilders. The total money supply is the sum of currency in the hands of the public (500,000 guilders) and bank deposits (5,000,000 guilders), or 5,500,000 guilders.

We can write a general relationship that captures the reasoning of Example 10.1. First, let's write out the fact that the money supply equals currency plus bank deposits:

$$\text{Money supply} = \text{Currency held by the public} + \text{Bank deposits}.$$

We also know that bank deposits equal bank reserves divided by the reserve-deposit ratio that is desired by commercial banks (Equation 10.1). Using that relationship to substitute for bank deposits in the expression for the money supply, we get

$$\text{Money supply} = \text{Currency held by public} + \frac{\text{Bank reserves}}{\text{Desired reserve-deposit ratio}}. \quad (10.2)$$

We can use Equation 10.2 to confirm our answer to Example 10.1. In that example, currency held by the public is 500,000 guilders, bank reserves are 500,000 guilders, and the desired reserve-deposit ratio is 0.10. Plugging these values into Equation 10.2, we get that the money supply equals 500,000 + 500,000/0.10 = 5,500,000, the same answer we found before.

EXAMPLE 10.2

### The money supply at Christmas

During the Christmas season people choose to hold unusually large amounts of currency for shopping. With no action by the central bank, how would this change in currency holding affect the national money supply?

To illustrate with a numerical example, suppose that initially bank reserves are 500, the amount of currency held by the public is 500, and the desired reserve-deposit ratio in the banking system is 0.2. Inserting these values into Equation 10.2, we find that the money supply equals 500 + 500/0.2 = 3,000.

Now suppose that because of Christmas shopping needs, the public increases its currency holdings to 600 by withdrawing 100 from commercial banks. These withdrawals reduce bank reserves to 400. Using Equation 10.2 we find now that the money supply is 600 + 400/0.2 = 2,600. So the public's increased holdings of currency have caused the money supply to drop, from 3,000 to 2,600. The reason for the drop is that with a reserve-deposit ratio of 20 percent, every dollar in the vaults of banks can "support" $5 of deposits and hence $5 of money supply. However, the same dollar in the hands of the public becomes $1 of currency, contributing only $1 to the total money supply. So when the public withdraws cash from the banks, the overall money supply declines. (We will see in the next section, however, that in practice the central bank has means to offset the impact of the public's actions on the money supply.)

## COMMERCIAL BANKS AND THE CREATION OF MONEY

Part of the money supply consists of deposits in private commercial banks. Hence the behavior of commercial banks and their depositors help to determine the money supply.

Cash or similar assets held by banks are called *bank reserves.* In modern economies, banks' reserves are less than their deposits, a situation called *fractional-reserve banking.* The ratio of bank reserves to deposits is called the *reserve-deposit ratio;* in a fractional-reserve banking system, this ratio is less than 1.

The portion of deposits not held as reserves can be lent out by the banks to earn interest. Banks will continue to make loans and accept deposits as long as the reserve-deposit ratio exceeds its desired level. This process stops only when the actual and desired reserve-deposit ratios are equal. At that point, total bank deposits equal bank reserves divided by the desired reserve-deposit ratio, and the money supply equals the currency held by the public plus bank deposits (see Equation 10.2).

# THE FEDERAL RESERVE SYSTEM

***Federal Reserve System (or Fed)*** the central bank of the United States

For participants in financial markets and the average citizen as well, one of the most important branches of the government is the **Federal Reserve System,** often called the **Fed.** The Fed is the *central bank* of the United States. Like central banks in other countries, the Fed has two main responsibilities.

First, it is responsible for monetary policy, which means that the Fed determines how much money circulates in the economy. As we will see in later chapters, changes in the supply of money can affect many important macroeconomic variables, including interest rates, inflation, unemployment, and exchange rates. Because of its ability to affect key variables, particularly financial variables such as interest rates, financial market participants pay close attention to Fed actions and announcements. As a necessary first step in understanding how Fed policies have the effects that they do, in this chapter we will focus on the basic question of how the Fed affects the supply of money, leaving for later the explanation of why changes in the money supply affect the economy.

Second, along with other government agencies, the Federal Reserve bears important responsibility for the oversight and regulation of financial markets. The Fed also plays a major role during periods of crisis in financial markets. To lay the groundwork for discussing how the Fed carries out its responsibilities, we first briefly review the history and structure of the Federal Reserve System.

## THE HISTORY AND STRUCTURE OF THE FEDERAL RESERVE SYSTEM

The Federal Reserve System was created by the Federal Reserve Act, passed by Congress in 1913, and began operations in 1914. Like all central banks, the Fed is a government agency. Unlike commercial banks, which are private businesses whose principal objective is making a profit, central banks like the Fed focus on promoting public goals such as economic growth, low inflation, and the smooth operation of financial markets.

*"I'm sorry, sir, but I don't believe you know us well enough to call us the Fed."*

The Federal Reserve Act established a system of 12 regional Federal Reserve banks, each associated with a geographical area called a Federal Reserve district. Congress hoped that the establishment of Federal Reserve banks around the country would ensure that different regions were represented in the national policymaking process. In fact, the regional Feds regularly assess economic conditions in their districts and report this information to policymakers in Washington. Regional Federal Reserve banks also provide various services, such as check-clearing services, to the commercial banks in their district.

At the national level, the leadership of the Federal Reserve System is provided by its **Board of Governors.** The Board of Governors, together with a large professional staff, is located in Washington, D.C. The Board consists of seven governors, who are appointed by the president of the United States to 14-year terms. The terms are staggered so that one governor comes up for reappointment every other year. The president also appoints one of these Board members to serve as chairman of the Board of Governors for a term of four years. The Fed chairman, along with the secretary of the Treasury, is probably one of the two most powerful economic policymakers in the United States government, after the president. Recent chairmen, such as Paul Volcker and Alan Greenspan, have been highly regarded and influential.

***Board of Governors*** the leadership of the Fed, consisting of seven governors appointed by the president to staggered 14-year terms

Decisions about monetary policy are made by a 12-member committee called the **Federal Open Market Committee** (or **FOMC**). The FOMC consists of the seven Fed governors, the president of the Federal Reserve Bank of New York, and four of the presidents of the other regional Federal Reserve banks, who serve on a rotating basis. The FOMC meets approximately eight times a year to review the state of the economy and to determine monetary policy.

***Federal Open Market Committee (or FOMC)*** the committee that makes decisions concerning monetary policy

## CONTROLLING THE MONEY SUPPLY: OPEN-MARKET OPERATIONS

The Fed's primary responsibility is making monetary policy, which involves decisions about the appropriate size of the nation's money supply. As we saw in the previous section, central banks in general, and the Fed in particular, do not control the money supply directly. However, they can control the money supply indirectly by changing the supply of reserves held by commercial banks.

The Fed has several ways of affecting the supply of bank reserves. By far the most important of these are *open-market operations*. Suppose that the Fed wants to increase bank reserves, with the ultimate goal of increasing bank deposits and the money supply. To accomplish this the Fed buys financial assets, usually government bonds, from the public. To simplify the actual procedure a bit, think of the Fed as paying for the bonds it acquires with newly printed money. Assuming that the public is already holding all the currency that it wants, they will deposit the cash they receive as payment for their bonds in commercial banks. Thus, the reserves of the commercial banking system will increase by an amount equal to the value of the bonds purchased by the Fed. The increase in bank reserves will lead in turn, through the process of lending and redeposit of funds described in the previous section, to an expansion of bank deposits and the money supply, as summarized by Equation 10.2. The Fed's purchase of government bonds from the public, with the result that bank reserves and the money supply are increased, is called an **open-market purchase.**

***open-market purchase*** the purchase of government bonds from the public by the Fed for the purpose of increasing the supply of bank reserves and the money supply

To reduce bank reserves and hence the money supply, the Fed reverses the procedure. It sells some of the government bonds that it holds (acquired in previous open-market purchases) to the public. Assume that the public pays for the bonds by writing checks on their accounts in commercial banks. Then, when the Fed presents the checks to the commercial banks for payment, reserves equal in value to the government bonds sold by the Fed are transferred from the commercial banks to the Fed. The Fed retires these reserves from circulation, lowering the supply of bank reserves and, hence, the overall money supply. The sale of government bonds by the Fed to the public for the purpose of reducing bank reserves and hence the money supply is called an **open-market sale.** Open-market purchases and sales together are called **open-market operations.** Open-market operations are the most convenient and flexible tool that the Federal Reserve has for affecting the money supply and are employed on a regular basis.

***open-market sale*** the sale by the Fed of government bonds to the public for the purpose of reducing bank reserves and the money supply

***open-market operations*** open-market purchases and open-market sales

**EXAMPLE 10.3**

### Increasing the money supply by open-market operations

In a particular economy, currency held by the public is 1,000 shekels, bank reserves are 200 shekels, and the desired reserve-deposit ratio is 0.2. What is the money supply? How is the money supply affected if the central bank prints 100 shekels and uses this new currency to buy government bonds from the public? Assume that the public does not wish to change the amount of currency it holds.

As bank reserves are 200 shekels and the reserve-deposit ratio is 0.2, bank deposits must equal 200 shekels/0.2, or 1,000 shekels. The money supply, equal to the sum of currency held by the public and bank deposits, is therefore 2,000 shekels, a result you can confirm using Equation 10.2.

The open-market purchase puts 100 more shekels into the hands of the public. We assume that the public continues to want to hold 1,000 shekels in currency, so they will deposit the additional 100 shekels in the commercial banking system, raising bank reserves from 200 to 300 shekels. As the desired reserve-deposit ratio is 0.2, multiple rounds of lending and redeposit will eventually raise the level of bank deposits to 300 shekels/0.2, or 1,500 shekels. The money supply, equal to 1,000 shekels held by the public plus bank deposits of 1,500 shekels, equals 2,500 shekels. So the open-market purchase of 100 shekels, by raising bank reserves by 100 shekels, has increased the money supply by 500 shekels. Again, you can confirm this result using Equation 10.2.

**EXERCISE 10.3**

**Continuing Example 10.3, suppose that instead of an open-market purchase of 100 shekels the central bank conducts an open-market sale of 50 shekels' worth of government bonds. What happens to bank reserves, bank deposits, and the money supply?**

## CONTROLLING THE MONEY SUPPLY: DISCOUNT WINDOW LENDING

A second means by which the Fed can affect bank reserves and thus the money supply is by what is called *discount window lending.* When individual commercial banks are short of reserves, they may choose to borrow reserves from the Fed. For historical reasons, lending of reserves by the Federal Reserve to commercial banks is called **discount window lending.** The interest rate that the Fed charges commercial banks that borrow reserves is called the **discount rate.** Loans of reserves by the Fed directly increase the quantity of reserves in the banking system, leading ultimately to increases in bank deposits and the money supply.

***discount window lending*** the lending of reserves by the Federal Reserve to commercial banks

***discount rate*** the interest rate that the Fed charges commercial banks to borrow reserves

## CONTROLLING THE MONEY SUPPLY: CHANGING RESERVE REQUIREMENTS

As Equation 10.2 shows, the economy's money supply depends on three factors: the amount of currency the public chooses to hold, the supply of bank reserves, and the reserve-deposit ratio maintained by commercial banks. For given quantities of currency held by the public and of reserves held by the banks, an increase in the reserve-deposit ratio reduces the money supply, as you can see from Equation 10.2. A higher reserve-deposit ratio implies that banks lend out a smaller share of their deposits in each of the rounds of lending and redeposit described earlier, limiting the overall expansion of loans and deposits.

Within a certain range, commercial banks are free to set the reserve-deposit ratio they want to maintain. However, Congress granted the Fed the power to set minimum values of the reserve-deposit ratio for commercial banks. The legally required values of the reserve-deposit ratio set by the Fed are called **reserve requirements.**

***reserve requirements*** set by the Fed, the minimum values of the ratio of bank reserves to bank deposits that commercial banks are allowed to maintain

Changes in reserve requirements can be used to affect the money supply, although the Fed does not usually use them in this way. For example, suppose that commercial banks are maintaining a 3 percent reserve-deposit ratio, and the Fed wants to contract the money supply. By raising required reserves to, say, 5 percent of deposits, the Fed could force commercial banks to raise their reserve-deposit ratio, at least until it reached 5 percent. As you can see from Equation 10.2, an increase in the reserve-deposit ratio lowers deposits and the money supply. Similarly, a reduction in required reserves by the Fed might allow at least some banks to lower their ratio of reserves to deposits. A decline in the economywide reserve-deposit ratio would in turn cause the money supply to rise.

## THE FED'S ROLE IN STABILIZING FINANCIAL MARKETS: BANKING PANICS

Besides controlling the money supply, the Fed also has the responsibility (together with other government agencies) of ensuring that financial markets operate smoothly. Indeed, the creation of the Fed in 1913 was prompted by a series of financial market crises that disrupted both the markets themselves and the U.S. economy as a whole. The hope of the Congress was that the Fed would be able to eliminate or at least control such crises.

***banking panic*** an episode in which depositors, spurred by news or rumors of the imminent bankruptcy of one or more banks, rush to withdraw their deposits from the banking system

Historically, in the United States, *banking panics* were perhaps the most disruptive type of recurrent financial crisis. In a **banking panic,** news or rumors of the imminent bankruptcy of one or more banks leads bank depositors to rush to withdraw their funds. Next, we will discuss banking panics and the Fed's attempts to control them.

Why do banking panics occur? An important factor that helps make banking panics possible is the existence of fractional-reserve banking. In a fractional-reserve banking system, like that of the United States and all other industrialized countries, bank reserves are less than deposits, which means that banks do not keep enough cash on hand to pay off their depositors if they were all to decide to withdraw their deposits. Normally this is not a problem, as only a small percentage of depositors attempt to withdraw their funds on any given day. But if a rumor circulates that one or more banks are in financial trouble and may go bankrupt, depositors may panic, lining up to demand their money. Since bank reserves are less than deposits, a sufficiently severe panic could lead even financially healthy banks to run out of cash, forcing them into bankruptcy and closure.

The Federal Reserve was established in response to a particularly severe banking panic that occurred in 1907. The Fed was equipped with two principal tools to try to prevent or moderate banking panics. First, the Fed was given the power to supervise and regulate banks. It was hoped that the public would have greater confidence in banks, and thus be less prone to panic, if people knew that the Fed was keeping a close watch on bankers' activities. Second, the Fed was allowed to make loans to banks through the discount window, discussed earlier. The idea was that, during a panic, banks could borrow cash from the Fed with which to pay off depositors, avoiding the need to close.

No banking panics occurred between 1914, when the Fed was established, and 1930. However, between 1930 and 1933 the United States experienced the worst and most protracted series of banking panics in its history. Economic historians agree that much of the blame for this panic should be placed on the Fed, which neither appreciated the severity of the problem nor acted aggressively enough to contain it.

ECONOMIC NATURALIST 10.2

### The banking panics of 1930–1933 and the money supply

The worst banking panics ever experienced in the United States occurred during the early stages of the Great Depression, between 1930 and 1933. During this period approximately one-third of the banks in the United States were forced to close. This near-collapse of the banking system was probably an important reason that the Depression was so severe. With many fewer banks in operation it was very difficult for small businesses and consumers during the early 1930s to obtain credit. Another important effect of the banking panics was to greatly reduce the nation's money supply. Why should banking panics reduce the national money supply?

During a banking panic people are afraid to keep deposits in a bank because of the risk that the bank will go bankrupt and their money will be lost (this was prior to the introduction of federal deposit insurance, discussed below). During the 1930–1933 period, many bank depositors withdrew their money from banks, holding currency instead. These withdrawals reduced bank reserves. Each extra dollar of currency held by the public adds $1 to the money supply; but each extra dollar of bank reserves translates into several dollars of money supply, because in a fractional-reserve banking system each dollar of reserves can "support" several dollars in bank deposits. Thus the public's withdrawals from banks, which increased currency holdings by the public but reduced bank reserves by an equal amount, led to a net decrease in the total money supply (currency plus deposits).

In addition, fearing banking panics and the associated withdrawals by depositors, banks increased their reserve-deposit ratios, which reduced the quantity of deposits that could be supported by any given level of bank reserves. This change in reserve-deposit ratios also tended to reduce the money supply.

Data on currency holdings by the public, the reserve-deposit ratio, bank reserves, and the money supply for selected dates are shown in Table 10.7. Notice the increase over the period in the amount of currency held by the public and in the reserve-deposit ratio, as well as the decline in bank reserves after 1930. The last column shows that the U.S. money supply dropped by about one-third between December 1929 and December 1933.

TABLE 10.7
**Key U.S. Monetary Statistics, 1929–1933**

| | Currency held by public | Reserve-deposit ratio | Bank reserves | Money supply |
|---|---|---|---|---|
| December 1929 | 3.85 | 0.075 | 3.15 | 45.9 |
| December 1930 | 3.79 | 0.082 | 3.31 | 44.1 |
| December 1931 | 4.59 | 0.095 | 3.11 | 37.3 |
| December 1932 | 4.82 | 0.109 | 3.18 | 34.0 |
| December 1933 | 4.85 | 0.133 | 3.45 | 30.8 |

NOTE: Data on currency, the monetary base, and the money supply are in billions of dollars.

SOURCE: Milton Friedman and Anna J. Schwartz, *A Monetary History of the United States, 1863–1960,* Princeton, N.J.: Princeton University Press, 1963, Table A-1.

Using Equation 10.2, we can see that increases in currency holdings by the public and increases in the reserve-deposit ratio both tend to reduce the money supply. These effects were so powerful in 1930–1933 that the nation's money supply, shown in the fourth column of Table 10.7, dropped precipitously, even though currency holdings and bank reserves, taken separately, actually rose during the period.

**EXERCISE 10.4**

**Using the data from Table 10.7, confirm that the relationship between the money supply and its determinants is consistent with Equation 10.2. Would the money supply have fallen in 1931–1933 if the public had stopped withdrawing deposits after December 1930 so that currency held by the public had remained at its December 1930 level?**

**EXERCISE 10.5**

**According to Table 10.7, the U.S. money supply fell from $44.1 billion to $37.3 billion over the course of 1931. The Fed did use open-market purchases during 1931 to replenish bank reserves in the face of depositor withdrawals. Find (a) the quantity of reserves that the Fed injected into the economy in 1931, and (b) the quantity of reserves the Fed would have had to add to the economy to keep the money supply unchanged from 1930, assuming that public currency holdings and reserve-deposit ratios for each year remained as reported in the table. Why has the Fed been criticized for being too timid in 1931?**

When the Fed failed to stop the banking panics of the 1930s, policymakers decided to look at other strategies for controlling panics. In 1934 Congress instituted a system of deposit insurance. Under a system of **deposit insurance,** the government guarantees depositors—specifically, under current rules, those with deposits of less than $100,000—that they will get their money back even if the bank goes bankrupt. Deposit insurance eliminates the incentive for people to withdraw their deposits when rumors circulate that the bank is in financial trouble,

***deposit insurance*** a system under which the government guarantees that depositors will not lose any money even if their bank goes bankrupt which nips panics in the bud. Indeed, since deposit insurance was instituted, the United States has had no significant banking panics.

Unfortunately, deposit insurance is not a perfect solution to the problem of banking panics. An important drawback is that when deposit insurance is in force, depositors know they are protected no matter what happens to their bank, and they become completely unconcerned about whether their bank is making prudent loans. This situation can lead to reckless behavior by banks or other insured intermediaries. For example, during the 1980s many savings and loan associations in the United States went bankrupt, in part because of reckless lending and financial investments. Like banks, savings and loans have deposit insurance, so the U.S. government had to pay savings and loans depositors the full value of their deposits. This action ultimately cost U.S. taxpayers hundreds of billions of dollars.

## MONEY AND PRICES

From a macroeconomic perspective, a major reason that control of the supply of money is important is that, *in the long run, the amount of money circulating in an economy and the general level of prices are closely linked.* Indeed, it is virtually unheard of for a country to experience high, sustained inflation without a comparably rapid growth in the amount of money held by its citizens. The economist Milton Friedman summarized the inflation–money relationship by saying, "Inflation is always and everywhere a monetary phenomenon." We will see in Part 4 that, over short periods, inflation can arise from sources other than an increase in the supply of money. But over a longer period, and particularly for more severe inflations, Friedman's dictum is certainly correct: The rate of inflation and the rate of growth of the money supply are closely related.

The existence of a close link between money supply and prices should make intuitive sense. Imagine a situation in which the available supply of goods and services is approximately fixed. Then the more cash (say, dollars) that people hold, the more they will be able to bid up the prices of the fixed supply of goods and services. Thus a large money supply relative to the supply of goods and services (too much money chasing too few goods) tends to result in high prices. Likewise, a rapidly *growing* supply of money will lead to quickly *rising* prices—that is, inflation.

### VELOCITY

***velocity*** a measure of the speed at which money circulates, or, equivalently, the value of transactions completed in a period of time divided by the stock of money required to make those transactions; numerically, $V = (P \times Y)/M$, where $V$ is velocity, $P \times Y$ is nominal GDP, and $M$ is the money supply whose velocity is being measured

To explore the relationship of money growth and inflation in a bit more detail, it is useful to introduce the concept of *velocity.* In economics, **velocity** is a measure of the speed at which money circulates. For example, a given dollar bill might pass from your hand to the grocer's when you buy a quart of milk. The same dollar may then pass from the grocer to his supplier, from the supplier to the dairy farmer who produced the milk, from the farmer to the feed supply store owner, and so on. The more quickly money circulates from one person to the next, the higher its velocity. More formally, velocity is defined as the value of transactions completed in a period of time divided by the stock of money required to make those transactions. The higher this ratio, the faster the "typical" dollar is circulating.

As a practical matter, we usually do not have precise measures of the total value of transactions taking place in an economy; so, as an approximation, economists often measure the total value of transactions in a given period by nominal GDP for that period. A numerical value of velocity can then be obtained from the following formula:

$$\text{Velocity} = \frac{\text{Value of transactions}}{\text{Money stock}} = \frac{\text{Nominal GDP}}{\text{Money stock}}.$$

Let $V$ stand for velocity and let $M$ stand for the particular money stock being considered (for example, M1 or M2). Nominal GDP (a measure of the total value of transactions) equals the price level $P$ times real GDP $Y$. Using this notation, we can write the definition of velocity as

$$V = \frac{P \times Y}{M} \qquad (10.3)$$

**EXAMPLE 10.4**

**The velocity of money in the U.S. economy**

In the United States in 2001, M1 was $1,177.9 billion, M2 was $5,449.1 billion, and nominal GDP was $10,082.2 billion. Find the velocity of M1 and of M2 for that year.

Using Equation 10.3, velocity for M1 is given by

$$V = \frac{\$10{,}082.2 \text{ billion}}{\$1{,}177.9 \text{ billion}} = 8.56.$$

Similarly, velocity for M2 was

$$V = \frac{\$10{,}082.2 \text{ billion}}{\$5{,}449.1 \text{ billion}} = 1.85.$$

You can see that the velocity of M1 is higher than that of M2. This makes sense: Because the components of M1, such as cash and checking accounts, are used more frequently for transactions, each dollar of M1 "turns over" more often than the average dollar of M2.

A variety of factors determine velocity. A leading example is advances in payment technologies, such as the introduction of credit cards and debit cards or the creation of networks of automatic teller machines (ATMs). These new technologies and payment methods have allowed people to carry out their daily business while holding less cash, and thus have tended to increase velocity over time.

## MONEY AND INFLATION IN THE LONG RUN

We can use the definition of velocity to see how money and prices are related in the long run. First, rewrite the definition of velocity, Equation 10.3, by multiplying both sides of the equation by the money stock $M$. This yields

$$M \times V = P \times Y. \qquad (10.4)$$

Equation 10.4, a famous relationship in economics, is called for historical reasons the *quantity equation*. The **quantity equation** states that money times velocity equals nominal GDP. Because the quantity equation is simply a rewriting of the definition of velocity, Equation 10.3, it always holds exactly.

***quantity equation*** money times velocity equals nominal GDP; $M \times V = P \times Y$

The quantity equation is historically important because late nineteenth and early twentieth-century monetary economists, such as Yale's Irving Fisher, used this relationship to theorize about the relationship between money and prices. We can do the same thing here. To keep things simple, imagine that velocity $V$ is determined by current payments technologies and thus is approximately constant over the period we are considering. Likewise, suppose that real output $Y$ is approximately constant. If we use a bar over a variable to indicate that the variable is constant, we can rewrite the quantity equation as

$$M \times \overline{V} = P \times \overline{Y}, \qquad (10.5)$$

where we are treating $\overline{V}$ and $\overline{Y}$ as fixed numbers.

Now look at Equation 10.5 and imagine that for some reason the Federal Reserve increases the money supply $M$ by 10 percent. Because $\overline{V}$ and $\overline{Y}$ are assumed to be fixed, Equation 10.5 can continue to hold only if the price level $P$ also rises by 10 percent. That is, according to the quantity equation, a 10 percent increase in the money supply $M$ should cause a 10 percent increase in the price level $P$, that is, an inflation of 10 percent.

The intuition behind this conclusion is the one we mentioned at the beginning of this section. If the quantity of goods and services $Y$ is approximately constant (and assuming also that velocity $V$ is also constant), an increase in the supply of money will lead people to bid up the prices of the available goods and services. Thus high rates of money growth will tend to be associated with high rates of inflation. Figure 10.1 shows this relationship for 10 countries in Latin America during the period 1995–2001. You can see that countries with higher rates of money growth tend also to have higher rates of inflation. The relationship between money growth and inflation is not exact, in part because—contrary to the simplifying assumption we made earlier—velocity and output are not constant but vary over time.

If high rates of money growth lead to inflation, why do countries allow their money supplies to rise quickly? Usually, rapid rates of money growth are the result of large government budget deficits. Particularly in developing countries or countries suffering from war or political instability, governments sometimes find that they cannot raise sufficient taxes or borrow enough from the public to cover their expenditures. In this situation the government's only recourse may be to print new money and use this money to pay its bills. If the resulting increase in the amount of money in circulation is large enough, the result will be inflation.

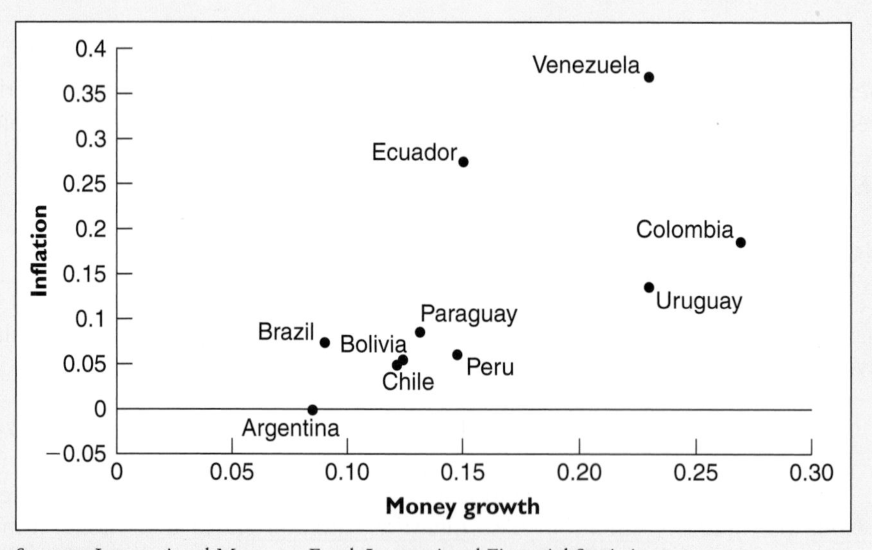


**FIGURE 10.1**
**Inflation and Money Growth in Latin America, 1995–2001.**
Latin American countries with higher rates of growth in their money supplies also tended to have higher rates of inflation between 1995 and 2001. (The data for Argentina and Uruguay end in 2000 and the data for Ecuador end in 1997. In 1997 Ecuador abandoned its currency, the sucre, and began using U.S. dollars instead.)

SOURCE: International Monetary Fund, International Financial Statistics

## RECAP MONEY AND PRICES

A high rate of money growth generally leads to inflation. The larger the amount of money in circulation, the higher the public will bid up the prices of available goods and services.

*Velocity* measures the speed at which money circulates; equivalently, it is the value of transactions completed in a period of time divided by the stock of money required to make those transactions. A numerical value for velocity can be obtained from the equation $V = (P \times Y) / M$, where $V$ is velocity, $P \times Y$

is nominal GDP (a measure of the total value of transactions), and $M$ is the money supply.

The *quantity equation* states that money times velocity equals nominal GDP, or, in symbols, $M \times V = P \times Y$. The quantity equation is a restatement of the definition of velocity and thus always holds. If velocity and output are approximately constant, the quantity equation implies that a given percentage increase in the money supply leads to the same percentage increase in the price level. In other words, the rate of growth of the money supply equals the rate of inflation.

## SUMMARY

- *Money* is any asset that can be used in making purchases, such as currency and checking account balances. Money has three main functions: It is a *medium of exchange,* which means that it can be used in transactions. It is a *unit of account,* in that economic values are typically measured in units of money (e.g., dollars). And it is a *store of value,* a means by which people can hold wealth. In practice it is difficult to measure the money supply, since many assets have some moneylike features. A relatively narrow measure of money is M1, which includes currency and checking accounts. A broader measure of money, M2, includes all the assets in M1 plus additional assets that are somewhat less convenient to use in transactions than those included in M1.

- Because bank deposits are part of the money supply, the behavior of commercial banks and of bank depositors affects the amount of money in the economy. A key factor is the *reserve-deposit ratio* chosen by banks. *Bank reserves* are cash or similar assets held by commercial banks, for the purpose of meeting depositor withdrawals and payments. The reserve-deposit ratio is bank reserves divided by deposits in banks. A banking system in which all deposits are held as reserves practices *100 percent reserve banking.* Modern banking systems have reserve-deposit ratios less than 100 percent, and are called *fractional-reserve banking systems.*

- Commercial banks create money through multiple rounds of lending and accepting deposits. This process of lending and increasing deposits comes to an end when banks' reserve-deposit ratios equal their desired levels. At that point, bank deposits equal bank reserves divided by the desired reserve-deposit ratio. The money supply equals currency held by the public plus deposits in the banking system.

- The central bank of the United States is called the *Federal Reserve System,* or the Fed for short. The Fed's two main responsibilities are making monetary policy, which means determining how much money will circulate in the economy, and overseeing and regulating financial markets, especially banks. Created in 1914, the Fed is headed by a *Board of Governors* made up of seven governors appointed by the president. One of these seven governors is appointed chairman. The *Federal Open Market Committee,* which meets about eight times a year to determine monetary policy, is made up of the seven governors and five of the presidents of the regional Federal Reserve banks.

- The Fed can affect the money supply indirectly through its control of the supply of bank reserves. The Fed has two ways to change bank reserves: through *open-market operations,* in which the Fed buys or sells government securities in exchange for currency held by banks or the public, or through *discount window lending,* which is the Fed's lending of reserves to commercial banks. In addition the Fed can affect the money supply by changing legal *reserve requirements,* which influences banks' reserve-deposit ratios and hence deposits.

- One of the original purposes of the Federal Reserve was to help eliminate or control banking panics. A *banking panic* is an episode in which depositors, spurred by news or rumors of the imminent bankruptcy of one or more banks, rush to withdraw their deposits from the banking system. Because banks do not keep enough reserves on hand to pay off all depositors, even a financially healthy bank can run out of cash during a panic and be forced to close. The Federal Reserve failed to contain banking panics during the Great Depression, which led to sharp declines in the money supply. The adoption of a system of *deposit insurance* in the United States eliminated banking panics. A disadvantage of deposit insurance is that if banks or other insured intermediaries make bad loans or financial investments, the tax-payers may be responsible for covering the losses.

- In the long run, the rate of growth of the money supply and the rate of inflation are closely linked because a larger amount of money in circulation allows people to bid up the prices of existing goods and services. *Velocity* measures the speed at which money circulates; equivalently, it is the value of transactions completed in a period of time, divided by the stock of money required to make those transactions. Velocity is defined by the equation $V = (P \times Y) / M$, where $V$ is velocity, $P \times Y$ is nominal GDP (a measure of the total value of transactions), and $M$ is the money supply. The definition of velocity can be rewritten as the *quantity equation,* $M \times V = P \times Y$. The quantity equation shows that, if velocity and output are constant, a given percentage increase in the money supply will lead to the same percentage increase in the price level.

## KEY TERMS

bank reserves (264)
banking panic (272)
barter (260)
Board of Governors of the Federal Reserve System (269)
deposit insurance (273)
discount rate (271)
discount window lending (271)
Federal Open Market Committee (FOMC) (269)
Federal Reserve System (the Fed) (268)
fractional-reserve banking system (264)
M1 (262)
M2 (262)
medium of exchange (260)
money (260)
100 percent reserve banking (264)
open-market operations (270)
open-market purchase (270)
open-market sale (270)
quantity equation (275)
reserve-deposit ratio (264)
reserve requirements (271)
store of value (261)
unit of account (261)
velocity (274)

## REVIEW QUESTIONS

1. What is *money?* Why do people hold money even though it pays a lower return than other financial assets?
2. Suppose that the public switches from doing most of its shopping with currency to using checks instead. If the Fed takes no action, what will happen to the national money supply? Explain.
3. The Fed wants to reduce the U.S. money supply. Describe the various actions it might take, and explain how each action would accomplish the Fed's objective.
4. What is a *banking panic?* Prior to the introduction of deposit insurance, why might even a bank that had made sound loans have reason to fear a panic?
5. Define *velocity.* How has the introduction of new payments technologies, such as ATM machines, affected velocity? Explain.
6. Use the quantity equation to explain why money growth and inflation tend to be closely linked.

## PROBLEMS

1. During World War II, an Allied soldier named Robert Radford spent several years in a large German prisoner-of-war camp. At times more than 50,000 prisoners were held in the camp, with some freedom to move about within the compound. Radford later wrote an account of his experiences. He described how an economy developed in the camp, in which prisoners traded food, clothing, and other items. Services, such as barbering, were also exchanged. Lacking paper money, the prisoners began to use cigarettes (provided monthly by the Red Cross) as money. Prices were quoted, and payments made, using cigarettes.
   a. In Radford's POW camp, how did cigarettes fulfill the three functions of money?
   b. Why do you think the prisoners used cigarettes as money, as opposed to other items of value such as squares of chocolate or pairs of boots?
   c. Do you think a nonsmoking prisoner would have been willing to accept cigarettes in exchange for a good or service in Radford's camp? Why or why not?
2. Obtain recent data on M1, M2, and their components. (For an online source, see Table 10.1). By what percentage have the two monetary aggregates grown over the past year? Which components of the two aggregates have grown the most quickly?
3. Redo the example of Gorgonzola in the text (see Tables 10.2 to 10.6), assuming that (a) initially, the Gorgonzolan central bank puts 5,000,000 guilders into circulation, and (b) commercial banks desire to hold reserves of 20 percent of deposits. As in the text, assume that the public holds no currency. Show the consolidated balance sheets of Gorgonzolan commercial banks after the initial deposits (compare to Table 10.2), after one round of loans (compare to Table 10.3), after the first redeposit of guilders (compare to Table 10.4), and after two rounds of loans and redeposits (Table 10.5). What are the final values of bank reserves, loans, deposits, and the money supply?
4. a. Bank reserves are 100, the public holds 200 in currency, and the desired reserve-deposit ratio is 0.25. Find deposits and the money supply.

b. The money supply is 500, and currency held by the public equals bank reserves. The desired reserve-deposit ratio is 0.25. Find currency held by the public and bank reserves.

c. The money supply is 1,250, of which 250 is currency held by the public. Bank reserves are 100. Find the desired reserve-deposit ratio.

5. When a central bank increases bank reserves by $1, the money supply rises by more than $1. The amount of extra money created when the central bank increases bank reserves by $1 is called the *money multiplier.*

   a. Explain why the money multiplier is generally greater than 1. In what special case would it equal 1?

   b. The initial money supply is $1,000, of which $500 is currency held by the public. The desired reserve-deposit ratio is 0.2. Find the increase in money supply associated with increases in bank reserves of $1, $5, and $10. What is the money multiplier in this economy?

   c. Find a general rule for calculating the money multiplier.

   d. Suppose the Fed wanted to reduce the money multiplier, perhaps because it believes that change would give it more precise control over the money supply. What action could the Fed take to achieve its goal?

6. Refer to Table 10.7. Suppose that the Fed had decided to set the U.S. money supply in December 1932 and in December 1933 at the same value as in December 1930. Assuming that the values of currency held by the public and the reserve-deposit ratio had remained as given in the table, by how much more should the Fed have increased bank reserves at each of those dates to accomplish that objective?

7. Real GDP is $8 trillion, nominal GDP is $10 trillion, M1 is $2 trillion, and M2 is $5 trillion.

   a. Find velocity for M1 and for M2.

   b. Show that the quantity equation holds for both M1 and M2.

8. You are given the following data for 2004 and 2005:

| | 2004 | 2005 |
|---|---|---|
| Money supply | 1,000 | 1,050 |
| Velocity | 8.0 | 8.0 |
| Real GDP | 12,000 | 12,000 |

   a. Find the price level for 2004 and 2005. What is the rate of inflation between the two years?

   b. What is the rate of inflation between 2004 and 2005 if the money supply in 2005 is 1,100 instead of 1050?

   c. What is the rate of inflation between 2004 and 2005 if the money supply in 2005 is 1,100 and output in 2005 is 12,600?

## ANSWERS TO IN-CHAPTER EXERCISES

10.1 Table 10.5 shows the balance sheet of banks after two rounds of lending and redeposits. At that point deposits are 2,710,000 guilders and reserves are 1,000,000 guilders. Since banks have a desired reserve-deposit ratio of 10 percent, they will keep 271,000 guilders (10 percent of deposits) as reserves and lend out the remaining 729,000 guilders. Loans to farmers are now 2,439,000 guilders. Eventually the 729,000 guilders lent to the farmers will be redeposited into the banks, giving the banks deposits of 3,439,000 guilders and reserves of 1,000,000 guilders. The balance sheet is as shown in the accompanying table.

| Assets | | Liabilities | |
|---|---|---|---|
| Currency (= reserves) | 1,000,000 guilders | Deposits | 3,439,000 guilders |
| Loans to farmers | 2,439,000 guilders | | |

Notice that assets equal liabilities. The money supply equals deposits, or 3,439,000 guilders. Currency held in the banks as reserves does not count in the money supply.

10.2 Because the public holds no currency, the money supply equals bank deposits, which in turn equal bank reserves divided by the reserve-deposit ratio (Equation 10.1). If bank reserves are 1,000,000 and the reserve-deposit ratio is 0.05, then deposits equal 1,000,000/0.05 = 20,000,000 guilders, which is also the money supply. If bank reserves are 2,000,000 guilders and the reserve-deposit ratio is 0.10, then the money supply and deposits are again equal to 20,000,000 guilders, or 2,000,000/0.10.

10.3 If the central bank sells 50 shekels of government bonds in exchange for currency, the immediate effect is to reduce the amount of currency in the hands of the public by 50 shekels. To restore their currency holding to the desired level of 1,000 shekels, the public will withdraw 50 shekels from commercial banks, reducing bank reserves from 200 shekels to 150 shekels. The desired reserve-deposit ratio is 0.2, so ultimately deposits must equal 150 shekels in reserves divided by 0.2, or 750 shekels. (Note that to contract deposits, the commercial banks will have to "call in" loans, reducing their loans outstanding.) The money supply equals 1,000 shekels in currency held by the public plus 750 shekels in deposits, or 1,750 shekels. Thus the open-market purchase has reduced the money supply from 2,000 to 1,750 shekels.

10.4 Verify directly for each date in Table 10.7 that

$$\text{Money supply} = \text{Currency} + \frac{\text{Bank reserves}}{\text{Desired reserve-deposit ratio}}.$$

For example, for December 1929 we can check that 45.9 = 3.85 + 3.15/0.075.

Suppose that the currency held by the public in December 1933 had been 3.79, as in December 1930, rather than 4.85, and that the difference (4.85 − 3.79 = 1.06) had been left in the banks. Then bank reserves in December 1933 would have been 3.45 + 1.06 = 4.51, and the money supply would have been 3.79 + 4.51/0.133 = 37.7. So the money supply would still have fallen between 1930 and 1933 if people had not increased their holdings of currency, but only by about half as much.

10.5 Over the course of 1931, currency holdings by the public rose by $0.80 billion but bank reserves fell overall by only $0.20 billion. Thus the Fed must have replaced $0.60 billion of lost reserves during the year through open-market purchases or discount window lending.

Currency holdings at the end of 1931 were $4.59 billion. To have kept the money supply at the December 1930 value of $44.1 billion, the Fed would have had to ensure that bank deposits equaled $44.1 billion − $4.59 billion, or $39.51 billion. As the reserve-deposit ratio in 1931 was 0.095, this would have required bank reserves of 0.095 × $39.51 billion, or $3.75 billion, compared to the actual value in December 1931 of $3.11 billion. Thus, to keep the money supply from falling, the Fed would have had to increase bank reserves by $0.64 billion more than it did. The Fed has been criticized for increasing bank reserves by only about half what was needed to keep the money supply from falling.

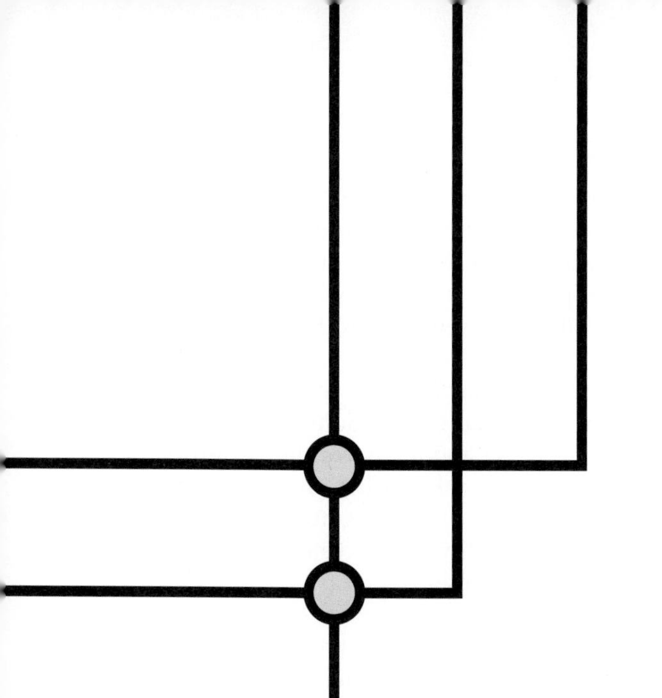

# CHAPTER 11

# FINANCIAL MARKETS AND INTERNATIONAL CAPITAL FLOWS

A recent television ad for an online trading company showed an office worker, call him Ed, sitting in front of his computer. Instead of working, Ed is checking the prices of the stocks he bought over the Internet. Suddenly his eyes widen, as on the computer screen a graph shows the price of his stock shooting up like a rocket. With a whoop Ed heads down to his boss's office, delivers a few well-chosen insults, and quits his job. Unfortunately, when Ed returns to his desk to pack up his belongings, the computer screen shows that the price of his stock has fallen as quickly as it rose. The last we hear of Ed are his futile attempts to convince his boss that he was only kidding.

Ed's story is not a bad metaphor for the behavior of U.S. financial markets in the years around the turn of the millennium. The 1990s were a boom period in the United States, with strong economic growth, record low unemployment rates, and almost nonexistent inflation. This prosperity was mirrored in the spectacular performance of the stock market. In 1999 the Dow Jones index, a popular measure of stock prices, broke 10,000 and then 11,000 for the first time—almost triple the value of the index five years earlier. Prices of stocks of start-up companies, particularly high-technology and Internet companies, rose to stunning levels, making billionaires of some entrepreneurs still in their twenties. Nor were the benefits of rising stock prices confined to the very rich or the very nerdy. Stock ownership among the general population—including both direct ownership of stocks and indirect ownership through mutual funds and pension plans—reached record levels. The stock market boom, however, was not sustained. High-tech companies did not produce the profits that their investors had hoped for, and stock prices began to drop. Many start-up companies saw their stock become nearly worthless, and the values of even established companies fell by one-third or more.

Would investors have spent $5.3 billion to build the Shoreham Nuclear Power Plant if they had known that regulators would never certify it as safe enough to operate?

During periods such as the 1990s, people often begin to think of the stock market as a place to gamble and, maybe, to strike it rich. Some people do get rich playing the market, and some people, like Ed and many real-life investors in stocks in recent years, lose everything. But the stock market, as well as other financial markets, plays a crucial role in the economy that is not shared by the gambling establishments in Las Vegas or Atlantic City. That role is to ensure that national saving is devoted to the most productive uses.

In Chapter 9 we discussed the importance of national saving. Under most circumstances, a high rate of national saving permits a high rate of capital formation, which both economic logic and experience predict will tend to increase labor productivity and living standards. However, creating new capital does not *guarantee* a richer and more productive economy. History is full of examples of "white elephants," capital projects in which millions, and even billions, of dollars were invested with little economic benefit: nuclear power plants that were never opened, massive dams whose main effect was to divert water supplies and disrupt the local agriculture, new technologies that just didn't work as planned.

A healthy economy not only saves adequately but also invests those savings in a productive way. In market economies, like that of the United States, channeling society's savings into the best possible capital investments is the role of the financial system: banks, stock markets, bond markets, and other financial markets and institutions. For this reason, many economists have argued that the development of well-functioning financial markets is a crucial precursor to sustained economic growth. In the first part of this chapter we discuss some major financial markets and institutions and their role in directing saving to productive uses.

Many of the people who purchased stock in U.S. companies during the roller-coaster years of the late 1990s and early 2000s were foreigners, looking to the United States for investment opportunities. More broadly, in the modern world saving often flows across national boundaries, as savers purchase financial assets in countries other than their own and borrowers look abroad for sources of financing. Flows of funds between lenders and borrowers located in different countries are referred to as *international capital flows*. We discuss the international dimension of saving and capital formation in the second part of the chapter. As we will see, for many countries, including the United States, foreign savings provide an important supplement to domestic savings as a means of financing the formation of new capital.

## THE FINANCIAL SYSTEM AND THE ALLOCATION OF SAVING TO PRODUCTIVE USES

We have emphasized the importance of high rates of saving and capital formation for economic growth and increased productivity. High rates of saving and investment by themselves are not sufficient, however. A case in point is the former Soviet Union, which had very high rates of saving and investment but often used its resources very inefficiently, for example, by constructing massive but poorly designed factories that produced inferior goods at high cost. A successful economy not only saves but also uses its savings wisely by applying these limited funds to the investment projects that seem likely to be the most productive.

In the Soviet Union, a centralized bureaucracy made decisions about the allocation of saving to alternative uses. Because the bureaucrats in Moscow had relatively poor information and because they allowed themselves to be influenced by noneconomic considerations such as political favoritism, they often made poor decisions. In a market economy like that of the United States, in contrast, savings are allocated by means of a decentralized, market-oriented financial system. The U.S. financial system consists both of financial institutions, like banks, and financial markets, such as bond markets and stock markets.

The financial system of countries like the United States improves the allocation of savings in at least two distinct ways. First, the financial system provides *information* to savers about which of the many possible uses of their funds are likely to prove most productive and hence pay the highest return. By evaluating the potential productivity of alternative capital investments, the financial system helps to direct savings to its best uses. Second, financial markets help savers to *share the risks* of individual investment projects. Sharing of risks protects individual savers from bearing excessive risk, while at the same time making it possible to direct savings to projects, such as the development of new technologies, which are risky but potentially very productive as well.

In this section we briefly discuss three key components of the financial system: the banking system, the bond market, and the stock market. In doing so we elaborate on the role of the financial system as a whole in providing information about investment projects and in helping savers to share the risks of lending.

## THE BANKING SYSTEM

The banking system consists of commercial banks, of which there are thousands in the United States. Commercial banks are privately owned firms that accept deposits from individuals and businesses and use those deposits to make loans. Banks are the most important example of a class of institutions called **financial intermediaries**, firms that extend credit to borrowers using funds raised from savers. Other examples of financial intermediaries are savings and loan associations and credit unions.

***financial intermediaries*** firms that extend credit to borrowers using funds raised from savers

Why are financial intermediaries such as banks, which "stand between" savers and investors, necessary? Why don't individual savers just lend directly to borrowers who want to invest in new capital projects? The main reason is that, through specialization, banks and other intermediaries develop a *comparative advantage* in evaluating the quality of borrowers—the information-gathering function that we referred to a moment ago. Most savers, particularly small savers, do not have the time or the knowledge to determine for themselves which borrowers are likely to use the funds they receive most productively. In contrast, banks and other intermediaries have gained expertise in performing the information-gathering activities necessary for profitable lending, including checking out the borrower's background, determining whether the borrower's business plans make sense, and monitoring the borrower's activities during the life of the loan. Because banks specialize in evaluating potential borrowers, they can perform this function at a much lower cost, and with better results, than individual savers could on their own. Banks also reduce the costs of gathering information about potential borrowers by pooling the savings of many individuals to make large loans. Each large loan needs to be evaluated only once, by the bank, rather than separately by each of the hundreds of individuals whose savings may be pooled to make the loan.

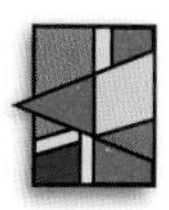

Banks help savers by eliminating their need to gather information about potential borrowers and by directing their savings toward higher-return, more productive investments. Banks help borrowers as well, by providing access to credit that might otherwise not be available. Unlike a *Fortune* 500 corporation, which typically has many ways to raise funds, a small business that wants to buy a copier or remodel its offices will have few options other than going to a bank. Because the bank's lending officer has developed expertise in evaluating small-business loans, and may even have an ongoing business relationship with the

*"O.K., folks, let's move along. I'm sure you've all seen someone qualify for a loan before."*

small-business owner, the bank will be able to gather the information it needs to make the loan at a reasonable cost. Likewise, consumers who want to borrow to finish a basement or add a room to a house will find few good alternatives to a bank. In sum, banks' expertise at gathering information about alternative lending opportunities allows them to bring together small savers, looking for good uses for their funds, and small borrowers with worthwhile investment projects.

In addition to being able to earn a return on their savings, a second reason that people hold bank deposits is to make it easier to make payments. Most bank deposits allow the holder to write a check against them or draw on them using a debit card or ATM card. For many transactions, paying by check or debit card is more convenient than using cash. For example, it is safer to send a check through the mail than to send cash, and paying by check gives you a record of the transaction, whereas a cash payment does not. We discussed the role of banks in the creation of money in the previous chapter.

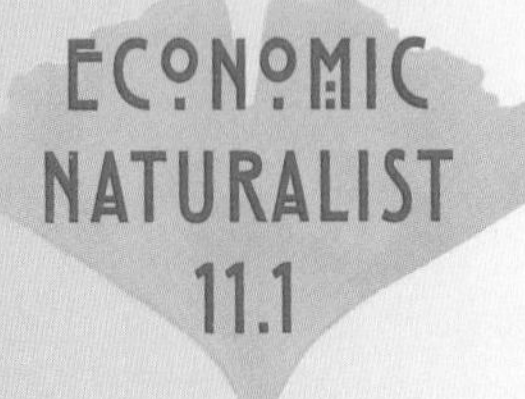


**How has the banking crisis in Japan affected the Japanese economy?**

During the 1980s, real estate and stock prices soared in Japan. Japanese banks made many loans to real estate developers, and the banks themselves acquired stock in corporations. (Unlike in the United States, in Japan it is legal for commercial banks to own stock.) However, in the early 1990s land prices plummeted in Japan, leading many bank borrowers to default on their loans. Stock prices also came down sharply, reducing the value of banks' shareholdings. The net result was that most Japanese banks fell into severe financial trouble, with many large banks near bankruptcy. What has been the effect of this crisis, which has lasted more than a decade, on the Japanese economy?

Relative to the United States, which has more developed stock and bond markets, Japan has traditionally relied very heavily on banks to allocate its savings. Thus, when the severe financial problems of the banks prevented them from operating normally, many borrowers found it unusually difficult to obtain credit—a situation known as a "credit crunch." Smaller borrowers, such as small- and medium-sized businesses, had been particularly dependent on banks for credit and thus suffered disproportionately.

The Japanese economy, after many years of robust growth, suffered a severe recession throughout the 1990s. Many factors contributed to this sharp slowdown. However, the virtual breakdown of the banking system certainly did not help the situation, as credit shortages interfered with smaller firms' ability to make capital investments and, in some cases, to purchase raw materials and pay workers. The Japanese government recognized the problem but has responded very slowly, in large part out of reluctance to bear the high costs of returning the banks to a healthy financial condition. Meanwhile, the Japanese economy has continued to perform poorly.

## BONDS AND STOCKS

Large and well-established corporations that wish to obtain funds for investment will sometimes go to banks. Unlike the typical small borrower, however, a larger firm usually has alternative ways of raising funds, notably through the corporate bond market and the stock market. We first discuss some of the mechanics of bonds and stocks, then return to the role of bond and stock markets in allocating saving.

**Bonds** A **bond** is a legal promise to repay a debt, usually including both the **principal amount,** which is the amount originally lent, and regular interest payments. The promised interest rate when a bond is issued is called the **coupon rate.** The regular interest payments made to the bondholder are called coupon payments. The **coupon payment** of a bond that pays interest annually equals the coupon rate times the principal amount of the bond. For example, if the principal amount of a bond is $1,000,000 and its coupon rate is 5 percent, then the annual coupon payment made to the holder of the bond is (0.05)($1,000,000), or $50,000.

***bond*** a legal promise to repay a debt, usually including both the principal amount and regular interest payments

***principal amount*** the amount originally lent

***coupon rate*** the interest rate promised when a bond is issued

***coupon payments*** regular interest payments made to the bondholder

Corporations and governments frequently raise funds by issuing bonds and selling them to savers. The coupon rate that a newly issued bond has to promise in order to be attractive to savers depends on a number of factors, including the bond's term, its credit risk, and its tax treatment. The *term* of a bond is the length of time before the debt it represents is fully repaid, a period that can range from 30 days to 30 years or more. Generally lenders will demand a higher interest rate to lend for a longer term. *Credit risk* is the risk that the borrower will go bankrupt and thus not repay the loan. A borrower that is viewed as risky will have to pay a higher interest rate to compensate lenders for taking the chance of losing all or part of their financial investment. For example, so-called high-yield bonds, less formally known as "junk bonds," are bonds issued by firms judged to be risky by credit-rating agencies; these bonds pay higher interest rates than bonds issued by companies thought to be less risky.

Bonds also differ in their *tax treatment.* For example, interest paid on bonds issued by local governments, called municipal bonds, is exempt from federal taxes, whereas interest on other types of bonds is treated as taxable income. Because of this tax advantage, lenders are willing to accepting a lower interest rate on municipal bonds.

Bondholders are not required to hold bonds until *maturity,* the time at which they are supposed to be repaid by the issuer, but are always free to sell their bonds in the *bond market,* an organized market run by professional bond traders. The market value of a particular bond at any given point in time is called the *price* of the bond. As it turns out, there is a close relationship between the price of a bond at a given point of time and the interest rate prevailing in financial markets at that time, illustrated by Example 11.1.

**EXAMPLE 11.1**

**Bond prices and interest rates**

On January 1, 2003, Tanya purchases a newly issued, two-year government bond with a principal amount of $1,000. The coupon rate on the bond is 5 percent, paid annually. Hence Tanya, or whoever owns the bond at the time, will receive a coupon payment of $50 (5 percent of $1,000) on January 1, 2004, and $1,050 (a $50 coupon payment plus repayment of the original $1,000 lent) on January 1, 2005.

On January 1, 2004, after receiving her first year's coupon payment, Tanya decides to sell her bond to raise the funds to take a vacation. She offers her bond for sale in the bond market. How much can she expect to get for her "used" bond if the prevailing interest rate in the bond market is 6 percent? If the prevailing interest rate is 4 percent?

As we mentioned, the price of a "used" bond at any point in time depends on the prevailing interest rate. Suppose first that, on January 1, 2004, when Tanya takes her bond to the bond market, the prevailing interest rate on newly issued one-year bonds is 6 percent. Would another saver be willing to pay Tanya the full $1,000 principal amount of her bond? No, because the purchaser of Tanya's bond will receive $1,050 in one year, when the bond matures; whereas if he uses his $1,000 to buy a new one-year bond paying 6 percent interest, he will receive $1,060 ($1,000 principal repayment plus $60 interest) in one year. So Tanya's bond is not worth $1,000 to another saver.

How much would another saver be willing to pay for Tanya's bond? Since newly issued one-year bonds pay a 6 percent return, he will buy Tanya's bond only at a price that allows him to earn at least that return. As the holder of Tanya's bond will receive $1,050 ($1,000 principal plus $50 interest) in one year, the price for her bond that allows the purchaser to earn a 6 percent return must satisfy the equation

$$\text{Bond price} \times 1.06 = \$1{,}050.$$

Solving the equation for the bond price, we find that Tanya's bond will sell for $1,050/1.06, or just under $991. To check this result, note that in one year the purchaser of the bond will receive $1,050, or $59 more than he paid. His rate of return is $59/$991, or 6 percent, as expected.

What if the prevailing interest rate had been 4 percent rather than 6 percent? Then the price of Tanya's bond would satisfy the relationship bond price $\times$ 1.04 = $1,050, implying that the price of her bond would be $1,050/1.04, or almost $1,010.

What happens if the interest rate when Tanya wants to sell is 5 percent, the same as it was when she originally bought the bond? You should show that in this case the bond would sell at its face value of $1,000.

This example illustrates a general principle, that *bond prices and interest rates are inversely related.* When the interest rate being paid on newly issued bonds rises, the price financial investors are willing to pay for existing bonds falls, and vice versa.

**EXERCISE 11.1**

**Three-year government bonds are issued at a face value (principal amount) of 100 and a coupon rate of 7 percent, interest payable at the end of each year. One year prior to the maturation of these bonds, a newspaper headline reads, "Bad Economic News Causes Prices of Bonds to Plunge," and the story reveals that these three-year bonds have fallen in price to 96. What has happened to interest rates? What is the one-year interest rate at the time of the newspaper story?**

Issuing bonds is one means by which a corporation or a government can obtain funds from savers. Another important way of raising funds, but one restricted to corporations, is by issuing stock to the public.

**Stocks** A share of **stock** (or *equity*) is a claim to partial ownership of a firm. For example, if a corporation has 1 million shares of stock outstanding, ownership of one share is equivalent to ownership of one-millionth of the company. Stockholders receive returns on their financial investment in two forms. First, stockholders receive a regular payment called a **dividend** for each share of stock they own. Dividends are determined by the firm's management and usually depend on the firm's recent profits. Second, stockholders receive returns in the form of *capital gains* when the price of their stock increases (we discussed capital gains and losses in Chapter 9).

***stock (or equity)*** a claim to partial ownership of a firm

***dividend*** a regular payment received by stockholders for each share that they own

Prices of stocks are determined through trading on a stock exchange, such as the New York Stock Exchange. A stock's price rises and falls as the demand for the stock changes. Demand for stocks in turn depends on factors such as news about the prospects of the company. For example, the stock price of a pharmaceutical company that announces the discovery of an important new drug is likely to rise on the announcement, even if actual production and marketing of the drug is some time away, because financial investors expect the company to become more profitable in the future. Example 11.2 illustrates numerically some key factors that affect stock prices.

**EXAMPLE 11.2**

### How much should you pay for a share of FortuneCookie.com?

You have the opportunity to buy shares in a new company called FortuneCookie.com, which plans to sell gourmet fortune cookies over the Internet. Your stockbroker estimates that the company will pay $1.00 per share in dividends a year from now, and that in a year the market price of the company will be $80.00 per share. Assuming that you accept your broker's estimates as accurate, what is the most that you should be willing to pay today per share of FortuneCookie.com? How does your answer change if you expect a $5.00 dividend? If you expect a $1.00 dividend but an $84.00 stock price in one year?

Based on your broker's estimates, you conclude that in one year each share of FortuneCookie.com you own will be worth $81.00 in your pocket—the $1.00 dividend plus the $80.00 you could get by reselling the stock. Finding the maximum price you would pay for the stock today therefore boils down to asking how much would you invest today to have $81.00 a year from today. Answering this question in turn requires one more piece of information, which is the expected rate of return that you require in order to be willing to buy stock in this company.

How would you determine your required rate of return to hold stock in FortuneCookie.com? For the moment, let's imagine that you are not too worried about the potential riskiness of the stock, either because you think that it is a "sure thing" or because you are a devil-may-care type who is not bothered by risk. In that case, your required rate of return to hold FortuneCookie.com should be about the same as you can get on other financial investments, such as government bonds. The available return on other financial investments gives the *opportunity cost* of your funds. So, for example, if the interest rate currently being offered by government bonds is 6 percent, you should be willing to accept a 6 percent return to hold FortuneCookie.com as well. In that case, the maximum price you would pay today for a share of FortuneCookie satisfies the equation

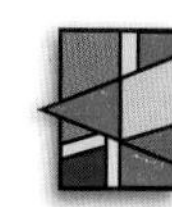

$$\text{Stock price} \times 1.06 = \$81.00.$$

This equation defines the stock price you should be willing to pay if you are willing to accept a 6 percent return over the next year. Solving this equation yields stock price = \$81.00/1.06 = \$76.42. If you buy FortuneCookie.com for \$76.42, then your return over the year will be (\$81.00 − \$76.42)/\$76.42 = \$4.58/\$71.42 = 6 percent, which is the rate of return you required to buy the stock.

If instead the dividend is expected to be \$5.00, then the total benefit of holding the stock in one year, equal to the expected dividend plus the expected price, is \$5.00 + \$80.00, or \$85.00. Assuming again that you are willing to accept a 6 percent return to hold FortuneCookie.com, the price you are willing to pay for the stock today satisfies the relationship stock price × 1.06 = \$85.00. Solving this equation for the stock price yields stock price = \$85.00/1.06 = \$80.19. Comparing with the previous case we see that a higher expected dividend in the future increases the value of the stock today. That's why good news about the future prospects of a company—such as the announcement by a pharmaceutical company that it has discovered a useful new drug—affects its stock price immediately.

If the expected future price of the stock is \$84.00, with the dividend at \$1.00, then the value of holding the stock in one year is once again \$85.00, and the calculation is the same as the previous one. Again, the price you should be willing to pay for the stock is \$80.19.

These examples show that an increase in the future dividend or in the future expected stock price raises the stock price today, whereas an increase in the return a saver requires to hold the stock lowers today's stock price. Since we expect required returns in the stock market to be closely tied to market interest rates, this last result implies that increases in interest rates tend to depress stock prices as well as bond prices.

Our examples also took the future stock price as given. But what determines the future stock price? Just as today's stock price depends on the dividend shareholders expect to receive this year and the stock price a year from now, the stock price a year from now depends on the dividend expected for next year and the stock price two years from now, and so on. Ultimately, then, today's stock price is affected not only by the dividend expected this year but future dividends as well. A company's ability to pay dividends depends on its earnings. Thus, as we noted in the example of the pharmaceutical company that announces the discovery of a new drug, news about future earnings—even earnings quite far in the future—is likely to affect a company's stock price immediately.

## EXERCISE 11.2

**As in Example 11.2, you expect a share of FortuneCookie.com to be worth \$80.00 per share in one year, and also to pay a dividend of \$1.00 in one year. What should you be willing to pay for the stock today if the prevailing interest rate, equal to your required rate of return, is 4 percent? What if the interest rate is 8 percent? In general, how would you expect stock prices to react if economic news arrives that implies that interest rates will rise in the very near future?**

In Example 11.2 we assumed that you were willing to accept a return of 6 percent to hold FortuneCookie.com, the same return that you could get on a government bond. However, financial investments in the stock market are quite risky in that returns to holding stocks can be highly variable and unpredictable. For example, although you expect a share of FortuneCookie.com to be worth \$80.00 in one year, you also realize that there is a chance it might sell as low as \$50.00 or as high as \$110.00 per share. Most financial investors dislike risk and unpredictability and thus have a higher required rate of return for holding risky assets like stocks than for holding relatively safe assets like government bonds. The difference between the required rate of return to hold risky assets and the rate of return on safe assets, like government bonds, is called the **risk premium.** Example 11.3 illustrates the effect of financial investors' dislike of risk on stock prices.

***risk premium*** the rate of return that financial investors require to hold risky assets minus the rate of return on safe assets

**Riskiness and stock prices**

**EXAMPLE 11.3**

Continuing Example 11.2, suppose that FortuneCookie.com is expected to pay a $1.00 dividend and have a market price of $80.00 per share in one year. The interest rate on government bonds is 6 percent per year. However, to be willing to hold a risky asset like a share of FortuneCookie.com, you require an expected return four percentage points higher than the rate paid by safe assets like government bonds (a risk premium of 4 percent). Hence you require a 10 percent expected return to hold FortuneCookie.com. What is the most you would be willing to pay for the stock now? What do you conclude about the relationship between perceived riskiness and stock prices?

As a share of FortuneCookie.com is expected to pay $81.00 in one year and the required return is 10 percent, we have stock price × 1.10 = $81.00. Solving for the stock price, we find the price to be $81.00/1.10 = $73.64, less than the price of $76.42 we found when there was no risk premium and the required rate of return was 6 percent (Example 11.2). We conclude that financial investors' dislike of risk, and the resulting risk premium, lowers the prices of risky assets like stocks.

**RECAP FACTORS AFFECTING STOCK PRICES**

1. An increase in expected future dividends or in the expected future market price of a stock raises the current price of the stock.
2. An increase in interest rates, implying an increase in the required rate of return to hold stocks, lowers the current price of stocks.
3. An increase in perceived riskiness, as reflected in an increase in the risk premium, lowers the current price of stocks.

## BOND MARKETS, STOCK MARKETS, AND THE ALLOCATION OF SAVINGS

Like banks, bond markets and stock markets provide a means of channeling funds from savers to borrowers with productive investment opportunities. For example, a corporation that is planning a capital investment but does not want to borrow from a bank has two other options: It can issue new bonds, to be sold to savers in the bond market, or it can issue new shares in itself, which are then sold in the stock market. The proceeds from the sales of new bonds or stocks are then available to the firm to finance its capital investment.

How do stock and bond markets help to ensure that available savings are devoted to the most productive uses? As we mentioned earlier, two important functions served by these markets are gathering information about prospective borrowers and helping savers to share the risks of lending.

**The informational role of bond and stock markets** Savers and their financial advisors know that to get the highest possible returns on their financial investments, they must find the potential borrowers with the most profitable opportunities. This knowledge provides a powerful incentive to scrutinize potential borrowers carefully.

For example, companies considering a new issue of stocks or bonds know that their recent performance and plans for the future will be carefully studied by professional analysts on Wall Street and other financial investors. If the analysts and other potential purchasers have doubts about the future profitability of the firm, they will offer a relatively low price for the newly issued shares or they will demand a high interest rate on newly issued bonds. Knowing this, a company will be reluctant to go to the bond or stock market for financing unless its management is confident that it can convince financial investors that the firm's planned use of the funds will be profitable. Thus the ongoing search by savers and their financial advisors for high returns leads the bond and stock markets to direct funds to the uses that appear most likely to be productive.

**Risk sharing and diversification** Many highly promising investment projects are also quite risky. The successful development of a new drug to lower cholesterol could create billions of dollars in profits for a drug company, for example; but if the drug turns out to be less effective than some others on the market, none of the development costs will be recouped. An individual who lent his or her life savings to help finance the development of the anticholesterol drug might enjoy a handsome return but also takes the chance of losing everything. Savers are generally reluctant to take large risks, so without some means of reducing the risk faced by each saver it might be very hard for the company to find the funds to develop the new drug.

Bond and stock markets help reduce risk by giving savers a means to *diversify* their financial investments. **Diversification** is the practice of spreading one's wealth over a variety of different financial investments to reduce overall risk. The idea of diversification follows from the adage that "you shouldn't put all your eggs in one basket." Rather than putting all of his or her savings in one very risky project, a financial investor will find it much safer to allocate a small amount of savings to each of a large number of stocks and bonds. That way, if some financial assets fall in value, there is a good chance that others will rise in value, with gains offsetting losses. Example 11.4 illustrates the benefits of diversification.

***diversification*** the practice of spreading of one's wealth over a variety of different financial investments to reduce overall risk

EXAMPLE 11.4

**The benefits of diversification**

Vikram has $1,000 to invest and is considering two stocks, the Smith Umbrella Company and the Jones Suntan Lotion Company. The price of Smith Umbrella stock will rise by 10 percent if it rains but will remain unchanged if the weather is sunny. The price of Jones Suntan stock is expected to rise by 10 percent if it is sunny but will remain unchanged if there is rain. The chance of rain is 50 percent, and the chance of sunshine is 50 percent. How should Vikram invest his $1,000?

If Vikram were to invest all his $1,000 in Smith Umbrella, he has a 50 percent chance of earning a 10 percent return, in the event that it rains, and a 50 percent chance of earning zero, if the weather is sunny. His average return is 50 percent times 10 percent plus 50 percent times zero, or 5 percent. Similarly, an investment in Jones Suntan yields 10 percent return half the time, when it's sunny, and 0 percent return the other half the time, when it rains, for an average return of 5 percent.

Although Vikram can earn an *average* return of 5 percent in either stock, investing in only one stock or the other is quite risky, since the actual return he receives varies widely depending on whether there is rain or shine. Can Vikram *guarantee* himself a 5 percent return, avoiding the uncertainty and risk? Yes, all he has to do is put $500 into each of the two stocks. If it rains, he will earn $50 on his Smith Umbrella stock and nothing on his Jones Suntan. If it's sunny, he will earn nothing on Smith Umbrella but $50 on Jones Suntan. Rain or shine, he is guaranteed to earn $50—a 5 percent return—without risk.

The existence of bond markets and stock markets make it easy for savers to diversify by putting a small amount of their savings into each of a wide variety of different financial assets, each of which represents a share of a particular company or investment project. From society's point of view, diversification makes it possible for risky but worthwhile projects to obtain funding, without individual savers having to bear too much risk.

***mutual fund*** a financial intermediary that sells shares in itself to the public, then uses the funds raised to buy a wide variety of financial assets

For the typical person, a particularly convenient way to diversify is to buy bonds and stocks indirectly through mutual funds. A **mutual fund** is a financial intermediary that sells shares in itself to the public, then uses the funds raised to buy a wide variety of financial assets. Holding shares in a mutual fund thus amounts to owning a little bit of many different financial assets, which helps to achieve diversification. The advantage of mutual funds is that it is usually less costly and time consuming to buy shares in one or two mutual funds than to buy many different stocks and bonds directly. Over the past decade mutual funds have become increasingly popular in the United States.

**Why did the U.S. stock market rise sharply in the 1990s, then fall in the new millennium?**

ECONOMIC NATURALIST 11.2

Stock prices soared during the 1990s in the United States. The Standard & Poor's 500 index, which summarizes the stock price performance of 500 major companies, rose 60 percent between 1990 and 1995, then more than doubled between 1995 and 2000. However, in the first two years of the new millennium this index lost nearly half its value. Why did the U.S. stock market boom in the 1990s and bust in the 2000s?

The prices of stocks depend on their purchasers' expectations about future dividends and stock prices and on the rate of return required by potential stockholders. The required rate of return in turn equals the interest rate on safe assets plus the risk premium. In principle, a rise in stock prices could be the result of increased optimism about future dividends, a fall in the required return, or some combination.

Probably both factors contributed to the boom in stock prices in the 1990s. Dividends grew rapidly in the 1990s, reflecting the strong overall performance of the U.S. economy. Encouraged by the promise of new technologies, many financial investors expected future dividends to be even higher.

There is also evidence that the risk premium that people required to hold stocks fell during the 1990s, thereby lowering the total required return and raising stock prices. One possible explanation for a decline in the risk premium in the 1990s is increased diversification. During that decade the number and variety of mutual funds available increased markedly. Millions of Americans invested in these funds, including many who had never owned stock before or had owned stock in only a few companies. This increase in diversification for the typical stock market investor may have lowered the perceived risk of holding stocks, which in turn reduced the risk premium and raised stock prices.

After 2000 both of these favorable factors reversed. The growth in dividends was disappointing to stockholders, in large part because many high-tech firms did not prove as profitable as had been hoped. An additional blow was a series of corporate accounting scandals in 2002, in which it became known that some large firms had taken illegal or unethical actions to make their profits seem larger than in fact they were. A number of factors, including a recession, a major terrorist attack, and the accounting scandals, also increased stockholders' concerns about the riskiness of stocks, so that the risk premium they required to hold stocks rose from its 1990s lows. The combination of lower expected dividends and a higher premium for risk sent stock prices sharply downward.

## INTERNATIONAL CAPITAL FLOWS

Our discussion thus far has focused on financial markets operating within a given country, such as the United States. However, economic opportunities are not necessarily restricted by national boundaries. The most productive use of a U.S. citizen's savings might be located far from U.S. soil, in helping to build a factory in Thailand or starting a small business in Poland. Likewise, the best way for a Brazilian saver to diversify her assets and reduce her risks could be to hold bonds and stocks from a number of different countries. Over time, extensive financial markets have developed to permit cross-border borrowing and lending. Financial markets in which borrowers and lenders are residents of different countries are called *international* financial markets.

International financial markets differ from domestic financial markets in at least one important respect: Unlike a domestic financial transaction, an international financial transaction is subject to the laws and regulations of at least two countries, the country that is home to the lender and the country that is home to the borrower. Thus the size and vitality of international financial markets depend on the degree of political and economic cooperation among countries. For example, during the relatively peaceful decades of the late nineteenth and early twentieth centuries, international financial markets were remarkably highly developed. Great Britain, at the time the world's dominant economic power, was a major international lender, dispatching its savings for use around the globe. However, during the turbulent years 1914–1945, two world wars and the Great Depression substantially reduced both international finance and international trade in goods and services. The extent of international finance and trade returned to the levels achieved in the late nineteenth century only in the 1980s.

In thinking about international financial markets, it is useful to understand that lending is economically equivalent to acquiring a real or financial asset, and borrowing is economically equivalent to selling a real or financial asset. For example, savers lend to companies by purchasing stocks or bonds, which are financial assets for the lender and financial liabilities for the borrowing firms. Similarly, lending to a government is accomplished in practice by acquiring a government bond—a financial asset for the lender, and a financial liability for the borrower, in this case the government. Savers can also provide funds by acquiring real assets such as land; if I purchase a parcel of land from you, though I am not making a loan in the usual sense, I am providing you with funds that you can use for consuming or investing. In lieu of interest or dividends from a bond or a stock, I receive the rental value of the land that I purchased.

***international capital flows*** purchases or sales of real and financial assets across international borders

***capital inflows*** purchases of domestic assets by foreign households and firms

***capital outflows*** purchases of foreign assets by domestic households and firms

Purchases or sales of real and financial assets across international borders (which are economically equivalent to lending and borrowing across international borders) are known as **international capital flows.** From the perspective of a particular country, say the United States, purchases of domestic (U.S.) assets by foreigners are called **capital inflows;** purchases of foreign assets by domestic (U.S.) households and firms are called **capital outflows.** To remember these terms, it may help to keep in mind that capital inflows represent funds "flowing in" to the country (foreign savers buying domestic assets), while capital outflows are funds "flowing out" of the country (domestic savers buying foreign assets). The difference between the two flows is expressed as *net capital inflows*—capital inflows minus capital outflows—or *net capital outflows*—capital outflows minus capital inflows. Note that capital inflows and outflows are *not* counted as exports or imports, because they refer to the purchase of existing real and financial assets rather than currently produced goods and services.

From a macroeconomic perspective, international capital flows play two important roles. First, they allow countries whose productive investment opportunities are greater than domestic savings to fill in the gap by borrowing from abroad. Second, they allow countries to run trade imbalances—situations in which the country's exports of goods and services do not equal its imports of goods and services. The rest of this chapter discusses these key roles. We begin by analyzing the important link between international capital flows and trade imbalances.

## CAPITAL FLOWS AND THE BALANCE OF TRADE

***trade balance (or net exports)*** the value of a country's exports less the value of its imports in a particular period (quarter or year)

In Chapter 5 we introduced the term *net exports (NX),* the value of a country's exports less the value of its imports. An equivalent term for the value of a country's exports less the value of its imports is the **trade balance.** Because exports need not equal imports in each quarter or year, the trade balance (or net exports) need not always equal zero. If the trade balance is positive in a particular period so that the value of exports exceeds the value of imports, a country is said to have a **trade surplus** for that period equal to the value of its exports minus the value of its imports. If the trade balance is negative, with imports greater than exports, the country is said to have a **trade deficit** equal to the value of its imports minus the value of its exports.

***trade surplus*** when exports exceed imports, the difference between the value of a country's exports and the value of its imports in a given period

***trade deficit*** when imports exceed exports, the difference between the value of a country's imports and the value of its exports in a given period

Figure 11.1 shows the components of the U.S. trade balance since 1960 (see Figure 4.5 for data extending back to 1900). The blue line represents U.S. exports as a percentage of GDP; the red line, U.S. imports as a percentage of GDP. When exports exceed imports, the vertical distance between the two lines gives the U.S. trade surplus as a percentage of GDP. When imports exceed exports, the vertical distance between the two lines represents the U.S. trade deficit. Figure 11.1 shows first that international trade has become an increasingly important part of the U.S. economy in the past several decades. In 1960, only 4.8 percent of U.S. GDP was exported, and the value of imports equaled 4.3 percent of U.S. GDP. In 2001, by comparison, 10.3 percent of U.S. production was sold abroad, and imports amounted to 13.7 percent of U.S. GDP. Second, the figure shows that since the late 1970s the United States has consistently run trade deficits, frequently equal to 2 percent or more of GDP. Why has the U.S. trade balance been in deficit for so long? We will answer that question later in this section.

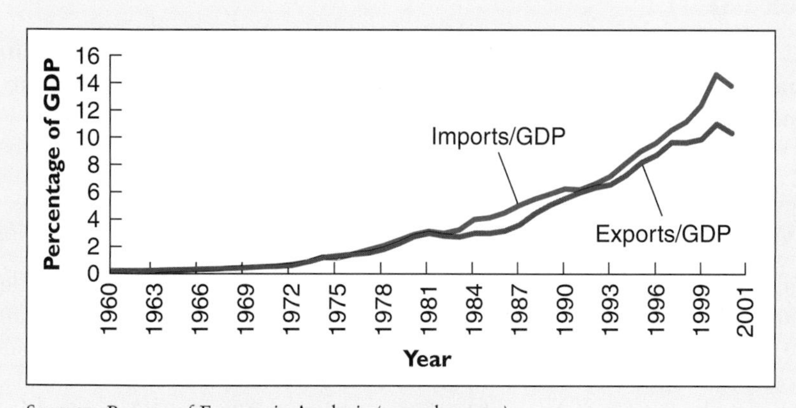


SOURCE: Bureau of Economic Analysis (www.bea.gov).

**FIGURE 11.1**
**The U.S. Trade Balance, 1960–2001.**
This figure shows U.S. exports and imports as a percentage of GDP. Since the late 1970s the United States has run a trade deficit, with imports exceeding exports.

The trade balance represents the difference between the value of goods and services exported by a country and the value of goods and services imported by the country. Net capital inflows represent the difference between purchases of domestic assets by foreigners and purchases of foreign assets by domestic residents. There is a precise and very important link between these two imbalances, which is that in any given period, *the trade balance and net capital inflows sum to zero*. For future reference, let's write this relationship as an equation:

$$NX + KI = 0. \tag{11.1}$$

where $NX$ is the trade balance (the same as net exports) and we use $KI$ to stand for net capital inflows. The relationship given by Equation 11.1 is an identity, meaning that it is true by definition.

To see why Equation 11.1 holds, consider what happens when (for example) a U.S. resident purchases an imported good, say a Japanese automobile priced at $20,000. Suppose the U.S. buyer pays by check so that the Japanese car manufacturer now holds $20,000 in an account in a U.S. bank. What will the Japanese manufacturer do with this $20,000? Basically, there are two possibilities.

First, the Japanese company may use the $20,000 to buy U.S.-produced goods and services, such as U.S.-manufactured car parts or Hawaiian vacations for its executives. In this case, the United States has $20,000 in exports to balance the $20,000 automobile import. Because exports equal imports, the U.S. trade balance is unaffected by these transactions (for these transactions, $NX = 0$). And because no assets are bought or sold, there are no capital inflows or outflows ($KI = 0$). So under this scenario, the condition that the trade balance plus net capital inflows equals zero, as stated in Equation 11.1, is satisfied.

Alternatively, the Japanese car producer might use the $20,000 to acquire U.S. assets, such as a U.S. Treasury bond or some land adjacent to its plant in Tennessee. In this case, the United States compiles a trade deficit of $20,000, because the $20,000 car import is not offset by an export ($NX = -\$20{,}000$). But there is a corresponding capital inflow of $20,000, reflecting the purchase of a U.S. asset by the Japanese ($KI = \$20{,}000$). So once again the trade balance and net capital inflows sum to zero, and Equation 11.1 is satisfied.[1]

In fact, there is a third possibility, which is that the Japanese car company might swap its dollars to some other party outside the United States. For example, the company might trade its dollars to another Japanese firm or individual in exchange for Japanese yen. However, the acquirer of the dollars would then have the same two options as the car company—to buy U.S. goods and services or acquire U.S. assets—so that the equality of net capital inflows and the trade deficit would continue to hold.

### EXERCISE 11.3

**A U.S. saver purchases a $20,000 Japanese government bond. Explain why Equation 11.1 is satisfied no matter what the Japanese government does with the $20,000 it receives for its bond.**

## THE DETERMINANTS OF INTERNATIONAL CAPITAL FLOWS

Capital inflows, recall, are purchases of domestic assets by foreigners, while capital outflows are purchases of foreign assets by domestic residents. For example, capital inflows into the United States include foreign purchases of items such as the stocks and bonds of U.S. companies, U.S. government bonds, and real assets such as land or buildings owned by U.S. residents. Why would foreigners want to acquire U.S. assets, and, conversely, why would Americans want to acquire assets abroad?

The basic factors that determine the attractiveness of any asset, either domestic or foreign, are *return* and *risk*. Financial investors seek high real returns; thus, with other factors (such as the degree of risk and the returns available abroad) held constant, a higher real interest rate in the home country promotes capital inflows by making domestic assets more attractive to foreigners. By the same token, a higher real interest rate in the home country reduces capital outflows by inducing domestic residents to invest their savings at home. Thus, all else being equal, a higher real interest rate at home leads to net capital inflows. Conversely, a low real interest rate at home tends to create net capital outflows, as financial investors look abroad for better opportunities. Figure 11.2 shows the relationship between a country's net capital inflows and the real rate of interest prevailing in that country. When the domestic real interest rate is high, net capital inflows are positive (foreign purchases of domestic assets exceed domestic purchases of foreign assets). But when the real interest rate is low, net capital inflows are negative (that is, the country experiences net capital outflows).

The effect of risk on capital flows is the opposite of the effect of the real interest rate. For a given real interest rate, an increase in the riskiness of domestic assets reduces net capital inflows, as foreigners become less willing to buy the home country's assets, and domestic savers become more inclined to buy foreign assets. For

[1] If the Japanese company simply left the $20,000 in the U.S. bank, it would still count as a capital inflow, since the deposit would still be a U.S. asset acquired by foreigners.

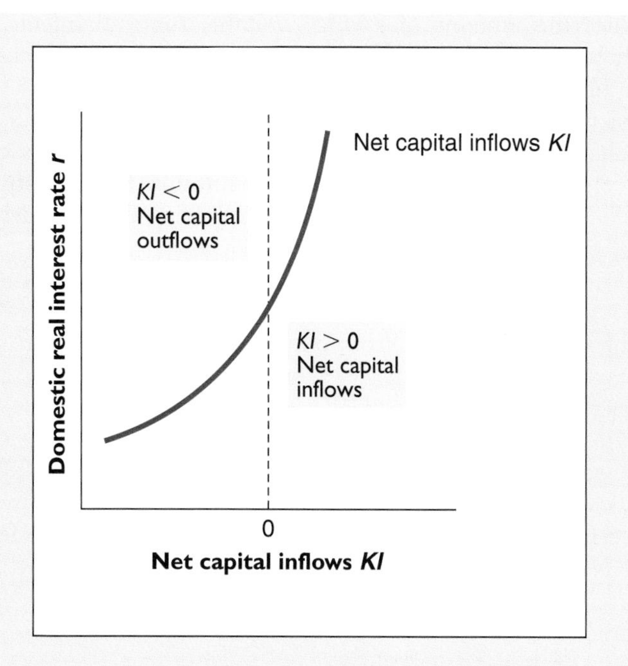


**FIGURE 11.2**
**Net Capital Inflows and the Real Interest Rate.**
Holding constant the degree of risk and the real returns available abroad, a high real interest rate in the home country will induce foreigners to buy domestic assets, increasing capital inflows. A high real rate in the home country also reduces the incentive for domestic savers to buy foreign assets, reducing capital outflows. Thus, all else being equal, the higher the domestic real interest rate *r*, the higher will be net capital inflows *KI*.

example, political instability, which increases the risk of investing in a country, tends to reduce net capital inflows. Figure 11.3 shows the effect of an increase in risk on capital flows: At each value of the domestic real interest rate, an increase in risk reduces net capital inflows, shifting the capital inflows curve to the left.

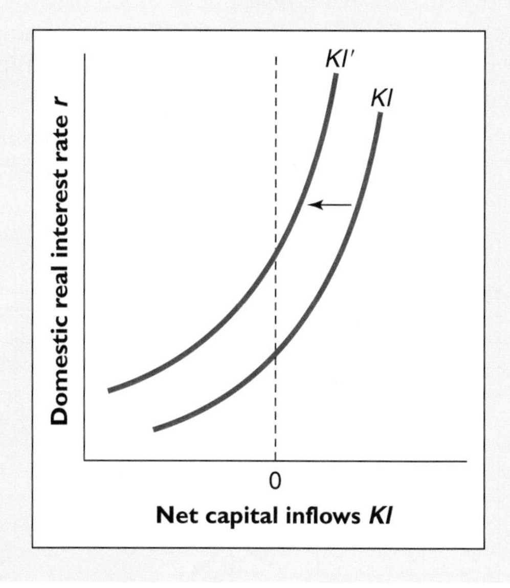


**FIGURE 11.3**
**An Increase in Risk Reduces Net Capital Inflows.**
An increase in the riskiness of domestic assets, arising, for example, from an increase in political instability, reduces the willingness of foreign and domestic savers to hold domestic assets. The supply of capital inflows declines at each value of the domestic real interest rate, shifting the *KI* curve to the left.

**EXERCISE 11.4**

**For given real interest rate and riskiness in the home country, how would you expect net capital inflows to be affected by an increase in real interest rates abroad? Show your answer graphically.**

## SAVING, INVESTMENT, AND CAPITAL INFLOWS

International capital flows have a close relationship to domestic saving and investment. As we will see next, capital inflows augment the domestic saving pool, increasing the funds available for investment in physical capital, while capital outflows reduce the amount of saving available for investment. Thus capital inflows can help to promote economic growth within a country, and capital outflows to restrain it.

To derive the relationship among capital inflows, saving, and investment, recall from Chapter 5 that total output or income $Y$ must always equal the sum of the four components of expenditure: consumption ($C$), investment ($I$), government purchases ($G$), and net exports ($NX$). Writing out this identity, we have

$$Y = C + I + G + NX.$$

Next, we subtract $C + G + NX$ from both sides of the identity to obtain

$$Y - C - G - NX = I.$$

In Chapter 9 we saw that national saving $S$ is equal to $Y - C - G$. Furthermore, Equation 11.1 above states that the trade balance plus capital inflows equals zero, or $NX + KI = 0$, which implies that $KI = -NX$. If we substitute $S$ for $Y - C - G$ and $KI$ for $-NX$ in the above equation, we find that

$$S + KI = I. \tag{11.2}$$

Equation 11.2, a key result, says that the sum of national saving $S$ and capital inflows from abroad $KI$ must equal domestic investment in new capital goods, $I$. In other words, in an open economy, the pool of saving available for domestic investment includes not only national saving (the saving of the domestic private and public sectors) but funds from savers abroad as well.

Chapter 9 introduced the saving-investment diagram, which shows that in a closed economy, the supply of saving must equal the demand for saving. A similar diagram applies to an open economy, except that the supply of saving in an open economy includes net capital inflows as well as domestic saving. Figure 11.4

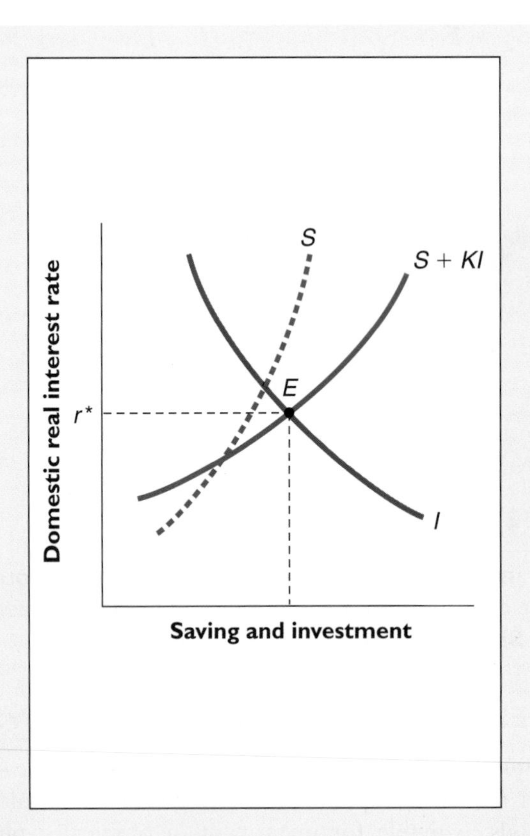


**FIGURE 11.4**
**The Saving-Investment Diagram for an Open Economy.**
The total supply of savings in an open economy is the sum of national saving $S$ and net capital inflows $KI$. The supply of domestic saving $S$ is shown for comparison. Because a low real interest rate prompts capital outflows ($KI < 0$), at low values of the domestic interest rate the total supply of saving $S + KI$ is smaller than national saving $S$. The domestic demand for saving for purposes of capital investment is shown by the curve labeled $I$. The equilibrium real interest rate $r^*$ sets the total supply of saving, including capital inflows, equal to the domestic demand for saving.

shows the open-economy version of the saving-investment diagram. The domestic real interest rate is shown on the vertical axis and saving and investment flows on the horizontal axis. As in a closed economy, the downward-sloping curve $I$ shows the demand for funds by firms that want to make capital investments. The solid upward-sloping curve, marked $S + KI$, shows the total supply of saving, including *both* domestic saving $S$ and net capital inflows from abroad $KI$. Also shown, for comparison, is the supply of domestic saving, marked $S$. You can see that for higher values of the domestic real interest rate, net capital inflows are positive, so the $S + KI$ curve falls to the right of the curve $S$ showing domestic saving only. But at low enough values of the real interst rate $r$, the economy sustains net capital outflows, as savers look abroad for higher returns on their financial investments. Thus, at low values of the domestic real interest rate, the net supply of savings is lower than it would be in a closed economy, and the $S + KI$ curve falls to the left of the domestic supply of saving curve $S$. As Figure 11.4 shows, the equilibrium real interest rate in an open economy, $r^*$, is the level that sets the total amount of saving supplied (including capital inflows from abroad) equal to the amount of saving demanded for purposes of domestic capital investment.

Figure 11.4 also indicates how net capital inflows can benefit an economy. A country that attracts significant amounts of foreign capital flows will have a larger pool of total saving and hence both a lower real interest rate and a higher rate of investment in new capital than it otherwise would. The United States and Canada both benefited from large inflows of capital in the early stages of their economic development, as do many developing countries today. Because capital inflows tend to react very sensitively to risk, an implication is that countries that are politically stable and safeguard the rights of foreign investors will attract more foreign capital and thus grow more quickly than countries without those characteristics.

Although capital inflows are generally beneficial to the countries that receive them, they are not costless. Countries that finance domestic capital formation primarily by capital inflows face the prospect of paying interest and dividends to the foreign financial investors from whom they have borrowed. A number of developing countries have experienced *debt crises,* arising because the domestic investments they made with foreign funds turned out poorly, leaving them insufficient income to pay what they owed their foreign creditors. An advantage to financing domestic capital formation primarily with domestic saving is that the returns from the country's capital investments accrue to domestic savers rather than flowing abroad.

**Why did the Argentine economy collapse in 2001–2002?**

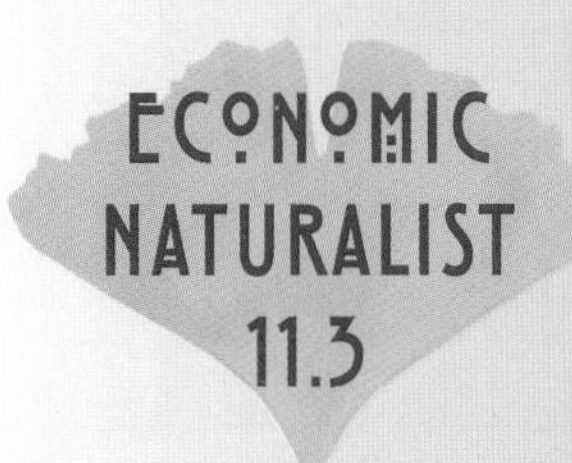


Argentina, with a wealth of natural resources and an educated population, has long been among the most prosperous economies of Latin America. However, in 2001–2002 the country faced a severe economic crisis. Political dissatisfaction reached so high a level that Argentines rioted in the streets of Buenos Aires, and the country had five different presidents within a span of a few months. Why did the Argentine economy collapse in 2001–2002?

Because Argentina is a developing economy with extensive human and natural resources, investments in new capital goods in that country could potentially be very profitable. However, Argentina's national saving rate is among the lowest in Latin America. To make up the difference between the demand for investment (new capital goods) and the domestic supply of saving, Argentina borrowed extensively from abroad, that is, capital inflows to Argentina were large. These capital inflows helped Argentina to invest more and grow more quickly than it otherwise might have (see Figure 11.4, which shows how capital inflows $KI$ augment the available pool of domestic saving $S$). However, the rapid capital inflows also implied that, over time, Argentina was building up a large debt to foreigners. Foreigners remained willing to lend to Argentina so long as they expected to earn good returns on their loans.

Unfortunately, in the late 1990s the situation in Argentina took a turn for the worse. In 1998, following a three-year growth boom, the Argentine economy slowed considerably. Moreover, partly as a result of the slowing economy, which reduced tax receipts and raised the public's demands for government services, the government budgetary situation worsened. The central government of Argentina, which had a budget surplus of over 2 percent of GDP in 1993, began to run large deficits. Free-spending provincial and city governments ran deficits of their own, which exacerbated the nation's fiscal problem.

The increased government budget deficits reduced Argentina's national saving, increasing the need to borrow abroad. But at the same time that Argentina's borrowing needs were rising, foreign lenders began to worry that the country—with its slowing economy, high debt burden, and worsening government budget deficits—was a much riskier location for investment than they had thought. Increased risk reduces the supply of capital inflows (see Figure 11.3) and thus also reduces the total pool of saving available; the result is a higher domestic interest rate, lower domestic investment, and hence a weakening economy. As the economy continued to weaken, and government budgets worsened, foreign lenders became so pessimistic about Argentina that they would lend only at very high interest rates, if at all. Ultimately Argentina was unable to repay even the interest on its foreign debt and was forced to default (refuse to pay). At that point the country became essentially unable to borrow abroad at any price. Investment in Argentina collapsed and real interest rates soared. As of this writing, Argentina is negotiating with public agencies, such as the International Monetary Fund, to try to obtain loans to help rebuild its economy.

## THE SAVING RATE AND THE TRADE DEFICIT

We have seen that a country's exports and imports do not necessarily balance in each period. Indeed, the United States has run a trade deficit, with its imports exceeding exports, for many years. What causes trade deficits? Stories in the media sometimes claim that trade deficits occur because a country produces inferior goods that no one wants to buy or because other countries impose unfair trade restrictions on imports. Despite the popularity of these explanations, however, there is little support for them in either economic theory or evidence. For example, the United States has a large trade deficit with China, but no one would claim U.S. goods are generally inferior to Chinese goods. And many developing countries have significant trade deficits even though they, rather than their trading partners, tend to impose the more stringent restrictions on trade.

Economists argue that, rather than the quality of a country's exports or the existence of unfair trade restrictions, *a low rate of national saving is the primary cause of trade deficits.*

To see the link between national saving and the trade deficit, recall the identity $Y = C + I + G + NX$. Subtracting $C + I + G$ from both sides of this equation and rearranging, we get $Y - C - G - I = NX$. Finally, recognizing that national saving $S$ equals $Y - C - G$, we can rewrite the relationship as

$$S - I = NX. \qquad (11.3)$$

Equation 11.3 can also be derived directly from Equations 11.1 and 11.2. According to Equation 11.3, if we hold domestic investment ($I$) constant, a high rate of national saving $S$ implies a high level of net exports $NX$, while a low level of national saving implies a low level of net exports. Furthermore, if a country's national saving is less than its investment, or $S < I$, then Equation 11.3 implies that net exports $NX$ will be negative. That is, the country will have a trade deficit. The conclusion from Equation 11.3 is that, holding domestic investment constant, low national saving tends to be associated with a trade deficit ($NX < 0$), and high national saving is associated with a trade surplus ($NX > 0$).

Why does a low rate of national saving tend to be associated with a trade deficit? A country with a low national saving rate is one in which households and the government have high spending rates, relative to domestic income and production. Since part of the spending of households and the government is devoted to imported goods, we would expect a low-saving, high-spending economy to have a high volume of imports. Furthermore, a low-saving economy consumes a large proportion of its domestic production, reducing the quantity of goods and services available for export. With high imports and low exports, a low-saving economy will experience a trade deficit.

A country with a trade deficit must also be receiving capital inflows, as we have seen. (Equation 11.1 tells us that if a trade deficit exists so that $NX < 0$, then it must be true that $KI > 0$—net capital inflows are positive.) Is a low national saving rate also consistent with the existence of net capital inflows? The answer is yes. A country with a low national saving rate will not have sufficient savings of its own to finance domestic investment. Thus there likely will be many good investment opportunities in the country available to foreign savers, leading to capital inflows. Equivalently, a shortage of domestic saving will tend to drive up the domestic real interest rate, which attracts capital flows from abroad.

We conclude that a low rate of national saving tends to create a trade deficit, as well as to promote the capital inflows that must accompany a trade deficit. Economic Naturalist 11.4 illustrates this effect for the case of the United States.

**Why is the U.S. trade deficit so large?**

ECONOMIC NATURALIST 11.4

As shown by Figure 11.1, U.S. trade was more or less in balance until the mid-1970s. Since the late 1970s, however, the United States has run large trade deficits, particularly in the mid-1980s and since the latter part of the 1990s. Indeed, in 2001 the trade deficit equaled 3.5 percent of U.S. GDP. Why is the U.S. trade deficit so large?

Figure 11.5 shows national saving, investment, and the trade balance for the United States from 1960 to 2001 (all measured relative to GDP). Note that the trade balance has been negative since the late 1970s, indicating a trade deficit. Note also that trade deficits correspond to periods in which investment exceeds national saving, as required by Equation 11.3.

U.S. national saving and investment were roughly in balance in the 1960s and early 1970s, and hence the U.S. trade balance was close to zero during that period.

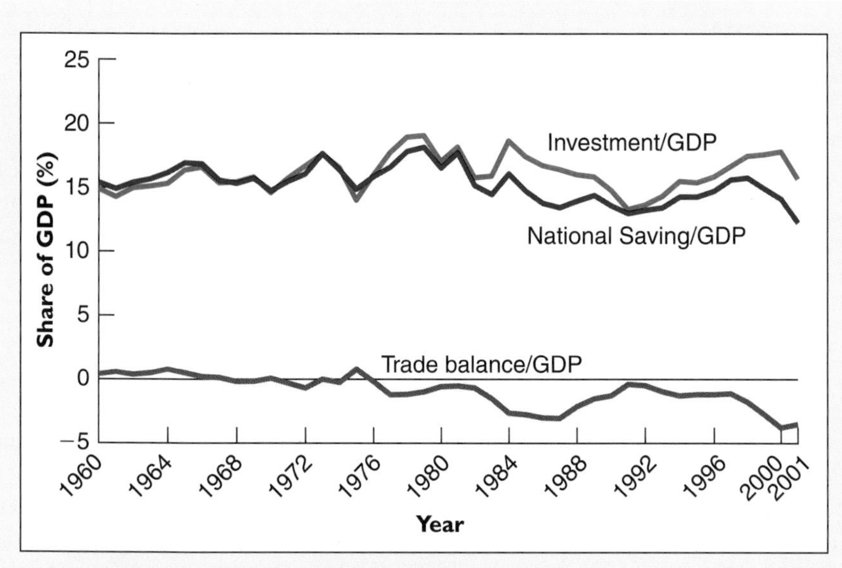


SOURCE: Bureau of Economic Analysis (www.bea.gov).

**FIGURE 11.5**
**National Saving, Investment, and the Trade Balance in the United States, 1960–2001.**
Since the 1970s U.S. national saving has fallen below domestic investment, implying a significant trade deficit.

However, U.S. national saving fell sharply during the late 1970s and 1980s. One factor that contributed to the decline in national saving was the large government deficits of the era (see Chapter 9). Because investment did not decline as much as saving, the U.S. trade deficit ballooned in the 1980s, coming under control only when investment fell during the recession of 1990–1991. Saving and investment both recovered during the 1990s, but in the latter part of the 1990s national saving dropped again. This time the federal government was not at fault, since its budget showed a healthy surplus. Rather, the fall in national saving reflected a decline in private saving, the result of a powerful upsurge in consumption spending. Much of the increase in consumption spending was for imported goods and services, which pushed the trade deficit to record levels.

Is the U.S. trade deficit a problem? The trade deficit implies that the United States is relying heavily on foreign savings to finance its domestic capital formation (net capital inflows). These foreign loans must ultimately be repaid with interest. If the foreign savings are well invested and the U.S. economy grows, repayment will not pose a problem. However, if economic growth in the United States slackens, repaying the foreign lenders will impose an economic burden in the future.

*"But we're not just talking about buying a car—we're talking about confronting this country's trade deficit with Japan."*

## RECAP INTERNATIONAL CAPITAL FLOWS AND THE BALANCE OF TRADE

- Purchases or sales of assets across borders are called international capital flows. If a person, firm, or government in (say) the United States borrows from abroad, we say that there is a capital inflow into the United States. In this case, foreign savers are acquiring U.S. assets. If a person, firm, or government in the United States lends to someone abroad, thereby acquiring a foreign asset, we say that there has been a capital outflow from the United States to the foreign country. Net capital inflows to a given country equal capital inflows minus outflows.
- If a country imports more goods and services than it exports, it must borrow abroad to cover the difference. Likewise, a country that exports more than it imports will lend the difference to foreigners. Thus, as a matter of accounting, the trade balance $NX$ and net capital inflows $KI$ must sum to zero in every period.

- The funds available for domestic investment in new capital goods equal the sum of domestic saving and net capital inflows from abroad. The higher the return and the lower the risk of investing in the domestic country, the greater will be the capital inflows from abroad. Capital inflows benefit an economy by providing more funds for capital investment, but they can become a burden if the returns from investing in new capital goods are insufficient to pay back the foreign lenders.
- An important cause of a trade deficit is a low national saving rate. A country that saves little and spends a lot will tend to import a greater quantity of goods and services than it is able to export. At the same time, the country's low saving rate implies a need for more foreign borrowing to finance domestic investment spending.

To this point in the book we have discussed a variety of issues relating to the long-run performance of the economy, including economic growth, the sources of increasing productivity and improved living standards, the determination of real wages, and the determinants of saving and capital formation. Beginning with the next chapter we will take a more short-run perspective, examining first the causes of recessions and booms in the economy and then turning to policy measures that can be used to affect these fluctuations.

## SUMMARY


- Besides balancing saving and investment in the aggregate, financial markets and institutions play the important role of allocating saving to the most productive investment projects. The financial system improves the allocation of saving in two ways: First, it provides information to savers about which of the many possible uses of their funds are likely to prove must productive, and hence pay the highest return. For example, *financial intermediaries* such as banks develop expertise in evaluating prospective borrowers, making it unnecessary for small savers to do that on their own. Similarly, stock and bond analysts evaluate the business prospects of a company issuing shares of stock or bonds, which determines the price the stock will sell for or the interest rate the company will have to offer on its bond. Second, financial markets help savers share the risks of lending by permitting them to *diversify* their financial investments. Individual savers often hold stocks through *mutual funds,* a type of financial intermediary that reduces risk by holding many different financial assets. By reducing the risk faced by any one saver, financial markets allow risky but potentially very productive projects to be funded.

- Corporations that do not wish to borrow from banks can obtain finance by issuing bonds or stocks. A *bond* is a legal promise to repay a debt, including both the *principal amount* and regular interest payments. The prices of existing bonds decline when interest rates rise. A share of *stock* is a claim to partial ownership of a firm. The price of a stock depends positively on the *dividend* the stock is expected to pay and on the expected future price of the stock and negatively on the rate of return required by financial investors to hold the stock. The required rate of return in turn is the sum of the return on safe assets and the additional return required to compensate financial investors for the riskiness of stocks, called the *risk premium.*

- The *trade balance,* or net exports, is the value of a country's exports less the value of its imports in a particular period. Exports need not equal imports in each period. If exports exceed imports, the difference is called a *trade surplus,* and if imports exceed exports, the difference is called a *trade deficit.* Trade takes place in assets as well as goods and services. Purchases of domestic assets (real or financial) by foreigners are called *capital inflows,* and purchases of foreign assets by domestic savers are called *capital outflows.* Because imports that are not financed by sales of exports must be financed by sales of assets, the trade balance and net capital inflows sum to zero.

- The higher the real interest rate in a country, and the lower the risk of investing there, the higher its capital inflows. The availability of capital inflows expands a country's pool of saving, allowing for more domestic investment and increased growth. A drawback to using capital inflows to finance domestic capital formation is that the returns to capital (interest and dividends) accrue to foreign financial investors rather than domestic residents.

- A low rate of national saving is the primary cause of trade deficits. A low-saving, high-spending country is likely to import more than a high-saving country. It also consumes more of its domestic production, leaving less for export. Finally, a low-saving country is likely to have a high real interest rate, which attracts net capital inflows. Because the sum of the trade balance and capital inflows is zero, a high level of net capital inflows is consistent with a large trade deficit.

## KEY TERMS

bond (285)
capital inflows (292)
capital outflows (292)
coupon payments (285)
coupon rate (285)
diversification (290)
dividend (287)
financial intermediaries (283)
international capital flows (292)
mutual fund (290)
principal amount (285)
risk premium (288)
stock (or equity) (287)
trade balance (292)
trade deficit (293)
trade surplus (293)

## REVIEW QUESTIONS

1. Give two ways that the financial system helps to improve the allocation of savings. Illustrate with examples.
2. Arjay plans to sell a bond that matures in one year and has a principal value of $1,000. Can he expect to receive $1,000 in the bond market for the bond? Explain.
3. Suppose you are much less concerned about risk than the typical person. Are stocks a good financial investment for you? Why or why not?
4. Stock prices surge but the prices of government bonds remain stable. What can you infer from the behavior of bond prices about the possible causes of the increase in stock values?
5. From the point of view of a given country, say, Spain, give an example of a capital inflow and a capital outflow.
6. How are capital inflows or outflows related to domestic investment in new capital goods?
7. Explain with examples why, in any period, a country's net capital inflows equal its trade deficit.
8. How would increased political instability in a country likely affect capital inflows, the domestic real interest rate, and investment in new capital goods? Show graphically.

## PROBLEMS

1. Simon purchases a bond, newly issued by the Amalgamated Corporation, for $1,000. The bond pays $60 to its holder at the end of the first and second years and pays $1,060 upon its maturity at the end of the third year.
   a. What are the principal amount, the term, the coupon rate, and the coupon payment for Simon's bond?
   b. After receiving the second coupon payment (at the end of the second year), Simon decides to sell his bond in the bond market. What price can he expect for his bond if the one-year interest rate at that time is 3 percent? 8 percent? 10 percent?
   c. Can you think of a reason that the price of Simon's bond after two years might fall below $1,000, even though the market interest rate equals the coupon rate?
2. Shares in Brothers Grimm, Inc., manufacturers of gingerbread houses, are expected to pay a dividend of $5.00 in one year and to sell for $100 per share at that time. How much should you be willing to pay today per share of Grimm:
   a. If the safe rate of interest is 5 percent and you believe that investing in Grimm carries no risk?
   b. If the safe rate of interest is 10 percent and you believe that investing in Grimm carries no risk?
   c. If the safe rate of interest is 5 percent but your risk premium is 3 percent?
   d. Repeat parts a to c, assuming that Grimm is not expected to pay a dividend but the expected price is unchanged.
3. Your financial investments consist of U.S. government bonds maturing in ten years and shares in a start-up company doing research in pharmaceuticals. How would you expect each of the following news items to affect the value of your assets? Explain.
   a. Interest rates on newly issued government bonds rise.

b. Inflation is forecasted to be much lower than previously expected (*Hint:* Recall the Fisher effect from Chapter 5.) Assume for simplicity that this information does *not* affect your forecast of the dollar value of the pharmaceutical company's future dividends and stock price.

In parts c to f, interest rates on newly issued government bonds are assumed to remain unchanged.

c. Large swings in the stock market increase financial investors' concerns about market risk.
d. The start-up company whose stock you own announces the development of a valuable new drug. However, the drug will not come to market for at least five years.
e. The pharmaceutical company announces that it will not pay a dividend next year.
f. The federal government announces a system of price controls on prescription drugs.

4. You have $1,000 to invest and are considering buying some combination of the shares of two companies, DonkeyInc and ElephantInc. Shares of DonkeyInc will pay a 10 percent return if the Democrats are elected, an event you believe to have a 40 percent probability; otherwise the shares pay a zero return. Shares of ElephantInc will pay 8 percent if the Republicans are elected (a 60 percent probability), zero otherwise. Either the Democrats or the Republicans will be elected.
   a. If your only concern is maximizing your average expected return, with no regard for risk, how should you invest your $1,000?
   b. What is your expected return if you invest $500 in each stock? (*Hint:* Consider what your return will be if the Democrats win and if the Republicans win, then weight each outcome by the probability that event occurs.)
   c. The strategy of investing $500 in each stock does *not* give the highest possible average expected return. Why might you choose it anyway?
   d. Devise an investment strategy that guarantees at least a 4.4 percent return, no matter which party wins.
   e. Devise an investment strategy that is riskless, that is, one in which the return on your $1,000 does not depend at all on which party wins.

5. From the web site of the Bureau of Economic Analysis, www.bea.gov, find data on the components of nominal GDP for the most recent quarter available and for the previous two complete years. Find net exports, national saving (which can be derived from the relationship $S = Y - C - G$), and gross private domestic investment for the period, and verify that they satisfy the relationship $S - I = NX$. How has the U.S. trade balance changed over the past two years as a percentage of GDP? Are the changes attributable to changes in the national saving rate, the rate of investment, or both?

6. How do each of the following transactions affect (1) the trade surplus or deficit and (2) capital inflows or outflows for the United States? Show that in each case the identity that the trade balance plus net capital inflows equals zero applies.
   a. A U.S. exporter sells software to Israel. She uses the Israeli shekels received to buy stock in an Israeli company.
   b. A Mexican firm uses proceeds from its sale of oil to the United States to buy U.S. government debt.
   c. A Mexican firm uses proceeds from its sale of oil to the United States to buy oil drilling equipment from a U.S. firm.
   d. A Mexican firm receives U.S. dollars from selling oil to the United States. A French firm accepts the dollars as payment for drilling equipment. The French firm uses the dollars to buy U.S. government debt.
   e. A British financial investor writes a check on his bank account in New York to purchase shares of General Motors stock (GM is a U.S. company).

7. Use a diagram like Figure 11.4 (solid lines only) to show the effects of each of the following on the real interest rate and capital investment of a country that is a net borrower from abroad.
   a. Investment opportunities in the country improve owing to new technologies.
   b. The government budget deficit rises.
   c. Domestic citizens decide to save more.
   d. Foreign investors believe that the riskiness of lending to the country has increased.

8. A country's domestic supply of saving, domestic demand for saving for purposes of capital formation, and supply of net capital inflows are given by the following equations:

$$S = 1{,}500 + 2{,}000r,$$
$$I = 2{,}000 - 4{,}000r,$$
$$KI = -100 + 6{,}000r.$$

   a. Assuming that the market for saving and investment is in equilibrium, find national saving, capital inflows, domestic investment, and the real interest rate.
   b. Repeat part a, assuming that desired national saving declines by 120 at each value of the real interest rate. What effect does a reduction in domestic saving have on capital inflows?
   c. Concern about the economy's macroeconomic policies causes capital inflows to fall sharply so that now $KI = -700 + 6{,}000r$. Repeat part a. What does a reduction in capital inflows do to domestic investment and the real interest rate?

## ANSWERS TO IN-CHAPTER EXERCISES

11.1 Since bond prices fell, interest rates must have risen. To find the interest rate, note that bond investors are willing to pay only 96 today for a bond that will pay back 107 (a coupon payment of 7 plus the principal amount of 100) in one year. To find the one-year return, divide 107 by 96 to get 1.115. Thus the interest rate must have risen to 11.5 percent.

11.2 The share of stock will be worth $81.00 in one year—the sum of its expected future price and the expected dividend. At an interest rate of 4 percent, its value today is $81.00/1.04 = $77.88. At an interest rate of 8 percent, the stock's current value is $81.00/1.08 = $75.00. Recall from Example 11.2 that when the interest rate is 6 percent, the value of a share of FortuneCookie.com is $76.42. Since higher interest rates imply lower stock values, news that interest rates are about to rise should cause the stock market to fall.

11.3 The purchase of the Japanese bond is a capital outflow for the United States, or $KI = -\$20{,}000$. The Japanese government now holds $20,000. What will it do with these funds? There are basically three possibilities. First, it might use the funds to purchase U.S. goods and services (military equipment, for example). In that case the U.S. trade balance equals +$20,000, and the sum of the trade balance and capital inflows is zero. Second, the Japanese government might acquire U.S. assets, for example, deposits in U.S. banks. In that case a capital inflow to the United States of $20,000 offsets the original capital outflow. Both the trade balance and net capital outflows individually are zero, and so their sum is zero.

Finally, the Japanese government might use the $20,000 to purchase non-U.S. goods, services, or assets—oil from Saudi Arabia, for example. But then the non-U.S. recipient of the $20,000 is holding the funds, and it has the same options that the Japanese government did. Eventually, the funds will be used to purchase U.S. goods, services, or assets, satisfying Equation 11.1. Indeed, even if the recipient holds onto the funds (in cash, or as a U.S. bank deposit), they would still count as a capital inflow to the United States, as U.S. dollars or accounts in a U.S. bank are U.S. assets acquired by foreigners.

11.4 An increase in the real interest rate abroad increases the relative attractiveness of foreign financial investments to both foreign and domestic savers. Net capital inflows to the home country will fall at each level of the domestic real interest rate. The supply curve of net capital inflows shifts left, as in Figure 11.3.

# PART 4

# THE ECONOMY IN THE SHORT RUN

A sign in Redwood City, California, boasts that the Bay Area town has the world's best climate. Redwood City's mean annual temperature and rainfall are similar to that of many other U.S. cities, so on what basis do Redwood City's boosters make their claim? The weather in Redwood City is attractive to many people because it varies so little over the year, being almost equally comfortable and temperate in winter and in summer. A city with the same average yearly temperature as Redwood City, but where the winters are freezing and the summers unbearably hot, would not be nearly so pleasant a place to live.

An analogous idea applies to the performance of the economy. As we saw, over a period of decades or more, the economy's average rate of growth is the crucial determinant of average living standards. But short-term fluctuations of the economy's growth rate around its long-run average matter for economic welfare as well. In particular, periods of slow or negative economic growth, known as *recessions,* may create significant economic hardship and dissatisfaction. In Part 4 we will explore the causes of short-term fluctuations in key economic variables, including output, unemployment, and inflation, and we will discuss the options available to government policymakers for stabilizing the economy.

Chapter 12 provides some necessary background for our study of short-term fluctuations by describing their key characteristics and reviewing the historical record of fluctuations in the U.S. economy. In Chapters 13 through 15 we develop a framework for the analysis of short-term fluctuations and the alternative policy responses. Chapter 13 shows how fluctuations in spending, or *aggregate demand,* may lead to short-run fluctuations in output and employment. That chapter also explains how changes in fiscal policy—policies relating to government spending and taxation—can be used to stabilize spending and output. Chapter 14 focuses on monetary policy, a second tool for stabilizing output and employment. Finally, Chapter 15 incorporates inflation into the analysis discussing both the sources of inflation and the policies that can be used to control it.

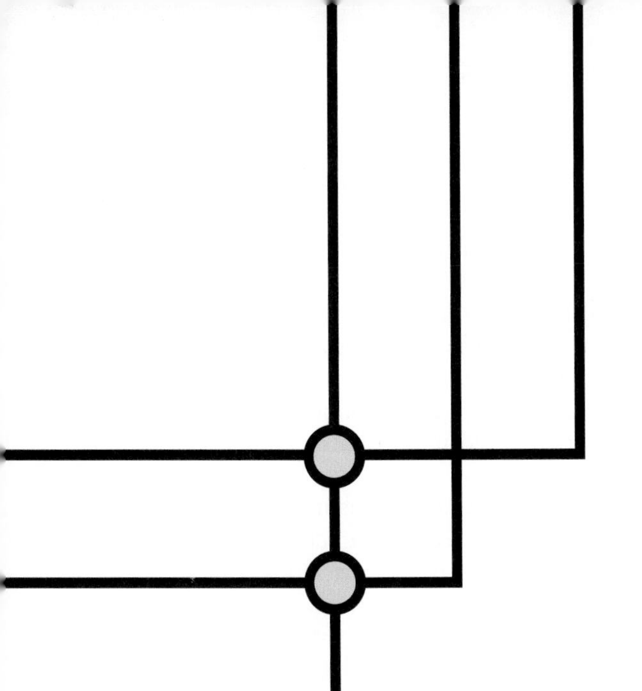

CHAPTER 12

# SHORT-TERM ECONOMIC FLUCTUATIONS: AN INTRODUCTION

In early 1991, following the defeat of Iraq in the Gulf War by the United States and its allies, polls showed that 90 percent of the American public approved of the job George H. W. Bush was doing as president. The Gulf War victory followed a number of other popular developments in the foreign policy sphere, including the ouster of the corrupt leader General Manuel Noriega from Panama in December 1989; improved relations with China; apparent progress in Middle East peace talks; and the end of apartheid in South Africa. The collapse of the Soviet Union in December 1991—a stunning event that signaled the end of the Cold War—also occurred during Bush's term. Yet despite these political pluses, in the months following the Gulf War Bush's sky-high approval rating declined sharply. According to one poll, by the time of the Republican national convention in the summer of 1992, only 29 percent of the public approved of Bush's performance. Although the president's ratings improved during the campaign, Bush and his running mate, Dan Quayle, lost the 1992 general election to Bill Clinton and Al Gore, receiving only 39 million of the 104 million votes cast. A third-party candidate, Ross Perot, received nearly 20 million votes.

What caused this turnaround in (the first) President Bush's political fortunes? Despite his high marks from voters in foreign policy, the president's domestic economic policies were widely viewed as ineffective. Bush received much criticism for breaking his campaign pledge not to raise taxes. More important, the economy weakened significantly in 1990–1991, then recovered only slowly. Although inflation was low, by mid-1992 unemployment had reached 7.8 percent of the labor force—2.5 percentage points higher than in the first year of Bush's term, and the highest level since 1984. A sign in Democratic candidate Bill Clinton's campaign headquarters summarized Clinton's strategy for winning the White House: "It's the economy, stupid." Clinton realized the importance of the nation's economic problems and pounded away at the Republican administration's inability to pull the country out of the doldrums. Clinton's focus on the economy was the key to his election.

Clinton's ability to parlay criticism of economic conditions into electoral success is not unusual in U.S. political history. Weakness in the economy played a decisive role in helping Franklin D. Roosevelt to beat Herbert Hoover in 1932, John F. Kennedy to best Richard Nixon in 1960, and Ronald Reagan to defeat Jimmy Carter in 1980. (And in an echo of his father's experience, President George W. Bush found the political popularity he enjoyed after ousting the Taliban from Afghanistan in 2001 eroded by an economic downturn and a slow subsequent recovery.) On the other hand, strong economic conditions have often helped incumbent presidents (or the incumbent's party) to retain office, including Nixon in 1972, Reagan in 1984, and Clinton in 1996. Indeed, a number of empirical studies have suggested that economic performance in the year preceding the election is among the most important determinants of whether an incumbent president is likely to win reelection.

Herbert Hoover
© Bettmann/Corbis

Jimmy Carter
© Bettmann/Corbis

George Bush
© Wally McNamee/Corbis

Victims of recession.

In Part 3 we discussed the factors that determine long-run economic growth. Over the broad sweep of history, those factors determine the economic success of a society. Indeed, over a span of 30, 50, or 100 years, relatively small differences in the rate of economic growth can have an enormous effect on the average person's standard of living. But even though the economic "climate" (long-run economic conditions) is the ultimate determinant of living standards, changes in the economic "weather" (short-run fluctuations in economic conditions) are also important. A good long-run growth record is not much consolation to a worker who has lost her job due to a recession. The bearing that short-term macroeconomic performance has on election results is one indicator of the importance the average person attaches to it.

In this part of the book we study short-term fluctuations in economic activity, commonly known as *recessions* and *expansions*. We will start, in this chapter, with some background on the history and characteristics of these economic ups and downs. However, the main focus of this part is the *causes* of short-term fluctuations, as well as the available *policy responses*. Because the analysis of short-term economic fluctuations can become complex and even controversial, we will proceed in a step-by-step fashion. In Chapter 12 we will introduce a basic—and oversimplified—model of booms and recessions, which we will refer to as the *basic Keynesian model* in honor of its principal originator, the British economist John Maynard Keynes. The basic Keynesian model focuses on the components of aggregate spending, such as consumption spending by households and investment spending by firms, and the effects of changes in spending on total real GDP.

Though this model is a useful starting point, it does not address some key issues. First, and perhaps most important, the basic Keynesian model has little to say about the determinants of inflation. Second, because it focuses on the very short run, this model does not give adequate attention to the economy's own natural tendency to eliminate deviations from full employment over the longer run. Because the basic Keynesian model does not take into account the "self-correcting" tendencies of the economy, it tends to overstate the need for government intervention to offset fluctuations. In Chapters 13 and 14 we will add new features to the model to make it more realistic. By the end of this part, we will have discussed the major causes of short-term economic fluctuations, as well as the options policymakers have in responding to them.

## RECESSIONS AND EXPANSIONS

As background to the study of short-term economic fluctuations, let's review the historical record of the fluctuations in the U. S. economy. Figure 12.1 shows the path of real GDP in the United States since 1920. (Figure 4.1, page 94, provides an even longer data series.) As you can see, the growth path of real GDP is not always smooth; the bumps and wiggles correspond to short periods of faster or slower growth.

A period in which the economy is growing at a rate significantly below normal is called a **recession** or a *contraction*. An extremely severe or protracted recession is called a **depression.** You should be able to pick out the Great Depression in Figure 12.1, particularly the sharp initial decline between 1929 and 1933. But you can also see that the U.S. economy was volatile in the mid-1970s and the early 1980s, with serious recessions in 1973–1975 and 1981–1982. A moderate recession (but not moderate enough for the first President Bush) occurred in 1990–1991. The most recent recession began in March 2001, exactly 10 years after the end of the 1990–1991 recession, which was declared over as of March 1991. This 10-year period without a recession was the longest such period in U.S. history.

***recession (or contraction)*** a period in which the economy is growing at a rate significantly below normal

***depression*** a particularly severe or protracted recession

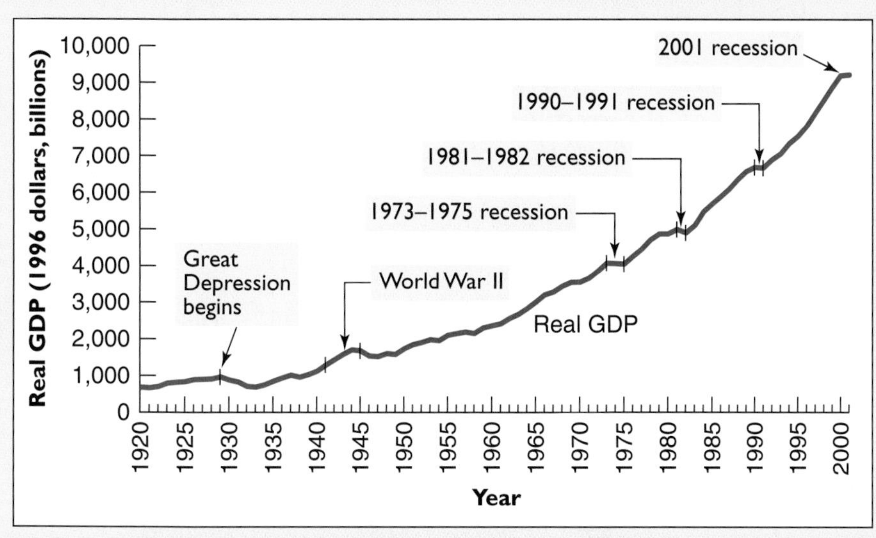


SOURCE: Bureau of Economic Analysis (www.bea.gov)

**FIGURE 12.1**
**Fluctuations in U.S. Real GDP, 1920–2001.**
Real GDP does not grow smoothly but has speedups (expansions or booms) and slowdowns (recessions or depressions). Note the contraction from 1929 to 1933 (the first phase of the Great Depression), the boom in 1941–1945 (World War II), and the recessions of 1973–1975, 1981–1982, 1990–1991, and 2001.

A more informal definition of a recession, often cited by reporters, is a period during which real GDP falls for at least two consecutive quarters. This definition is not a bad rule of thumb, as real GDP usually does fall during recessions. However, many economists would argue that periods in which real GDP growth is well below normal, though not actually negative, should be counted as recessions. Another problem with relying on GDP figures for dating recessions is that GDP data can be substantially revised, sometimes years after the fact. In practice, when trying to determine whether a recession is in progress, economists look at a variety of economic data, not just GDP.

Table 12.1 lists the beginning and ending dates of U.S. recessions since 1929, as well as the *duration* (length, in months) of each. The table also gives the highest unemployment rate recorded during each recession and the percentage change in real GDP. (Ignore the last column of the table for now.) The beginning of a recession is called the **peak,** because it represents the high point of economic activity prior to a downturn. The end of a recession, which marks the low point of economic activity prior to a recovery, is called the **trough.** The dates of peaks and troughs reported in Table 12.1 were determined by the National Bureau of Economic Research (NBER), a nonprofit organization of economists that has been a major source of research on short-term economic fluctuations since its founding

***peak*** the beginning of a recession, the high point of economic activity prior to a downturn

***trough*** the end of a recession, the low point of economic activity prior to a recovery in 1920 (see Economic Naturalist 12.1). The NBER is not a government agency, but it is usually treated by the news media and the government as the "official" arbiter of the dates of peaks and troughs.

*"Please stand by for a series of tones. The first indicates the official end of the recession, the second indicates prosperity, and the third the return of the recession."*

**TABLE 12.1**
**U.S. Recessions since 1929**

| Peak date (beginning) | Trough date (end) | Duration (months) | Highest unemployment rate (%) | Change in real GDP (%) | Duration of subsequent expansion (months) |
|---|---|---|---|---|---|
| Aug. 1929 | Mar. 1933 | 43 | 24.9 | −28.8 | 50 |
| May 1937 | June 1938 | 13 | 19.0 | −5.5 | 80 |
| Feb. 1945 | Oct. 1945 | 8 | 3.9 | −8.5 | 37 |
| Nov. 1948 | Oct. 1949 | 11 | 5.9 | −1.4 | 45 |
| July 1953 | May 1954 | 10 | 5.5 | −1.2 | 39 |
| Aug. 1957 | Apr. 1958 | 8 | 6.8 | −1.7 | 24 |
| Apr. 1960 | Feb. 1961 | 10 | 6.7 | 2.3 | 106 |
| Dec. 1969 | Nov. 1970 | 11 | 5.9 | 0.1 | 36 |
| Nov. 1973 | Mar. 1975 | 16 | 8.5 | −1.1 | 58 |
| Jan. 1980 | July 1980 | 6 | 7.6 | −0.3 | 12 |
| July 1981 | Nov. 1982 | 16 | 9.7 | −2.1 | 92 |
| July 1990 | Mar. 1991 | 8 | 7.5 | −0.9 | 120 |
| Mar. 2001 | Dec. 2001* | 9* | 6.0* | 0.2* | |

*Unofficial, based on authors' estimate at time of publication

SOURCES: Peak and trough dates, National Bureau of Economic Research; unemployment and real GDP, *Historical Statistics of the United States* and *Economic Report of the President.*

NOTES: Unemployment rate is the annual rate. Peak and trough dates from the National Bureau of Economic Research. Unemployment and real GDP data from *Historical Statistics of the United States* and the *Economic Report of the President.* Dates and figures with asterisks are unofficial and represent the authors' best guesses as of the time of publication. Unemployment rate is the annual rate for the trough year or the subsequent year, whichever is higher. Change in annual real GDP is measured from the peak year to the trough year, except that the entry for the 1945 recession is the 1945–1946 change in real GDP, the entry for the 1980 recession is the 1979–1980 change, and the estimated entry for 2001 is the 2000–2001 change.

Table 12.1 shows that since 1929, by far the longest and most severe recession in the United States was the 43-month economic collapse that began in August 1929 and lasted until March 1933, initiating what became known as the Great Depression. Between 1933 and 1937 the economy grew fairly rapidly, so technically the period was not a recession, although unemployment remained very high at close to 20 percent of the workforce. In 1937–1938 the nation was hit by another significant recession. Full economic recovery from the Depression did not come until U.S. entry into World War II at the end of 1941. The economy boomed from 1941 to 1945 (see Figure 12.1), reflecting the enormous wartime production of military equipment and supplies.

In sharp contrast to 1930s, U.S. recessions since World War II have generally been short—between 6 and 16 months, from peak to trough. As Table 12.1 shows, the two most severe postwar recessions, 1973–1975 and 1981–1982, lasted just 16 months. And, though unemployment rates during those two recessions were quite high by today's standards, they were low compared to the Great Depression. Since 1982 the U.S. economy has experienced only two relatively mild recessions, in 1990–1991 and in 2001. Although recent recessions have not been among the worst that Americans have experienced, they warn us to guard against overconfidence. Prosperity and economic stability can never be guaranteed.

The opposite of a recession is an **expansion**—a period in which the economy is growing at a rate that is significantly *above* normal. A particularly strong and protracted expansion is called a **boom.** In the United States, strong expansions occurred during 1933–1937, 1961–1969, 1982–1990, and in 1991–2001, with exceptionally strong growth during 1995–2000 (see Figure 12.1). On average, expansions have been much longer than recessions. The final column of Table 12.1 shows the duration, in months, of U.S. expansions since 1929. As you can see in the table, the 1961–1969 expansion lasted 106 months; the 1982–1990 expansion, 92 months. The longest expansion of all began in March 1991, at the trough of the 1990–1991 recession. This expansion lasted 120 months, a full 10 years, until a new recession began in March 2001.

***expansion*** a period in which the economy is growing at a rate significantly above normal

***boom*** a particularly strong and protracted expansion

ECONOMIC NATURALIST 12.1

### Calling the 2001 recession

The Business Cycle Dating Committee of the National Bureau of Economic Research determined that a recession, the first in 10 years, began in March 2001. What led the committee to choose that date?

As of the beginning of 2001, there had been no "official" recession in the United States since the one that began in July 1990 and ended in March 1991. As mentioned above, this 10-year period without a recession was the longest expansion in U.S. history. However, the economy weakened considerably during the latter part of 2000 and in the spring and summer of 2001. A further blow was the terrorist attacks of September 11, 2001, which caused the loss of many jobs both in the affected areas and (because people became afraid to travel) in industries such as airlines and hotels. Because of the increased likelihood that a recession had begun, the six economists who form the Business Cycle Dating Committee—the group within the National Bureau of Economic Research that actually determines recession dates—found themselves called upon for the first time in a decade.

The determination of whether and when a recession has begun involves intensive statistical analysis, mixed in with a significant amount of human judgment. The Business Cycle Dating Committee typically relies heavily on a small set of statistical indicators that measure the overall strength of the economy. The committee prefers indicators that are available monthly, because they are available quickly and may provide relatively precise information about the timing of peaks and troughs. Four of the most important indicators used by the committee are

- Industrial production, which measures the output of factories and mines.
- Total sales in manufacturing, wholesale trade, and retail trade.

- Nonfarm employment (the number of people at work outside of agriculture).
- Real after-tax income received by households, excluding transfers like Social Security payments.

Each of these indicators measures a different aspect of the economy. Because their movements tend to coincide with the overall movements in the economy, they are called *coincident indicators.*

Normally the coincident indicators move more or less together, during the 2001 recession they did not. Industrial production and sales in the manufacturing sector began to decline as early as September 2000. This early weakness in manufacturing reflected slow sales of information technology (computers, software, communications devices, and the like) following the collapse of the "dot-com bubble" in the stock market during 2000. (The values of high-tech stocks fell by two-thirds or more during the year.) However, the weakness in manufacturing was not immediately reflected in the economy as a whole, as both employment and personal income grew strongly in the fall of 2000. Employment did not start to decline until around March 2001, and real personal income continued to grow into the fall of 2001.

The failure of the coincident indicators to move closely together made the committee's job difficult. The committee ultimately chose the date of the peak, or beginning of the recession, to be March 2001, the month in which nonfarm employment began to decline. A rationale for their choice is that recessions are supposed to reflect declines in the entire economy, not just a few specific sectors. It might be argued that the decline in nonfarm employment in March 2001 (which counts people at work in the entire economy, outside of agriculture) was the first clear indication that the slowdown which had begun in the high-tech sector had spread to the broader economy.

### EXERCISE 12.1

**Update Table 12.1 using the National Bureau of Economic Research website (go to http://www.nber.org and click on *business cycle dates*). According to the NBER, is the U.S. economy currently in recession or expansion? How much time has elapsed since the last peak or trough? Explore the NBER website to find additional useful information about current conditions in the U.S. economy.**

## SOME FACTS ABOUT SHORT-TERM ECONOMIC FLUCTUATIONS

Recessions are very difficult to forecast.

Although Figure 12.1 and Table 12.1 show only twentieth-century data, periods of expansion and recession have been a feature of industrial economies since at least the late eighteenth century. Karl Marx and Friedrich Engels referred to these fluctuations, which they called "commercial crises," in their Communist Manifesto of 1848. In the United States, economists have been studying short-term fluctuations for at least a century. The traditional term for these fluctuations is *business cycles,* and they are still often referred to as *cyclical fluctuations.* Neither term is accurate though; as Figure 12.1 shows, economic fluctuations are not "cyclical" at all in the sense that they recur at predictable intervals, but instead are *irregular in their length and severity.* This irregularity makes the dates of peaks and troughs extremely hard to predict, despite the fact that professional forecasters have devoted a great deal of effort and brainpower to the task.

Expansions and recessions usually are not limited to a few industries or regions but, as noted in Economic Naturalist 12.1, are *felt throughout the economy.* Indeed, the largest fluctuations may have a *global impact.* For instance, the Great Depression of the 1930s affected nearly all the world's economies, and the 1973–1975 and 1981–1982 recessions were also widely felt outside the United States. When

East Asia suffered a major slowdown in the late 1990s, the effects of that slowdown spilled over into many other regions (although not so much the United States).

Even a relatively moderate recession, like the one that occurred in 2001, can have global effects. Figure 12.2, which shows growth rates of real GDP over the period 1999–2002 for Canada, Germany, Japan, the United Kingdom, and the United States, illustrates this point. You can see that—except for Japan, which has performed poorly since the early 1990s—all the countries experienced relatively strong growth in 1999 and 2000. The 2001 recession lowered growth rates in all of the countries, with some recovery apparent in early 2002.

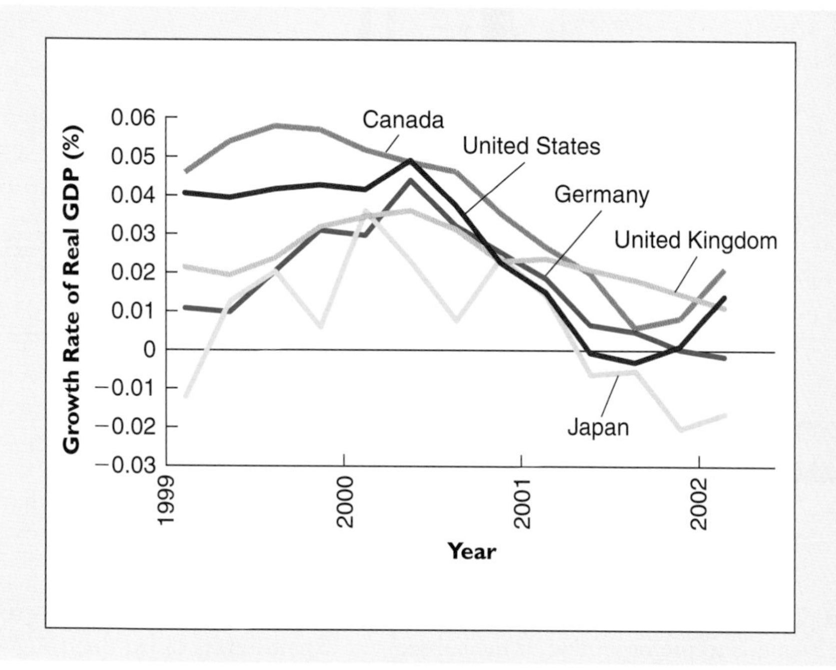


**FIGURE 12.2**
**Real GDP Growth in Five Major Countries, 1999–2002.**
Quarterly growth rates (measured as the change in real GDP over the past four quarters) for five major industrialized countries show that all the countries except Japan enjoyed rapid GDP growth in 1999 and 2000. (Japan has been growing slowly for more than a decade.) The 2001 recession was reflected in falling growth rates in all of the major countries, with some recovery apparent in 2002.

*Unemployment* is a key indicator of short-term economic fluctuations. The unemployment rate typically rises sharply during recessions and recovers (although more slowly) during expansions. Figure 5.3 on page 132 shows the U.S. unemployment rate since 1960. You should be able to identify the recessions that began in 1960, 1969, 1973, 1981, 1990, and 2001 by noting the sharp peaks in the unemployment rate in those years. Recall from Chapter 8 that the part of unemployment that is associated with recessions is called *cyclical unemployment.* Beyond this increase in unemployment, labor market conditions generally worsen during recessions. For example, during recessions real wages grow more slowly, workers are less likely to receive promotions or bonuses, and new entrants to the labor force (such as college graduates) have a much tougher time finding attractive jobs.

Generally, industries that produce *durable goods,* such as cars, houses, and capital equipment, are more affected than others by recessions and booms. In contrast, industries that provide *services* and *nondurable goods* like food are much less sensitive to short-term fluctuations. Thus an automobile worker or a construction worker is far more likely to lose his or her job in a recession than is a barber or a baker.

Like unemployment, *inflation* follows a typical pattern in recessions and expansions, though it is not so sharply defined. Figure 12.3 shows the U.S. inflation rate since 1960; in the figure, periods of recession are indicated by shaded vertical bars. As you can see, recessions tend to be followed soon after by a decline in the rate of inflation. For example, the recession of 1981–1982 was followed by a sharp reduction in inflation. Furthermore, many—though not all—postwar recessions have been preceded by increases in inflation, as Figure 12.3 shows. The behavior of inflation during expansions and recessions will be discussed more fully in Chapter 15.

Unemployment among construction workers rises substantially during recessions.

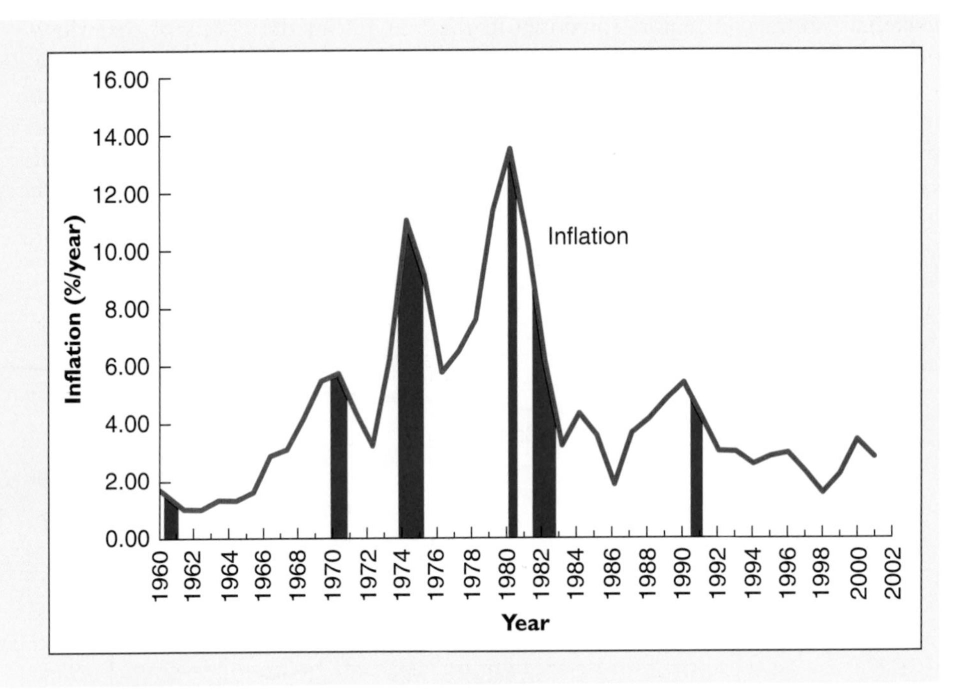


**FIGURE 12.3**
**U.S. Inflation, 1960–2001.**
U.S. inflation since 1960 is measured by the change in the CPI, and periods of recession are indicated by the shaded vertical bars. Note that inflation declined following the recessions of 1960–1961, 1969–1970, 1973–1975, 1981–1982, and 1990–1991 and rose prior to many of those recessions.

## RECAP RECESSIONS, BOOMS, AND THEIR CHARACTERISTICS

- A recession is a period in which output is growing more slowly than normal. An expansion, or boom, is a period in which output is growing more quickly than normal.
- The beginning of a recession is called the peak, and its end (which corresponds to the beginning of the subsequent expansion) is called the trough.
- The sharpest recession in the history of the United States was the initial phase of the Great Depression in 1929–1933. Severe recessions also occurred in 1973–1975 and 1981–1982. Two relatively mild recessions occurred in 1990–1991 and 2001.
- Short-term economic fluctuations (recessions and expansions) are irregular in length and severity, and thus are difficult to predict.
- Expansions and recessions have widespread (and sometimes global) impacts, affecting most regions and industries.
- Unemployment rises sharply during a recession and falls, usually more slowly, during an expansion.
- Durable goods industries are more affected by expansions and recessions than other industries. Services and nondurable goods industries are less sensitive to ups and downs in the economy.
- Recessions tend to be followed by a decline in inflation and are often preceded by an increase in inflation.

# OUTPUT GAPS AND CYCLICAL UNEMPLOYMENT

If policymakers are to respond appropriately to recessions and expansions, and economists are to study them, knowing whether a particular economic fluctuation is "big" or "small" is essential. Intuitively, a "big" recession or expansion is one in which output and the unemployment rate deviate significantly from their normal or trend levels. In this section we will attempt to be more precise about this idea by introducing the concept of the *output gap,* which measures how far output is from its normal level at a particular time. We will also revisit the idea of *cyclical unemployment,* or the deviation of unemployment from its normal level. Finally, we will examine how these two concepts are related.

## POTENTIAL OUTPUT AND THE OUTPUT GAP

The concept of potential output is a useful starting point for thinking about the measurement of expansions and recessions. **Potential output,** also called *potential GDP* or *full-employment output,* is the amount of output (real GDP) that an economy can produce when using its resources, such as capital and labor, at normal rates.[1] Potential output is not a fixed number but grows over time, reflecting increases in both the amounts of available capital and labor and their productivity. We discussed the sources of growth in potential output (the economy's productive capacity) in Chapter 7. We will use the symbol $Y^*$ to signify the economy's potential output at a given point in time.

***potential output, $Y^*$ (or potential GDP or full-employment output)*** the amount of output (real GDP) that an economy can produce when using its resources, such as capital and labor, at normal rates

Why does a nation's output sometimes grow quickly and sometimes slowly, as shown for the United States in Figure 12.1? Logically, there are two possibilities: First, changes in the rate of output growth may reflect *changes in the rate at which the country's potential output is increasing.* For example, unfavorable weather conditions, such as a severe drought, would reduce the rate of potential output growth in an agricultural economy, and a decline in the rate of technological innovation might reduce the rate of potential output growth in an industrial economy. Under the assumption that the country is using its resources at normal rates, so that actual output equals potential output, a significant slowdown in potential output growth would tend to result in recession. Similarly, new technologies, increased capital investment, or a surge in immigration that swells the labor force could produce unusually brisk growth in potential output, and hence an economic boom.

Undoubtedly, changes in the rate of growth of potential output are part of the explanation for expansions and recessions. In the United States, for example, the economic boom of the second half of the 1990s was propelled in part by new information technologies, such as the Internet. And the severe slowdown in Japan during the decade of the 1990s reflected in part a reduction in the growth of potential output, arising from factors such as slower growth in the Japanese labor force and capital stock. When changes in the rate of GDP growth reflect changes in the growth rate of potential output, the appropriate policy responses are those discussed in Chapter 7. In particular, when a recession results from slowing growth in potential output, the government's best response is to try to promote saving, investment, technological innovation, human capital formation, and other activities that support growth.

A second possible explanation for short-term economic fluctuations is that *actual output does not always equal potential output.* For example, potential output may be growing normally, but for some reason the economy's capital and labor resources may not be fully utilized, so that actual output is significantly below the level of potential output. This low level of output, resulting from underutilization of economic resources, would generally be interpreted as a recession. Alternatively, capital and labor may be working much harder than normal—firms may put workers on overtime, for example—so that actual output expands beyond potential output, creating a boom.

***output gap, $Y^* - Y$*** the difference between the economy's potential output and its actual output at a point in time

At any point in time, the difference between potential output and actual output is called the **output gap.** Recalling that $Y^*$ is the symbol for potential output

[1]The term "potential output" is slightly misleading, in that *potential* output is not the same as *maximum* output. Because capital and labor can be utilized at greater-than-normal rates, at least for a time, a country's actual output can exceed its potential output.

and that $Y$ stands for actual output (real GDP), we can express the output gap as $Y^* - Y$. A positive output gap—when actual output is below potential, and resources are not being fully utilized—is called a **recessionary gap.** A negative output gap—when actual output is above potential, and resources are being utilized at above-normal rates—is referred to as an **expansionary gap.**

***recessionary gap*** a positive output gap, which occurs when potential output exceeds actual output ($Y^* > Y$)

***expansionary gap*** a negative output gap, which occurs when actual output is higher than potential output ($Y > Y^*$)

Policymakers generally view both recessionary gaps and expansionary gaps as problems. It is not difficult to see why a recessionary gap is bad news for the economy: When there is a recessionary gap, capital and labor resources are not being fully utilized, and output and employment are below normal levels. This is the sort of situation that poses problems for politicians' reelection prospects, as discussed in this chapter's introduction. An expansionary gap is considered a problem by policymakers for a more subtle reason: What's wrong, after all, with having higher output and employment than normal? A prolonged expansionary gap is problematic because, when faced with a demand for their products that significantly exceeds their normal capacity, firms tend to raise prices. Thus an expansionary gap typically results in increased inflation, which reduces the efficiency of the economy in the longer run. (We discuss the genesis of inflation in more detail in Chapter 15.)

*"Ed, this is Art Simbley over at Hollis, Bingham, Cotter & Krone. What did you get for thirty-four across, 'Persian fairy,' four letters?"*

Thus, whenever an output gap exists, whether it is recessionary or expansionary, policymakers have an incentive to try to eliminate the gap by returning actual output to potential. In the next three chapters we will discuss both how output gaps arise and the tools that policymakers have for *stabilizing* the economy—that is, bringing actual output into line with potential output.

## THE NATURAL RATE OF UNEMPLOYMENT AND CYCLICAL UNEMPLOYMENT

Whether recessions arise because of slower growth in potential output or because actual output falls below potential, they bring bad times. In either case output falls (or at least grows more slowly), implying reduced living standards. Recessionary output gaps are particularly frustrating for policymakers, however, because they imply that the economy has the *capacity* to produce more, but for some reason available resources are not being fully utilized. Recessionary gaps are *inefficient* in that they unnecessarily reduce the total economic pie, making the typical person worse off.

An important indicator of the low utilization of resources during recessions is the unemployment rate. In general, a *high* unemployment rate means that labor resources are not being fully utilized, so that output has fallen below potential (a recessionary gap). By the same logic, an unusually *low* unemployment rate suggests that labor is being utilized at a rate greater than normal, so that actual output exceeds potential output (an expansionary gap).

To better understand the relationship between the output gap and unemployment, recall from Chapter 8 the three broad types of unemployment: frictional unemployment, structural unemployment, and cyclical unemployment. *Frictional unemployment* is the short-term unemployment that is associated with the matching of workers and jobs. Some amount of frictional unemployment is necessary for the labor market to function efficiently in a dynamic, changing economy. *Structural unemployment* is the long-term and chronic unemployment that occurs even when the economy is producing at its normal rate. Structural unemployment often results when workers' skills are outmoded and do not meet the needs of employers—so, for example, steelworkers may become structurally unemployed as the steel industry goes into a long-term decline, unless those workers can retrain to find jobs in growing industries. Finally, *cyclical unemployment* is the extra unemployment that occurs during periods of recession. Unlike cyclical unemployment, which is present only during recessions, frictional unemployment and structural unemployment are always present in the labor market, even when the economy is operating normally. Economists call the part of the total unemployment rate that is attributable to frictional and structural unemployment the **natural rate of unemployment.** Put another way, the natural rate of unemployment is the unemployment rate that prevails when cyclical unemployment is zero, so that the economy has neither a recessionary nor an expansionary output gap. We will denote the natural rate of unemployment as $u^*$.

***natural rate of unemployment, $u^*$*** the part of the total unemployment rate that is attributable to frictional and structural unemployment; equivalently, the unemployment rate that prevails when cyclical unemployment is zero, so that the economy has neither a recessionary nor an expansionary output gap

Cyclical unemployment, which is the difference between the total unemployment rate and the natural rate, can thus be expressed as $u - u^*$, where $u$ is the actual unemployment rate and $u^*$ denotes the natural rate of unemployment. In a recession, the actual unemployment rate $u$ exceeds the natural unemployment rate $u^*$, so cyclical unemployment, $u - u^*$, is positive. When the economy experiences an expansionary gap, in contrast, the actual unemployment rate is lower than the natural rate, so that cyclical unemployment is negative. Negative cyclical unemployment corresponds to a situation in which labor is being used more intensively than normal, so that actual unemployment has dipped below its usual frictional and structural levels.

ECONOMIC NATURALIST 12.2

### Why has the natural rate of unemployment in the United States apparently declined?

According to the Congressional Budget Office, which regularly estimates the natural rate of unemployment in the United States, the natural rate has fallen steadily since about 1979, from 6.3 percent of the labor force to about 5.2 percent.[2] Some economists, noting that unemployment remained close to 4 percent for several years around the turn of the millennium, have argued for an even lower natural rate, perhaps 4.5 percent or lower. Why is the U.S. natural rate of unemployment apparently so much lower today than it was 20 years ago?

[2]Congressional Budget Office, *Economic and Budget Outlook: Fiscal Years 2001–2010,* Table F-1, available online at http://www.cbo.gov/.

The natural rate of unemployment may have fallen because of reduced frictional unemployment, reduced structural unemployment, or both. A variety of ideas have been advanced to explain declines in both types of unemployment. One promising suggestion is based on the changing age structure of the U.S. labor force.[3] The average age of U.S. workers is rising, reflecting the aging of the baby boom generation. Indeed, over the past 20 years the share of the labor force aged 16–24 has fallen from about 25 percent to about 16 percent. Since young workers are more prone to unemployment than older workers, the aging of the labor force may help to explain the overall decline in unemployment.

Why are young workers more likely to be unemployed? Compared to teenagers and workers in their twenties, older workers are much more likely to hold long-term, stable jobs. In contrast, younger workers tend to hold short-term jobs, perhaps because they are not ready to commit to a particular career, or because their time in the labor market is interrupted by schooling or military service. Because they change jobs more often, younger workers are more prone than others to frictional unemployment. They also have fewer skills, on average, than older workers, so they may experience more structural unemployment. As workers age and gain experience, however, their risk of unemployment declines.

Another possible explanation for the declining natural rate of unemployment is that labor markets have become more efficient at matching workers with jobs, thereby reducing both frictional and structural unemployment. For example, agencies that arrange temporary help have become much more commonplace in the United States in recent years. Although the placements these agencies make are intended to be temporary, they often become permanent when an employer and worker discover that a particularly good match has been made. Online job services, which allow workers to search for jobs nationally and even internationally, are also becoming increasingly important. By reducing the time people must spend in unemployment and by creating more lasting matches between workers and jobs, temporary help agencies, online job services, and similar innovations may have reduced the natural rate of unemployment.[4]

## OKUN'S LAW

What is the relationship between an output gap and the amount of cyclical unemployment in the economy? We have already observed that by definition, cyclical unemployment is positive when the economy has a recessionary gap, negative when there is an expansionary gap, and zero when there is no output gap. A more quantitative relationship between cyclical unemployment and the output gap is given by a rule of thumb called *Okun's law,* after Arthur Okun, one of President Kennedy's chief economic advisers. According to **Okun's law,** each extra percentage point of cyclical unemployment is associated with about a 2 percentage point increase in the output gap, measured in relation to potential output.[5] So, for example, if cyclical unemployment increases from 1 percent to 2 percent of the labor force, the recessionary gap will increase from 2 percent to 4 percent of potential GDP. Example 12.1 illustrates further.

***Okun's law*** each extra percentage point of cyclical unemployment is associated with about a 2 percentage point increase in the output gap, measured in relation to potential output

[3]See Robert Shimer, "Why Is the U.S. Unemployment Rate So Much Lower?" in B. Bernanke and J. Rotemberg, eds., *NBER Macroeconomics Annual,* 1998.

[4]For a detailed analysis of factors affecting the natural rate, see Lawrence Katz and Alan Krueger, "The High-Pressure U.S. Labor Market of the 1990s," *Brookings Papers on Economic Activity,* 1:1–88, 1999.

[5]This relationship between unemployment and output has weakened over time. When Arthur Okun first formulated his law in the 1960s, he suggested that each extra percentage point of unemployment was associated with about a 3 percentage point increase in the output gap.

EXAMPLE 12.1

**Okun's law and the output gap in the U.S. economy**

Following are the actual unemployment rate, the natural unemployment rate, and potential GDP (in billions of dollars) for the U.S. economy in three selected years. Using Okun's law, estimate the output gap in each year, in billions of dollars.

| Year | $u$ | $u^*$ | $Y^*$ |
|---|---|---|---|
| 1982 | 9.7% | 6.1% | 3,433 |
| 1991 | 6.8 | 5.8 | 6,093 |
| 1998 | 4.5 | 5.2 | 8,563 |

SOURCES: Unemployment rate, *Economic Report of the President*. Natural unemployment rate and potential GDP, Congressional Budget Office.

In 1982 cyclical unemployment, $u - u^*$, was 9.7% − 6.1%, or 3.6 percent of the labor force. According to Okun's law, the output gap for that year would be twice that percentage, or 7.2 percent of potential output. Since potential output in 1982 was $3,433 billion, the value of the output gap for that year was 7.2% × $3,433 billion, or $247 billion.

In 1991 cyclical unemployment was 6.8% − 5.8%, or 1.0 percent of the labor force. According to Okun's law, the output gap for 1991 would be twice 1.0 percent, or 2.0 percent of potential GDP. Since potential GDP in 1991 was $6,093 billion, the output gap in that year should have been 2.0% × $6,093, or $122 billion.

Both 1982 and 1991 were recession years, so the output gaps were recessionary gaps. In contrast, 1998 was a year of expansion, in which unemployment was below the natural rate and the economy experienced an expansionary gap. Cyclical unemployment in 1998 was 4.5% − 5.2%, or −0.7 percent. The output gap in 1998 should therefore have been about −1.4 percent of the potential GDP of $8,563 billion, or −$120 billion. In other words, in 1998 actual GDP was about $120 billion greater than potential GDP.

**EXERCISE 12.2**

**In the first half of 2002 the U.S. unemployment rate averaged about 6 percent. Assuming that the natural rate is 5.2 percent, by what percentage amount did actual GDP differ from potential GDP in the first six months of 2002?**

The output losses sustained in recessions, calculated according to Okun's law, can be quite significant. In Example 12.1 we found the U.S. output gap in 1982 to be $247 billion. The U.S. population in 1982 was about 230 million. Hence the output loss per person in that year equaled the total output gap of $247 billion divided by 230 million people, or about $1,074—about $4,300 for a family of four. Adjusting for inflation between 1982 and 2001, the equivalent cost in terms of 2001 dollars would have been about $7,100 for a family of four. This calculation implies that output gaps and cyclical unemployment may have significant costs—a conclusion that justifies the concern that the public and policymakers have about recessions.

ECONOMIC NATURALIST 12.3

**Why did the Federal Reserve take measures to slow down the economy in 1999 and 2000?**

As noted in Chapters 4 and 10, monetary policy decisions of the Federal Reserve—actions that change the level of the nation's money supply—affect the performance of the U.S. economy. Why did the Federal Reserve take measures to slow down the economy in 1999 and 2000?

Throughout the 1990s cyclical unemployment in the U.S. fell dramatically, becoming negative sometime in 1997, according to Congressional Budget Office estimates. Okun's law indicates that growing negative cyclical unemployment rates signal an increasing expansionary gap (illustrated in Example 12.1), and with it an increased risk of future inflation.

In 1997 and 1998 the Federal Reserve argued that the inflationary pressures typically caused by rapidly expanding output and falling unemployment rates were being offset by productivity gains and international competition, leaving inflation rates lower than expected. Because inflation remained low during this period—despite a small but growing expansionary gap—the Federal Reserve did little to eliminate the gap.

However, as the actual unemployment rate continued to fall throughout 1999 and early 2000 the expansionary gap continued to widen, causing the Federal Reserve to grow increasingly concerned about the growing imbalance between actual and potential GDP and the threat of increasing inflation. In response, the Federal Reserve took actions in 1999 and 2000 to slow the growth of output and bring actual and potential output closer into alignment (we will give more details about how the Fed can do this in Chapters 14 and 15). The Fed's actions helped to "promote overall balance in the economy"[6] and restrain inflation throughout 2000. By early 2001, however, the U.S. economy stalled and fell into recession (a topic analyzed in Economic Naturalist 13.2), leading the Federal Reserve to reverse course and take policy measures aimed at eliminating the growing *recessionary* gap.

### OUTPUT GAPS, CYCLICAL UNEMPLOYMENT, AND OKUN'S LAW

Potential output is the amount of output (real GDP) that an economy can produce when using its resources, such as capital and labor, at normal rates. The output gap, $Y^* - Y$, is the difference between potential output $Y^*$ and actual output $Y$. When actual output is below potential, the resulting output gap is called a recessionary gap. When actual output is above potential, the difference is called an expansionary gap. A recessionary gap reflects a waste of resources, while an expansionary gap threatens to ignite inflation; hence policymakers have an incentive to try to eliminate both types of output gaps.

The natural rate of unemployment $u^*$ is the sum of the frictional and structural unemployment rates. It is the rate of unemployment that is observed when the economy is operating at a normal level, with no output gap.

Cyclical unemployment, $u - u^*$, is the difference between the actual unemployment rate $u$ and the natural rate of unemployment $u^*$. Cyclical unemployment is positive when there is a recessionary gap, negative when there is an expansionary gap, and zero when there is no output gap.

Okun's law relates cyclical unemployment and the output gap. According to this rule of thumb, each percentage point increase in cyclical unemployment is associated with about a 2 percentage point increase in the output gap, measured in relation to potential output.

## WHY DO SHORT-TERM FLUCTUATIONS OCCUR? A PREVIEW AND A PARABLE

What causes periods of recession and expansion? In the preceding section we discussed two possible reasons for slowdowns and speedups in real GDP growth. First, growth in potential output itself may slow down or speed up, reflecting

[6]Testimony of Chairman Alan Greenspan, *The Federal Reserve's semiannual report on the economy and monetary policy,* Committee on Banking and Financial Services, U.S. House of Representatives February 17, 2000. Available online: http://www.federalreserve.gov/boarddocs/hh/2000/February/Testimony.htm

changes in the growth rates of available capital and labor and in the pace of technological progress. Second, even if potential output is growing normally, actual output may be higher or lower than potential output—that is, expansionary or recessionary output gaps may develop. Earlier in this book we discussed some of the reasons that growth in potential output can vary, and the options that policymakers have for stimulating growth in potential output. But we have not yet addressed the question of how output gaps can arise or what policymakers should do in response. The causes and cures of output gaps will be a major topic of the next three chapters. Here is a brief preview of the main conclusions of those chapters:

1. In a world in which prices adjusted immediately to balance the quantities supplied and demanded for all goods and services, output gaps would not exist. However, for many goods and services, the assumption that prices will adjust immediately is not realistic. Instead, many firms adjust the prices of their output only periodically. In particular, rather than changing prices with every variation in demand, firms tend to adjust to changes in demand in the short run by varying the quantity of output they produce and sell. This type of behavior is known as "meeting the demand" at a preset price.

2. Because in the short run firms tend to meet the demand for their output at preset prices, changes in the amount that customers decide to spend will affect output. When total spending is low for some reason, output may fall below potential output; conversely, when spending is high output may rise above potential output. In other words, *changes in economywide spending are the primary cause of output gaps.* Thus government policies can help to eliminate output gaps by influencing total spending. For example, the government can affect total spending directly simply by changing its own level of purchases.

3. Although firms tend to meet demand in the short run, they will not be willing to do so indefinitely. If customer demand continues to differ from potential output, firms will eventually adjust their prices to eliminate output gaps. If demand exceeds potential output (an expansionary gap), firms will raise their prices aggressively, spurring inflation. If demand falls below potential output (a recessionary gap), firms will raise their prices less aggressively or even cut prices, reducing inflation.

4. Over the longer run, price changes by firms eliminate any output gap and bring production back into line with the economy's potential output. Thus the economy is "self-correcting" in the sense that it operates to eliminate output gaps over time. Because of this self-correcting tendency, in the long run actual output equals potential output, so that output is determined by the economy's productive capacity rather than by the rate of spending. In the long run, total spending influences only the rate of inflation.

These ideas will become clearer as we proceed through the next chapters. Before plunging into the details of the analysis, though, let's consider an example that illustrates the links between spending and output in the short and long run.

Al's ice cream store produces gourmet ice cream on the premises and sells it directly to the public. What determines the amount of ice cream that Al produces on a daily basis? The productive capacity, or potential output, of the shop is one important factor. Specifically, Al's potential output of ice cream depends on the amount of capital (number of ice cream makers) and labor (number of workers) that he employs, and on the productivity of that capital and labor. Although Al's potential output usually changes rather slowly, on occasion it can fluctuate significantly—for example, if an ice cream maker breaks down or Al contracts the flu.

The main source of day-to-day variations in Al's ice cream production, however, is not changes in potential output but fluctuations in the demand for ice cream by the public. Some of these fluctuations in spending occur predictably over the course of the day (more demand in the afternoon than in the morning, for example), the week (more demand on weekends), or the year (more demand in the summer). Other changes in demand are less regular—more demand on a hot day than a cool one, or when a parade is passing by the store. Some changes in demand are hard for Al to interpret: For example, a surge in demand for rocky road ice cream on one particular Tuesday could reflect a permanent change in consumer tastes, or it might just be a random, one-time event.

How should Al react to these ebbs and flows in the demand for ice cream? The basic supply-and-demand model that we introduced in Chapter 3, if applied to the market for ice cream, would predict that the price of ice cream should change with every change in the demand for ice cream. For example, prices should rise just after the movie theater next door to Al's shop lets out on Friday night, and they should fall on unusually cold, blustery days, when most people would prefer a hot cider to an ice cream cone. Indeed, taken literally, the supply and demand model of Chapter 3 predicts that ice cream prices should change almost moment to moment. Imagine Al standing in front of his shop like an auctioneer, calling out prices in an effort to determine how many people are willing to buy at each price!

Of course, we do not expect to see this behavior by an ice cream store owner. Price setting by auction does in fact occur in some markets, such as the market for grain or the stock market, but it is not the normal procedure in most retail markets, such as the market for ice cream. Why this difference? The basic reason is that sometimes the economic benefits of hiring an auctioneer and setting up an auction exceed the costs of doing so, and sometimes they do not. In the market for grain, for example, many buyers and sellers gather together in the same place at the same time to trade large volumes of standardized goods (bushels of grain). In that kind of situation, an auction is an efficient way to determine prices and balance the quantities supplied and demanded. In an ice cream store, by contrast, customers come in by twos and threes at random times throughout the day. Some want shakes, some cones, and some sodas. With small numbers of customers and a low sales volume at any given time, the costs involved in selling ice cream by auction are much greater than the benefits of allowing prices to vary with demand.

So how does Al the ice cream store manager deal with changes in the demand for ice cream? Observation suggests that he begins by setting prices based on the best information he has about the demand for his product and the costs of production. Perhaps he prints up a menu or makes a sign announcing the prices. Then, over a period of time, he will keep his prices fixed and serve as many customers as want to buy (up to the point where he runs out of ice cream or room in the store at these prices). This behavior is what we call "meeting the demand" at preset prices, and it implies that *in the short run,* the amount of ice cream Al produces and sells is determined by the demand for his products.

However, *in the long run* the situation is quite different. Suppose, for example, that Al's ice cream earns a citywide reputation for its freshness and flavor. Day after day Al observes long lines in his store. His ice cream maker is overtaxed, as are his employees and his table space. There can no longer be any doubt that at current prices, the quantity of ice cream the public wants to consume exceeds what Al is able and willing to supply on a normal basis (his potential output). Expanding the store is an attractive possibility, but not one (we assume) that is immediately feasible. What will Al do?

Certainly one thing Al can do is raise his prices. At higher prices, Al will earn higher profits. Moreover, raising ice cream prices will bring the quantity of ice cream demanded closer to Al's normal production capacity—his potential output. Indeed, when the price of Al's ice cream finally rises to its equilibrium level, the shop's actual output will equal its potential output. Thus, over the long run, ice cream prices adjust to their equilibrium level, and the amount that is sold is determined by potential output.

This example illustrates in a simple way the links between spending and output—except, of course, that we must think of this story as applying to the whole economy, not to a single business. The key point is that there is an important difference between the short run and the long run. In the short run, producers often choose not to change their prices, but rather to meet the demand at preset prices. Because output is determined by demand, in the short run total spending plays a central role in determining the level of economic activity. Thus Al's ice cream store enjoys a boom on an unusually hot day, when the demand for ice cream is strong, while an unseasonably cold day brings an ice cream recession. But in the long run, prices adjust to their market-clearing levels, and output equals potential output. Thus the quantities of inputs and the productivity with which they are used are the primary determinants of economic activity in the long run, as we saw in Chapter 7. Although total spending affects output in the short run, in the long run its main effects are on prices.

**Why did the Coca-Cola Company test a vending machine that "knows" when the weather is hot?**

ECONOMIC NATURALIST 12.4

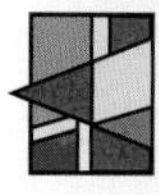

According to the *New York Times* (October 28, 1999, p. C1), the Coca-Cola Company has quietly tested a soda vending machine that includes a temperature sensor. Why would Coca-Cola want a vending machine that "knows" when the weather is hot?

When the weather is hot, the demand for refreshing soft drinks rises, increasing their market-clearing price. To take advantage of this variation in consumer demand, the vending machines that Coca-Cola tested were equipped with a computer chip that gave them the capability to raise soda prices automatically when the temperature climbs. The company's chairman and chief executive, M. Douglas Ivester, described in an interview how the desire for a cold drink increases during a sports championship final held in the summer heat. "So it is fair that it should be more expensive," Mr. Ivester was quoted as saying. "The machine will simply make this process automatic." Company officials suggested numerous other ways in which vending machine prices could be made dependent on demand. For example, machines could be programmed to reduce prices during off-peak hours or at low-traffic machines.

In traditional vending machines, cold drinks are priced in a way analogous to the way Al prices his ice cream: A price is set, and demand is met at the preset price, until the machine runs of soda. The weather-sensitive vending machine illustrates how technology may change pricing practices in the future. Indeed, increased computing power and access to the Internet have already allowed some firms, such as airline companies, to change prices almost continuously in response to variations in demand. Conceivably, the practice of meeting demand at a preset price may someday be obsolete.

On the other hand, Coca-Cola's experiments with "smart" vending machines also illustrate the barriers to fully flexible pricing in practice. First, the new vending machines are more costly than the standard model. In deciding whether to use them, the company must decide whether the extra profits from variable pricing justify the extra cost of the machines. Second, in early tests many consumers reacted negatively to the new machines, complaining that they take unfair advantage of thirsty customers. In practice, customer complaints and concerns about "fairness" make companies less willing to vary prices sensitively with changing demand.

## SUMMARY

- Real GDP does not grow smoothly. Periods in which the economy is growing at a rate significantly below normal are called *recessions;* periods in which the economy is growing at a rate significantly above normal are called *expansions.* A severe or protracted recession, like the long decline that occurred between 1929 and 1933, is called a *depression,* while a particularly strong expansion is called a *boom.*

- The beginning of a recession is called the *peak,* because it represents the high point of economic activity prior to a downturn. The end of a recession, which marks the low point of economic activity prior to a recovery, is called the *trough.* Since World War II, U.S. recessions have been much shorter on average than booms, lasting between 6 and 16 months. The longest boom period in U.S. history began with the end of the 1990–1991 recession in March 1991, ending exactly 10 years later in March 2001 when a new recession began.

- Short-term economic fluctuations are irregular in length and severity, and are thus hard to forecast. Expansions and recessions are typically felt throughout the economy and may even be global in scope. Unemployment rises sharply during recessions, while inflation tends to fall during or shortly after a recession. Durable goods industries tend to be particularly sensitive to recessions and booms, whereas services and nondurable goods industries are less sensitive.

- *Potential output,* also called potential GDP or full-employment output, is the amount of output (real GDP) that an economy can produce when it is using its resources, such as capital and labor, at normal rates. The difference between potential output and actual output is the *output gap.* When output is below potential, the gap is called a *recessionary gap*; when output is above potential, the difference is called an *expansionary gap.* Recessions can occur either because potential output is growing unusually slowly or because actual output is below potential. Because recessionary gaps represent wasted resources and expansionary gaps threaten to create inflation, policymakers have an incentive to try to eliminate both types of gap.

- The *natural rate of unemployment* is the part of the total unemployment rate that is attributable to frictional and structural unemployment. Equivalently, the natural rate of unemployment is the rate of unemployment that exists when the output gap is zero. Cyclical unemployment, the part of unemployment that is associated with recessions and expansions, equals the total unemployment rate less the natural unemployment rate. Cyclical unemployment is related to the output gap by *Okun's law,* which states that each extra percentage point of cyclical unemployment is associated with about a 2 percentage point increase in the output gap, measured in relation to potential output.

- In the next several chapters our study of recessions and expansions will focus on the role of economywide spending. If firms adjust prices only periodically, and in the meantime produce enough output to meet demand, then fluctuations in spending will lead to fluctuations in output over the short run. During that short-run period, government policies that influence aggregate spending may help to eliminate output gaps. In the long run, however, firms' price changes will eliminate output gaps—that is, the economy will "self-correct"—and total spending will influence only the rate of inflation.

## KEY TERMS

boom (311)
depression (309)
expansion (311)
expansionary gap (316)
natural rate of unemployment, $u^*$ (317)
Okun's law (318)
output gap, $Y^* - Y$ (315)
peak (309)
potential output, $Y^*$ (or potential GDP or full-employment output) (315)
recession (or contraction) (309)
recessionary gap (316)
trough (309)

## REVIEW QUESTIONS

1. Define *recession* and *expansion.* What are the beginning and ending points of a recession called? In the postwar United States, which have been longer on average, recessions or expansions?
2. Why is the traditional term *business cycles* a misnomer? How does your answer relate to the ease or difficulty of forecasting peaks and troughs?

3. Which firm is likely to see its profits reduced the most in a recession: an automobile producer, a manufacturer of boots and shoes, or a janitorial service? Which is likely to see its profits reduced the least? Explain.
4. How is each of the following likely to be affected by a recession: the natural unemployment rate, the cyclical unemployment rate, the inflation rate, the poll ratings of the president?
5. Define *potential output.* Is it possible for an economy to produce an amount greater than potential output? Explain.
6. True or false: All recessions are the result of output gaps. Explain.
7. True or false: When output equals potential output, the unemployment rate is zero. Explain.
8. If the natural rate of unemployment is 5 percent, what is the total rate of unemployment if output is 2 percent below potential output? What if output is 2 percent above potential output?

## PROBLEMS

1. Using Table 12.1, find the average duration, the minimum duration, and the maximum duration of expansions in the United States since 1929. Are expansions getting longer or shorter on average over time? Is there any tendency for long expansions to be followed by long recessions?
2. From the homepage of the Bureau of Economic Analysis (http://www.bea.gov/bea/dn1.htm) obtain quarterly data for U.S. real GDP for 2001 and 2002. Did the 2001 recession satisfy the informal criterion that a recession must have two consecutive quarters of negative GDP growth?
3. Given below are data on real GDP and potential GDP for the United States for the years 1988–1993, in billions of 1992 dollars. For each year calculate the output gap as a percentage of potential GDP, and state whether the gap is a recessionary gap or an expansionary gap. Also calculate the year-to-year growth rates of real GDP. Can you identify the recession that occurred during this period?

| Year | Real GDP | Potential GDP |
|---|---|---|
| 1988 | 5,844 | 5,788 |
| 1989 | 6,056 | 5,943 |
| 1990 | 6,172 | 6,102 |
| 1991 | 6,075 | 6,265 |
| 1992 | 6,214 | 6,432 |
| 1993 | 6,360 | 6,604 |

SOURCE: International Monetary Fund.

4. From the homepage of the Bureau of Labor Statistics (http://stats.bls.gov/), obtain the most recent available data on the unemployment rate for workers aged 16–19 and workers aged 20 or over. How do they differ? What are some of the reasons for the difference? How does this difference relate to the decline in the overall natural rate of unemployment since 1980?
5. Using Okun's law, fill in the four pieces of missing data in the table below. The data are hypothetical.

| Year | Real GDP | Potential GDP | Natural unemployment rate (%) | Actual unemployment rate (%) |
|---|---|---|---|---|
| 2001 | 7,840 | 8,000 | (a) | 6 |
| 2002 | 8,100 | (b) | 5 | 5 |
| 2003 | (c) | 8,200 | 4.5 | 4 |
| 2004 | 8,415 | 8,250 | 5 | (d) |

## ■ ANSWERS TO IN-CHAPTER EXERCISES ■

12.1 Answers will vary, depending on when the data is obtained.

12.2 The actual unemployment rate in the first half of 2002 exceeded the natural rate by 0.8 percent, so by Okun's law actual output fell below potential output by $2 \times 0.8\% = 1.6\%$ of potential output (a recessionary gap).

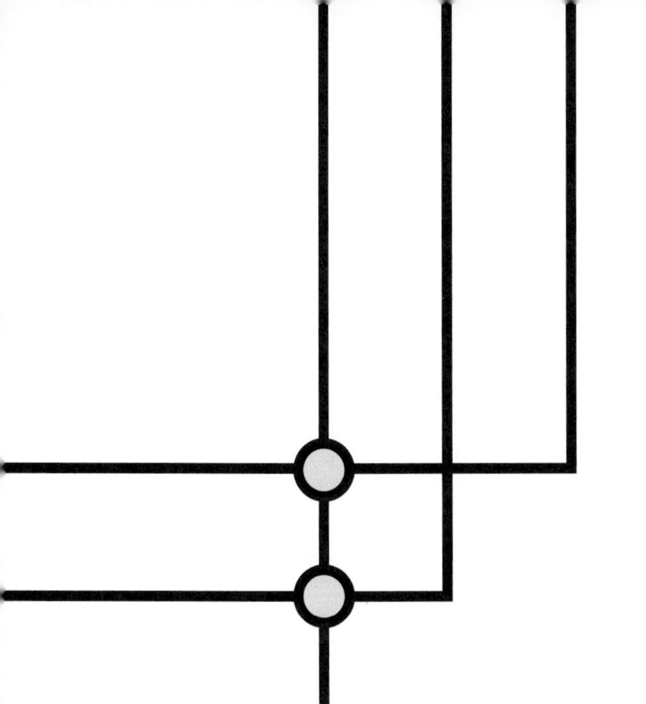

CHAPTER

13

# SPENDING AND OUTPUT IN THE SHORT RUN

When one of the authors of this book was a small boy, he used to spend some time every summer with his grandparents, who lived a few hours from his home. A favorite activity of his during these visits was to spend a summer evening on the front porch with his grandmother, listening to her stories. For some reason Grandma's recounting of her own life was particularly fascinating to her grandson.

Grandma had spent the early years of her marriage in New England, during the worst part of the Great Depression. In one of her reminiscences she remarked that at that time, in the mid-1930s, it had been a satisfaction to her to be able to buy her children a new pair of shoes every year. In the small town where she and her family lived, many children had to wear their shoes until they fell apart, and a few unlucky boys and girls went to school barefoot. Her grandson thought this was scandalous: "Why didn't their parents just buy them new shoes?" he demanded.

"They couldn't," said Grandma. "They didn't have the money. Most of the fathers had lost their jobs because of the Depression."

"What kind of jobs did they have?"

"They worked in the shoe factories, which had to close down."

"Why did the factories close down?"

"Because," Grandma explained, "nobody had any money to buy shoes."

The grandson was only six or seven years old at the time, but even he could see that there was something badly wrong with Grandma's logic. On the one side were boarded-up shoe factories and shoe workers with no jobs; on the other, children without shoes. Why couldn't the shoe factories just open

and produce the shoes the children so badly needed? He made his point quite firmly, but Grandma just shrugged and said it didn't work that way.

The story of the closed-down shoe factories illustrates in a microcosm the cost to society of a recessionary gap. In an economy with a recessionary gap, available resources, which could in principle be used to produce valuable goods and services, are instead allowed to lie fallow. This waste of resources lowers the economy's output and economic welfare, compared to its potential.

Grandma's account also suggests how such an unfortunate situation might come about. Suppose factory owners and other producers, being reluctant to accumulate unsold goods on their shelves, produce just enough output to satisfy the demand for their products. And suppose that, for some reason, the public's willingness or ability to spend declines. If spending declines, factories will respond by cutting their production (because they don't want to produce goods they can't sell) and by laying off workers who are no longer needed. And because the workers who are laid off will lose most of their income—a particularly serious loss in the 1930s, in the days before government-sponsored unemployment insurance was common—they must reduce their own spending. As their spending declines, factories will reduce their production again, laying off more workers, who in turn reduce their spending—and so on, in a vicious circle. In this scenario, the problem is not a lack of productive capacity—the factories have not lost their ability to produce—but rather *insufficient spending* to support the normal level of production.

The idea that a decline in aggregate spending may cause output to fall below potential output was one of the key insights of John Maynard Keynes, a highly influential British economist of the first half of the twentieth century. Box 13.1

© Bettmann/Corbis

### BOX 13.1: JOHN MAYNARD KEYNES AND THE KEYNESIAN REVOLUTION

John Maynard Keynes (1883–1946), perhaps the most influential economist of the twentieth century, was a remarkable individual who combined a brilliant career as an economic theorist with an active life in diplomacy, finance, journalism, and the arts. Keynes (pronouced "canes") first came to prominence at the end of World War I when he attended the Versailles peace conference as a representative of the British Treasury. He was appalled by the shortsightedness of the diplomats at the conference, particularly their insistence that the defeated Germans make huge compensatory payments (called reparations) to the victorious nations. In his widely read book *The Economic Consequences of the Peace* (1919), Keynes argued that the reparations imposed on Germany were impossibly large, and that attempts to extract the payments would prevent Germany's economic recovery and perhaps lead to another war. Unfortunately for the world, he turned out to be right.

In the period between the two world wars, Keynes held a professorship at Cambridge, where his father had taught economics. Keynes's early writings had been on mathematics and logic, but after his experience in Versailles he began to work primarily on economics, producing several well-regarded books. He developed an imposing intellectual reputation, editing Great Britain's leading scholarly journal in economics, writing articles for newspapers and magazines, advising the government, and playing a major role in the political and economic debates of the day. On the side, Keynes made fortunes both for himself and for King's College (a part of Cambridge University) by speculating in international currencies and commodities. He was also an active member of the Bloomsbury Group, a circle of leading artists, performers, and writers that included E. M. Forster and Virginia Woolf. In 1925 Keynes married the glamorous Russian ballerina Lydia Lopokova. Theirs was by all accounts a very successful marriage, and Keynes devoted significant energies to managing his wife's career and promoting the arts in Britain.

Like other economists of the time, Keynes struggled to understand the Great Depression that gripped the world in the 1930s. His work on the problem led to the publication in 1936 of *The General Theory of Employment, Interest, and Money.* In the *General Theory,* Keynes tried to explain how economies can remain at low levels of output and employment for protracted periods. He stressed a number of factors, most notably that aggregate spending may be too low to permit full employment during such periods. Keynes recommended increases in government spending as the most effective way to increase aggregate spending and restore full employment.

The *General Theory* is a difficult book, reflecting Keynes's own struggle to understand the complex causes of the Depression. In retrospect, some of the *General Theory*'s arguments seem unclear or even inconsistent. Yet the book is full of fertile ideas, many of which had a worldwide impact and eventually led to what has been called the *Keynesian revolution.* Over the years many economists have added to or modified Keynes's conception, to the point that Keynes himself, were he alive today, probably would not recognize much of what is now called Keynesian economics. But the ideas that insufficient aggregate spending can lead to recession and that government policies can help to restore full employment are still critical to Keynesian theory.

In 1937 a heart attack curtailed Keynes's activities, but he remained an important figure on the world scene. In 1944 he led the British delegation to the international conference in Bretton Woods, New Hampshire, which established the key elements of the postwar international monetary and financial system, including the International Monetary Fund and the World Bank. Keynes died in 1946.

gives a brief account of Keynes's life and ideas. The goal of this chapter is to present a theory, or model, of how recessions and expansions may arise from fluctuations in aggregate spending, along the lines first suggested by Keynes. This model, which we call the *basic Keynesian model,* is also known as the *Keynesian cross,* after the diagram that is used to illustrate the theory. In the body of the chapter we will emphasize a numerical and graphical approach to the basic Keynesian model. The appendix to this chapter provides a more general algebraic analysis.

We will begin with a brief discussion of the key assumptions of the basic Keynesian model. We will then turn to the important concept of total, or aggregate, *planned spending* in the economy. We will show how, in the short run, the rate of aggregate spending helps to determine the level of output, which can be greater than or less than potential output. In other words, depending on the level of spending, the economy may develop an output gap. "Too little" spending leads to a recessionary output gap, while "too much" creates an expansionary output gap.

An implication of the basic Keynesian model is that government policies that affect the level of spending can be used to reduce or eliminate output gaps. Policies used in this way are called *stabilization policies.* Keynes himself argued for the active use of fiscal policy—policy relating to government spending and taxes—to eliminate output gaps and stabilize the economy. In the latter part of this chapter we will show why Keynes thought fiscal policy could help to stabilize the economy, and we will discuss the usefulness of fiscal policy as a stabilization tool.

As we mentioned in Chapter 12, the basic Keynesian model is not a complete or entirely realistic model of the economy, since it applies only to the relatively short period during which firms do not adjust their prices but instead meet the demand forthcoming at preset prices. Furthermore, by treating prices as fixed, the basic Keynesian model presented in this chapter does not address the determination of inflation. Nevertheless, this model is an essential building block of leading current theories of short-run economic fluctuations and stabilization policies.

In subsequent chapters we will extend the basic Keynesian model to incorporate inflation and other important features of the economy.

## THE KEYNESIAN MODEL'S CRUCIAL ASSUMPTION: FIRMS MEET DEMAND AT PRESET PRICES

The basic Keynesian model is built on a key assumption, highlighted in Box 13.2. This assumption is that firms do not continuously change their prices as supply and demand conditions change; rather, over short periods, firms tend to keep their prices fixed and *meet the demand* that is forthcoming at those prices. As we will see, the assumption that firms vary their production in order to meet demand at preset prices implies that fluctuations in spending will have powerful effects on the nation's real GDP.

> **BOX 13.2: KEY ASSUMPTION OF THE BASIC KEYNESIAN MODEL**
>
> **In the short run, firms meet the demand for their products at preset prices.**
> Firms do not respond to every change in the demand for their products by changing their prices. Instead, they typically set a price for some period, then *meet the demand* at that price. By "meeting the demand," we mean that firms produce just enough to satisfy their customers at the prices that have been set.

The assumption that over short periods of time, firms meet the demand for their products at preset prices is generally realistic. Think of the stores where you shop. The price of a pair of jeans does not fluctuate from moment to moment according to the number of customers who enter the store or the latest news about the price of denim. Instead, the store posts a price and sells jeans to any customer who wants to buy at that price, at least until the store runs out of stock. Similarly, the corner pizza restaurant may leave the price of its large pie unchanged for months or longer, allowing its pizza production to be determined by the number of customers who want to buy at the preset price.

***menu costs*** the costs of changing prices

Firms do not normally change their prices frequently because doing so would be costly. Economists refer to the costs of changing prices as **menu costs.** In the case of the pizza restaurant, the menu cost is literally just that—the cost of printing up a new menu when prices change. Similarly, the clothing store faces the cost of remarking all its merchandise if the manager changes prices. But menu costs may also include other kinds of costs—for example, the cost of doing a market survey to determine what price to charge and the cost of informing customers about price changes. Economic Naturalist 13.1 discusses how technology may affect menu costs in the future.

Menu costs will not prevent firms from changing their prices indefinitely. As we saw in the case of Al's ice cream store (see Chapter 12), too great an imbalance between demand and supply, as reflected by a difference between sales and potential output, will eventually lead firms to change their prices. If no one is buying jeans, for example, at some point the clothing store will mark down its jeans prices. Or if the pizza restaurant becomes the local hot spot, with a line of customers stretching out the door, eventually the manager will raise the price of a large pie. Like many other economic decisions, the decision to change prices reflects a cost-benefit comparison: Prices should be changed if the benefit of doing so—the fact that sales will be brought more nearly in line with the firm's normal production capacity—outweighs the menu costs associated with making the change. As we have stressed, the basic Keynesian model developed in this chapter ignores the fact that prices will eventually adjust, and should therefore be interpreted as applying to the short run.

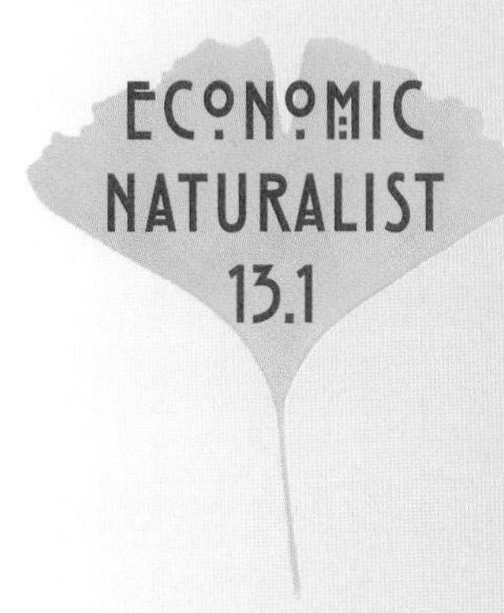


**Will new technologies eliminate menu costs?**

Thanks to new technologies, changing prices and informing customers about price changes is becoming increasingly less costly. Will technology eliminate menu costs as a factor in price setting?

Keynesian theory is based on the assumption that costs of changing prices, which economists refer to as menu costs, are sufficiently large to prevent firms from adjusting prices immediately in response to changing market conditions. However, in many industries, new technologies have eliminated or greatly reduced the direct costs of changing prices. For example, the use of bar codes to identify individual products, together with scanner technologies, allows a grocery store manager to change prices with just a few keystrokes, without having to change the price label on each can of soup or loaf of bread. Airlines use sophisticated computer software to implement complex pricing strategies, under which two travelers on the same flight to Milwaukee may pay very different fares, depending on whether they are business or vacation travelers and on how far in advance their flights were booked. Online retailers, such as booksellers, have the ability to vary their prices by type of customer and even by individual customer, while other Internet-based companies, such as Ebay and Priceline, allow for negotiation over the price of each individual purchase. As described in Economic Naturalist 12.4, the Coca-Cola company experimented with a vending machine that automatically varied the price of a soft drink according to the outdoor temperature, charging more when the weather is hot.

"You thought we would offer lower fares? How insensitive."

Will these reductions in the direct costs of changing prices make the Keynesian theory, which assumes that firms meet demand at preset prices, less relevant to the real world? This is certainly a possibility that macroeconomists must take into account. However, it is unlikely that new technologies will completely eliminate the costs of changing prices anytime soon. Gathering the information about market conditions needed to set the profit-maximizing price—including the prices charged by competitors, the costs of producing the good or service, and the likely demand for the product—will remain costly for firms. Another cost of changing prices is the use of valuable managerial time and attention needed to make informed pricing decisions. A more subtle cost of changing prices—particularly raising prices—is that doing so may lead regular customers to rethink their choice of suppliers and decide to search for a better deal elsewhere.

## PLANNED AGGREGATE EXPENDITURE

In the Keynesian theory discussed in this chapter, output at each point in time is determined by the amount that people throughout the economy want to spend—what we will refer to as *planned aggregate expenditure.* Specifically, **planned aggregate expenditure** *(PAE)* is total planned spending on final goods and services.

***planned aggregate expenditure (PAE)*** total planned spending on final goods and services

The four components of spending on final goods and services were introduced in Chapter 5:

1. *Consumer expenditure,* or simply *consumption* (*C*), is spending by households on final goods and services. Examples of consumer expenditure are spending on food, clothes, and entertainment and on consumer durable goods like automobiles and furniture.

2. *Investment* (*I*) is spending by firms on new capital goods, such as office buildings, factories, and equipment. Spending on new houses and apartment buildings (residential investment) and increases in inventories (inventory investment) are also included in investment.[1]

[1]In everyday conversations, people often use the term "investment" to mean *financial* investment, for example, the purchase of stocks or bonds. As we discussed earlier, we use "investment" here to mean spending on new capital goods, such as factories, housing, and equipment, which is not the same as financial investment. This distinction is important to keep in mind.

3. *Government purchases* ($G$) is spending by governments (federal, state, and local) on goods and services. Examples of government purchases include new schools and hospitals, military hardware, equipment for the space program, and the services of government employees, such as soldiers, police, and government office workers. Recall from Chapter 5 that *transfer payments,* such as social security benefits and unemployment insurance, and interest on the government debt are *not* included in government purchases. Transfer payments and interest contribute to aggregate expenditure only at the point when they are spent by their recipients (for example, when a recipient of a Social Security check uses the funds to buy food, clothing, or other consumption goods).
4. *Net exports* ($NX$) equals exports minus imports. Exports are sales of domestically produced goods and services to foreigners; imports are purchases by domestic residents of goods and services produced abroad. Net exports represents the net demand for domestic goods by foreigners.

Together these four types of spending—by households, firms, the government, and the rest of the world—sum to total, or aggregate, spending.

## PLANNED SPENDING VERSUS ACTUAL SPENDING

In the Keynesian model, output is determined by planned aggregate expenditure, or planned spending for short. Could *planned* spending ever differ from *actual* spending? The answer is yes. The most important case is that of a firm that sells either less or more of its product than expected. As was noted in Chapter 5, additions to the stocks of goods sitting in a firm's warehouse are treated in official government statistics as inventory investment by the firm. In effect, government statisticians assume that the firm buys its unsold output from itself; they then count those purchases as part of the firm's investment spending.[2]

Suppose, then, that a firm's actual sales are less than expected, so that part of what it had planned to sell remains in the warehouse. In this case, the firm's actual investment, including the unexpected increases in its inventory, is greater than its planned investment, which did not include the added inventory. Suppose we agree to let $I^p$ equal the firm's planned investment, including planned inventory investment. A firm that sells less of its output than planned, and therefore adds more to its inventory than planned, will find that its actual investment (including unplanned inventory investment) exceeds its planned investment, so that $I > I^p$.

What about a firm that sells more of its output than expected? In that case, the firm will add less to its inventory than it planned, so actual investment will be less than planned investment, or $I < I^p$. Example 13.1 gives a numerical illustration.

EXAMPLE 13.1

### Actual and planned investment

The Fly-by-Night Kite Company produces \$5,000,000 worth of kites during the year. It expects sales of \$4,800,000 for the year, leaving \$200,000 worth of kites to be stored in the warehouse for future sale. During the year, Fly-by-Night adds \$1,000,000 in new production equipment as part of an expansion plan. Find Fly-by-Night's actual investment $I$ and its planned investment $I^p$ if actual kite sales turn out to be \$4,600,000. What if sales are \$4,800,000? What if they are \$5,000,000?

Fly-by-Night's planned investment $I^p$ equals its purchases of new production equipment (\$1,000,000) plus its planned additions to inventory (\$200,000), for a total of \$1,200,000 in planned investment. The company's planned investment does not depend on how much it actually sells.

[2]For the purposes of measuring GDP, treating unsold output as being purchased by its producer has the advantage of ensuring that actual production and actual expenditure are equal.

If Fly-by-Night sells only \$4,600,000 worth of kites, it will add \$400,000 in kites to its inventory instead of the \$200,000 worth originally planned. In this case, actual investment equals the \$1,000,000 in new equipment plus the \$400,000 in inventory investment, so $I = \$1{,}400{,}000$. We see that when the firm sells less output than planned, actual investment exceeds planned investment ($I > I^p$).

If Fly-by-Night has \$4,800,000 in sales, then it will add \$200,000 in kites to inventory, just as planned. In this case, actual and planned investment are the same:

$$I = I^p = \$1{,}200{,}000.$$

Finally, if Fly-by-Night sells \$5,000,000 worth of kites, it will have no output to add to inventory. Its inventory investment will be zero, and its total actual investment (including the new equipment) will equal \$1,000,000, which is less than its planned investment of \$1,200,000 ($I < I^p$).

Because firms that are meeting the demand for their product or service at preset prices cannot control how much they sell, their actual investment (including inventory investment) may well differ from their planned investment. However, for households, the government, and foreign purchasers, we may reasonably assume that actual spending and planned spending are the same. Thus, from now on we will assume that, for consumption, government purchases, and net exports, actual spending equals planned spending.

With these assumptions, we can define planned aggregate expenditure by the following equation:

$$PAE = C + I^p + G + NX. \quad (13.1)$$

Equation 13.1 says that planned aggregate expenditure is the sum of planned spending by households, firms, governments, and foreigners. We use a superscript $p$ to distinguish planned investment spending by firms, $I^p$, from actual investment spending, $I$. However, because planned spending equals actual spending for households, the government, and foreigners, we do not need to use superscripts for consumption, government purchases, or net exports.

## HEY BIG SPENDER! CONSUMER SPENDING AND THE ECONOMY

The largest component of planned aggregate expenditure—nearly two-thirds of total spending—is consumption spending, denoted $C$. As already mentioned, consumer spending includes household purchases of goods, such as groceries and clothing; services, such as health care, concerts, and college tuition; and consumer durables, such as cars, furniture, and home computers. Thus consumers' willingness to spend affects sales and profitability in a wide range of industries. (Households' purchases of new homes are classified as investment, rather than consumption; but home purchases represent another channel through which household decisions affect total spending.)

What determines how much people plan to spend on consumer goods and services in a given period? While many factors are relevant, a particularly important determinant of the amount people plan to consume is their after-tax, or *disposable*, income. All else being equal, households and individuals with higher disposable incomes will consume more than those with lower disposable incomes. Keynes himself stressed the importance of disposable income in determining household consumption decisions, claiming a "psychological law" that people would tie their spending closely to their incomes.

Recall from Chapter 9 that the disposable income of the private sector is the total production of the economy, *Y*, less net taxes (taxes minus transfers), or *T*. So we can assume that consumption spending (*C*) increases as disposable income ($Y - T$) increases. As already mentioned, other factors may also affect consumption, such as the real interest rate, also discussed in Chapter 9. For now we will ignore those other factors, returning to some of them later.

A general equation that captures the link between consumption and the private sector's disposable income is

$$C = \overline{C} + c(Y - T). \quad (13.2)$$

This equation, which we will dissect in a moment, is known as the *consumption function*. The **consumption function** relates consumption spending to its determinants, in particular, disposable (after-tax) income.

***consumption function*** the relationship between consumption spending and its determinants, in particular, disposable (after-tax) income.

Let's look at the consumption function, Equation 13.2, more carefully. The right side of the equation contains two terms, $\overline{C}$ and $c(Y - T)$. The first term, $\overline{C}$, is a constant term in the equation that is intended to capture factors *other than disposable income* that affect consumption. For example, suppose consumers were to become more optimistic about the future, so that they desire to consume more and save less at any given level of their current disposable incomes. An increase in desired consumption at any given level of disposable income would be represented in the consumption function as an increase in the term $\overline{C}$.

We can imagine other factors that may affect the term $\overline{C}$ in the consumption function. Suppose, for example, that there is a boom in the stock market or a sharp increase in home prices, making consumers feel wealthier and hence more inclined to spend, for a given level of current disposable income. This effect could be captured by assuming that $\overline{C}$ increases. Likewise, a fall in home prices or stock prices that made consumers feel poorer and less inclined to spend would be represented by a decrease in $\overline{C}$. Economists refer to the effect of changes in asset prices on households' wealth and hence their consumption spending as the **wealth effect** of changes in asset prices.

***wealth effect*** the tendency of changes in asset prices to affect households' wealth and thus their spending on consumption goods

**What effect did the 2000–2002 decline in U.S. stock market values have on consumption spending?**

From March 2000 to October 2002, the U.S. stock market suffered a 49 percent drop in value as measured by the Standard and Poor's 500 stock index, a widely referenced benchmark of U.S. stock performance. What effect did this decline in U.S. stock market values have on consumption spending?

According to MIT economist James Poterba, U.S. households owned roughly 13.3 trillion dollars of corporate stock in 2000.[3] If households' stock market holdings reflect those of the Standard and Poor's stock index, the 49 percent drop in the value of the stock market wiped out approximately 6.5 trillion dollars of household wealth in two years. According to economic models based on historical experience, a dollar's decrease in household wealth reduces consumer spending by 3 to 7 cents per year, so the reduction in stock market wealth had the potential to reduce overall consumer spending by 195 billion to 455 billion dollars, a drop of approximately 3 percent to 7 percent. Yet, real consumption spending continued to rise from 2000 through 2002. Why did this happen?

Despite the start of a recession in March 2001, overall consumption spending remained strong during 2000–2002 for a variety of reasons. Housing prices rose dramatically during this period, increasing consumers' housing wealth and offsetting their decline in stock-related wealth. In addition, as Economic Naturalist 12.1 points out, consumers' real after-tax income continued to grow into

[3]"The Stock Market and the Consumer," Hoover Institution Weekly Essays, November 6, 2000 Online: www-hoover.stanford.edu/pubaffairs/we/current/poterba_1100.html.

the fall of 2001, helping to maintain strong consumer spending despite the drop in the stock market. Finally, throughout 2001 and into early 2002, the Federal Reserve significantly reduced interest rates (how the Federal Reserve does this will be discussed in Chapter 14). A reduction in interest rates helps to promote consumer spending, especially on "big-ticket" items such as automobiles, by reducing consumers' borrowing costs. Overall, while the drop in stock market values clearly had a negative effect on consumer wealth, other offsetting factors helped to keep the 2000–2002 stock market decline from dampening consumption spending during this period.

The second term on the right side of Equation 13.2, $c(Y - T)$, reflects the effect of disposable income, $Y - T$, on consumption. The parameter $c$, a fixed number, is called the *marginal propensity to consume.* The **marginal propensity to consume** (***MPC***) is the amount by which consumption rises when current disposable income rises by one dollar. Presumably, if people receive an extra dollar of income, they will consume part of the dollar and save the rest. In other words, their consumption will increase, but by less than the full dollar of extra income. Thus it is realistic to assume that the marginal propensity to consume is greater than 0 (an increase in income leads to an increase in consumption) but less than 1 (the increase in consumption will be less than the full increase in income). Mathematically, we can summarize these assumptions as $0 < c < 1$.

***marginal propensity to consume (MPC),*** or $c$ the amount by which consumption rises when disposable income rises by one dollar. We assume that $0 < c < 1$

Figure 13.1 shows a hypothetical consumption function, with consumption spending (C) on the vertical axis and disposable income ($Y - T$) on the horizontal axis. The intercept of the consumption function on the vertical axis equals exogenous consumption $\overline{C}$, and the slope of the consumption function equals the marginal propensity to consume, $c$.

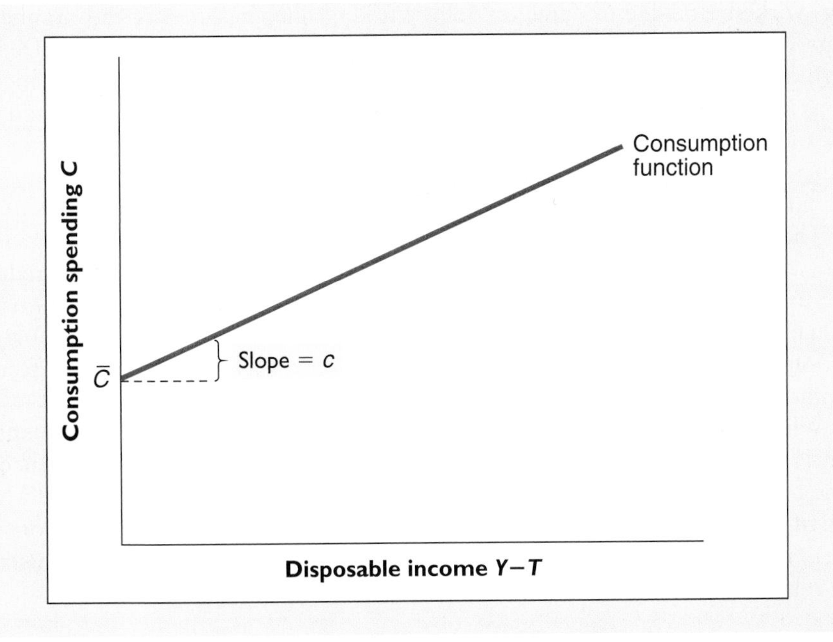


**FIGURE 13.1**
**A Consumption Function.** The consumption function relates households' consumption spending, $C$, to disposable income, $Y - T$. The vertical intercept of this consumption function is the exogenous component of consumption, $\overline{C}$, and the slope of the line equals the marginal propensity to consume, $c$.

To see how this consumption function fits reality, compare Figure 13.1 to Figure 13.2, which shows the relationship between aggregate real consumption expenditures and real disposable income in the United States for the period 1960–2001. Figure 13.2, a type of diagram called a *scatter plot,* shows aggregate real consumption on the vertical axis and aggregate real disposable income on the horizontal axis. Each point on the graph corresponds to a year between 1960 and 2001 (selected years are indicated in the figure). The position of each point is determined by the combination of consumption and disposable income associated with that year. As you can see, there is indeed a close relationship

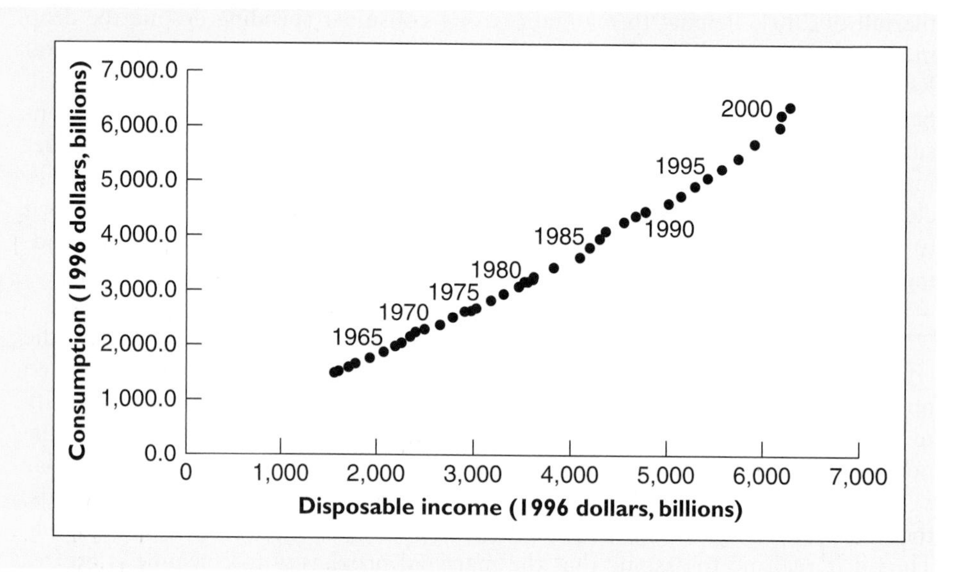


**FIGURE 13.2**
**The U.S. Consumption Function, 1960–2001.** Each point on this figure represents a combination of aggregate real consumption and aggregate real disposable income for a specific year between 1960 and 2001. Note the strong positive relationship between consumption and disposable income.

between aggregate consumption and disposable income: Higher disposable income implies higher consumption.

## PLANNED AGGREGATE EXPENDITURE AND OUTPUT

Thinking back to Grandma's reminiscences, recall that an important element of her story involved the links among production, income, and spending. As the shoe factories in Grandma's town reduced production, the incomes of both factory workers and factory owners fell. Workers' incomes fell as the number of hours of work per week were reduced (a common practice during the Depression), as workers were laid off, or as wages were cut. Factory owners' income fell as profits declined. Reduced incomes, in turn, forced both workers and factory owners to curtail their spending—which led to still lower production and further reductions in income. This vicious circle led the economy further and further into recession.

The logic of Grandma's story has two key elements: (1) declines in production (which imply declines in the income received by producers) lead to reduced spending; and (2) reductions in spending lead to declines in production and income. In this section we look at the first part of the story, the effects of production and income on spending. We return later in this chapter to the effects of spending on production and income.

Why do changes in production and income affect planned aggregate spending? The consumption function, which relates consumption to disposable income, is the basic source of this relationship. Because consumption spending $C$ is a large part of planned aggregate spending, and because consumption depends on output $Y$, aggregate spending as a whole depends on output. Example 13.2 illustrates this relationship numerically.

**EXAMPLE 13.2**

### Linking planned aggregate expenditure to output

In a particular economy, the consumption function is

$$C = 620 + 0.8(Y - T),$$

so that the intercept term in the consumption function $\overline{C}$, equals 620, and the marginal propensity to consume $c$ equals 0.8. Also, suppose that we are given that planned investment spending $I^P = 220$, government purchases $G = 300$, net exports $NX = 20$, and taxes $T = 250$.

Write a numerical equation linking planned aggregate expenditure *PAE* to output *Y*. How does planned spending change when output and hence income change?

Recall the definition of planned aggregate expenditure, Equation 13.1:

$$PAE = C + I^p + G + NX$$

To find a numerical equation for planned aggregate expenditure, we need to find numerical expressions for each of its four components. The first component of spending, consumption, is defined by the consumption function, $C = 620 + 0.8(Y - T)$. Since taxes $T = 250$, we can substitute for $T$ to write the consumption function as $C = 620 + 0.8(Y - 250)$. Now plug this expression for $C$ into the definition of planned aggregate expenditure above to get

$$PAE = [620 + 0.8(Y - 250)] + I^p + G + NX,$$

where we have just replaced $C$ by its value as determined by the consumption function. Similarly, we can substitute the given numerical values of planned investment $I^p$, government purchases $G$, and net exports $NX$ into the definition of planned aggregate expenditure to get

$$PAE = [620 + 0.8(Y - 250)] + 220 + 300 + 20.$$

To simplify this equation, first note that $0.8(Y - 250) = 0.8Y - 200$, then add all the terms that don't depend on output $Y$. The result is

$$\begin{aligned} PAE &= (620 - 200 + 220 + 300 + 20) + 0.8Y \\ &= 960 + 0.8Y. \end{aligned}$$

The final expression shows the relationship between planned aggregate expenditure and output in this numerical example. Note that, according to this equation, a \$1 increase in $Y$ leads to an increase in *PAE* of (0.8) (\$1), or 80 cents. The reason for this is that the marginal propensity to consume, $c$, in this example is 0.8. Hence a \$1 increase in income raises consumption spending by 80 cents. Since consumption is a component of total planned spending, total spending rises by 80 cents as well.

The solution to Example 13.2 illustrates a general point: Planned aggregate expenditure can be divided into two parts, a part that depends on output ($Y$) and a part that is independent of output. The portion of planned aggregate expenditure that is independent of output is called **autonomous expenditure.** In Example 13.2, autonomous expenditure is the constant term in the equation for planned aggregate expenditure, or 960. This portion of planned spending, being a fixed number, does not vary when output varies. By contrast, the portion of planned aggregate expenditure that depends on output ($Y$) is called **induced expenditure.** In Example 13.2, induced expenditure equals $0.8Y$, the second term in the expression for planned aggregate expenditure. Note that the numerical value of induced expenditure depends, by definition, on the numerical value taken by output. Autonomous expenditure and induced expenditure together equal planned aggregate expenditure.

***autonomous expenditure*** the portion of planned aggregate expenditure that is independent of output

***induced expenditure*** the portion of planned aggregate expenditure that depends on output $Y$

## RECAP PLANNED AGGREGATE EXPENDITURE

Planned aggregate expenditure (*PAE*) is total planned spending on final goods and services. The four components of planned spending are consumer expenditure ($C$), planned investment ($I^p$), government purchases ($G$), and net exports ($NX$). Planned investment differs from actual investment when firms' sales are different from what they expected, so that additions to inventory (a component of investment) are different from what firms anticipated.

The largest component of aggregate expenditure is consumer expenditure, or simply consumption. Consumption depends on disposable, or after-tax, income, according to a relationship known as the consumption function, stated algebraically as $C = \overline{C} + c(Y - T)$.

The constant term in the consumption function, $\overline{C}$, captures factors other than disposable income that affect consumer spending. For example, an increase in housing or stock prices that makes households wealthier and thus more willing to spend—an effect called the wealth effect—could be captured by an increase in $\overline{C}$. The slope of the consumption function equals the marginal propensity to consume, $c$, where $0 < c < 1$. This is the amount by which consumption rises when disposable income rises by one dollar.

Increases in output $Y$, which imply equal increases in income, cause consumption to rise. As consumption is part of planned aggregate expenditure, planned spending depends on output as well. The portion of planned aggregate expenditure that depends on output is called induced expenditure. The portion of planned aggregate expenditure that is independent of output is autonomous expenditure.

## SHORT-RUN EQUILIBRIUM OUTPUT

Now that we have defined planned aggregate expenditure and seen how it is related to output, the next task is to see how output itself is determined. Recall the assumption of the basic Keynesian model: In the short run, producers leave prices at preset levels and simply meet the demand that is forthcoming at those prices. In other words, during the short-run period in which prices are preset, firms produce an amount that is equal to planned aggregate expenditure. Accordingly, we define **short-run equilibrium output** as the level of output at which output $Y$ equals planned aggregate expenditure $PAE$:

***short-run equilibrium output*** the level of output at which output $Y$ equals planned aggregate expenditure $PAE$; short-run equilibrium output is the level of output that prevails during the period in which prices are predetermined

$$Y = PAE. \tag{13.3}$$

Short-run equilibrium output is the level of output that prevails during the period in which prices are predetermined.

We can find the short-run equilibrium output for the economy described in Example 13.2 using Table 13.1. Column 1 of the table gives some possible values for short-run equilibrium output. To find the correct value, we must compare

**TABLE 13.1**
**Numerical Determination of Short-Run Equilibrium Output**

| (1) Output $Y$ | (2) Planned aggregate expenditure $PAE = 960 + 0.8Y$ | (3) $Y - PAE$ | (4) $Y = PAE$? |
|---|---|---|---|
| 4,000 | 4,160 | −160 | No |
| 4,200 | 4,320 | −120 | No |
| 4,400 | 4,480 | −80 | No |
| 4,600 | 4,640 | −40 | No |
| 4,800 | 4,800 | 0 | **Yes** |
| 5,000 | 4,960 | 40 | No |
| 5,200 | 5,120 | 80 | No |

each to the value of planned aggregate expenditure at that output level. Column 2 shows the value of planned aggregate expenditure corresponding to the values of short-run equilibrium output in column 1. Recall that in Example 13.2, planned spending is determined by the equation

$$PAE = 960 + 0.8Y.$$

Because consumption rises with output, total planned spending (which includes consumption) rises also. But if you compare columns 1 and 2, you will see that when output rises by 200, planned spending rises by only 160. That is because the marginal propensity to consume in this economy is 0.8, so that each dollar in added income raises consumption and planned spending by 80 cents.

Again, short-run equilibrium output is the level of output at which $Y = PAE$, or, equivalently, $Y - PAE, = 0$. Looking at Table 13.1, we can see there is only one level of output that satisfies that condition, $Y = 4{,}800$. At that level, output and planned aggregate expenditure are precisely equal, so that producers are just meeting the demand for their goods and services.

In this economy, what would happen if output differed from its equilibrium value of 4,800? Suppose, for example, that output were 4,000. Looking at the second column of Table 13.1, you can see that when output is 4,000, planned aggregate expenditure equals 960 + 0.8(4,000), or 4,160. Thus if output is 4,000, firms are not producing enough to meet the demand. They will find that as sales exceed the amounts they are producing, their inventories of finished goods are being depleted by 160 per year, and that actual investment (including inventory investment) is less than planned investment. Under the assumption that firms are committed to meeting their customers' demand, firms will respond by expanding their production.

Would expanding production to 4,160, the level of planned spending firms faced when output was 4,000, be enough? The answer is no, because of induced expenditure. That is, as firms expand their output, aggregate income (wages and profits) rises with it, which in turn leads to higher levels of consumption. Indeed, if output expands to 4,160, planned spending will increase as well, to 960 + 0.8(4,160), or 4,288. So an output level of 4,160 will still be insufficient to meet demand. As Table 13.1 shows, output will not be sufficient to meet planned aggregate expenditure until it expands to its short-run equilibrium value of 4,800.

What if output were initially greater than its equilibrium value—say, 5,000? From Table 13.1 we can see that when output equals 5,000, planned spending equals only 4,960—less than what firms are producing. So at an output level of 5,000, firms will not sell all they produce, and they will find that their merchandise is piling up on store shelves and in warehouses (actual investment, including inventory investment, is greater than planned investment). In response, firms will cut their production runs. As Table 13.1 shows, they will have to reduce production to its equilibrium value of 4,800 before output just matches planned spending.

### EXERCISE 13.1

**Construct a table like Table 13.1 for an economy like the one we have been working with, assuming that the consumption function is $C = 820 + 0.7(Y - T)$ and that $I^P = 600$, $G = 600$, $NX = 200$, and $T = 600$.**

**What is short-run equilibrium output in this economy? (*Hint:* Try using values for output above 5,000.)**

Short-run equilibrium output can also be determined graphically, as Example 13.3 shows.

**EXAMPLE 13.3** **Finding short-run equilibrium output (graphical approach)**

Using a graphical approach, find short-run equilibrium output for the economy described in Example 13.2.

Figure 13.3 shows the graphical determination of short-run equilibrium output for the economy described in Example 13.2. Output $Y$ is plotted on the horizontal axis and planned aggregate expenditure $PAE$ on the vertical axis. The figure contains two lines, one of which is a 45° line extending from the origin. In general, a 45° line from the origin includes the points at which the variable on the vertical axis equals the variable on the horizontal axis. Hence, in this case, the 45° line represents the equation $Y = PAE$. Since short-run equilibrium output must satisfy the equation $Y = PAE$, the combination of output and spending that satisfies this condition must lie somewhere on the 45° line in Figure 13.3.

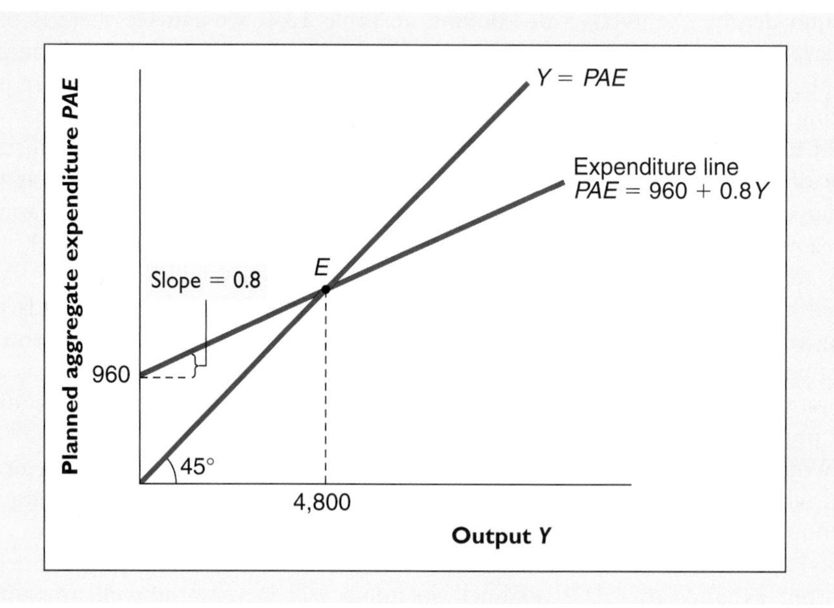


**FIGURE 13.3**
**Determination of Short-Run Equilibrium Output (Keynesian Cross).**
The 45° line represents the short-run equilibrium condition $Y = PAE$. The line $PAE = 960 + 0.8Y$, referred to as the expenditure line, shows the relationship of planned aggregate expenditure to output. Short-run equilibrium output (4,800) is determined at the intersection of the two lines, point $E$. This type of diagram is known as a Keynesian cross.

The second line in Figure 13.3, less steep than the 45° line, shows the relationship between planned aggregate expenditure $PAE$ and output $Y$. Because it summarizes how planned spending depends on output, we will call this line the *expenditure line*. In this example, we know that the relationship between planned aggregate expenditure and output (the equation for the expenditure line) is

$$PAE = 960 + 0.8Y.$$

According to this equation, when $Y = 0$, the value of $PAE$ is 960. Thus 960 is the intercept of the expenditure line, as shown in Figure 13.3. Notice that *the intercept of the expenditure line equals autonomous expenditure,* a result that will always hold. The slope of the line relating aggregate demand to output is 0.8, the value of the coefficient of output in the equation $PAE = 960 + 0.8Y$. Where does the number 0.8 come from? (*Hint*: What determines by how much aggregate spending increases when output rises by a dollar?)

Only one point in Figure 13.3 is consistent with *both* the definition of short-run equilibrium output, $Y = PAE$, and the given relationship between planned spending and output, $PAE = 960 + 0.8Y$. That point is the intersection of the two lines, point $E$. At point $E$, short-run equilibrium output equals 4,800, which is the same value that we obtained using Table 13.1 and by a direct numerical solution. At points to the right of $E$, output exceeds planned aggregate expenditure. Hence, to the right of point $E$, firms will be producing more than they can sell, which will lead them to reduce their rate of production. By contrast, to the left of point $E$, planned aggregate spending exceeds output. In that region, firms will not be producing enough to meet demand, and they will tend to increase their production. Only at point *E*, where output equals 4,800, will firms be producing enough to just satisfy planned spending on goods and services.

The diagram in Figure 13.3 is often called the *Keynesian cross*, after its characteristic shape. The Keynesian cross shows graphically how short-run equilibrium output is determined in a world in which producers meet demand at predetermined prices.

**EXERCISE 13.2**

**Use a Keynesian cross diagram to show graphically the determination of short-run equilibrium output for the economy described in Exercise 13.1. What are the intercept and the slope of the expenditure line?**

## PLANNED SPENDING AND THE OUTPUT GAP

We are now ready to use the basic Keynesian model to show how insufficient spending can lead to a recession. To illustrate the effects of spending changes on output, we will continue to work with the economy introduced in Example 13.2. We have shown that in this economy, short-run equilibrium output equals 4,800. Let's now make the additional assumption that potential output in this economy also equals 4,800, or $Y^* = 4{,}800$, so that initially there is no output gap. Starting from this position of full employment, Example 13.4 shows how a fall in planned aggregate expenditure can lead to a recession.

**EXAMPLE 13.4**

### A fall in planned spending leads to a recession

For the economy introduced in Example 13.2, we have found that short-run equilibrium output $Y$ equals 4,800. Assume also that potential output $Y^*$ equals 4,800, so that the output gap $Y^* - Y$ equals zero.

Suppose, though, that consumers become more pessimistic about the future, so that they begin to spend less at every level of current disposable income. We can capture this change by assuming that $\overline{C}$, the constant term in the consumption function, falls to a lower level. To be specific, suppose that $\overline{C}$ falls by 10 units, which in turn implies a decline in autonomous expenditure of 10 units. What is the effect of this reduction in planned spending on the economy?

We can see the effects of the decline in consumer spending on the economy using the Keynesian cross diagram. Figure 13.4 shows the original short-run

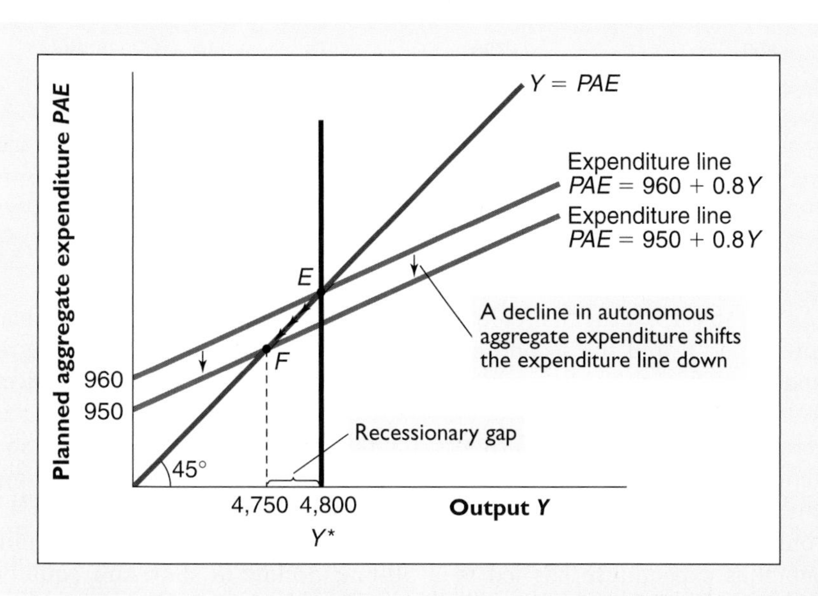


**FIGURE 13.4**
**A Decline in Planned Spending Leads to a Recession.**
A decline in consumers' willingness to spend at any current level of disposable income reduces planned autonomous expenditure and shifts the expenditure line down. The short-run equilibrium point drops from *E* to *F*, reducing output and opening up a recessionary gap.

equilibrium point of the model ($E$), at the intersection of the 45° line, along which $Y = PAE$, and the original expenditure line, representing the equation $PAE = 960 + 0.8Y$. As before, the initial value of short-run equilibrium output is 4,800, which we have now assumed also corresponds to potential output $Y^*$. But what happens $\overline{C}$ declines by 10, reducing autonomous expenditure by 10 as well?

Originally, autonomous expenditure in this economy was 960, so a decline of 10 units causes it to fall to 950. Instead of the economy's planned spending being described by the equation $PAE = 960 + 0.8Y$, as initially, it is now given by $PAE = 950 + 0.8Y$. What does this change imply for the graph in Figure 13.4? Since the intercept of the expenditure line (equal to autonomous expenditure) has decreased from 960 to 950, the effect of the decline in consumer spending will be to shift the expenditure line down in parallel fashion, by 10 units. Figure 13.4 indicates this downward shift in the expenditure line. The new short-run equilibrium point is at point $F$, where the new, lower expenditure line intersects the 45° line.

Point $F$ is to the left of the original equilibrium point $E$, so we can see that output and spending have fallen from their initial levels. Since output at point $F$ is lower than potential output, 4,800, we see that the fall in consumer spending has resulted in a recessionary gap in the economy. More generally, starting from a situation of full employment (where output equals potential output), any decline in autonomous expenditure leads to a recession.

Numerically, how large is the recessionary gap in Figure 13.4? To answer this question, we can use Table 13.2, which is in the same form as Table 13.1. The key difference is that in Table 13.2 planned aggregate expenditure is given by $PAE = 950 + 0.8Y$, rather than by $PAE = 960 + 0.8Y$, as in Table 13.1.

**TABLE 13.2**
**Determination of Short-Run Equilibrium Output after a Fall in Spending**

| (1) Output $Y$ | (2) Planned aggregate expenditure $PAE = 950 + 0.8Y$ | (3) $Y - PAE$ | (4) $Y = PAE$? |
|---|---|---|---|
| 4,600 | 4,630 | −30 | No |
| 4,650 | 4,670 | −20 | No |
| 4,700 | 4,710 | −10 | No |
| 4,750 | 4,750 | 0 | **Yes** |
| 4,800 | 4,790 | 10 | No |
| 4,850 | 4,830 | 20 | No |
| 4,900 | 4,870 | 30 | No |
| 4,950 | 4,910 | 40 | No |
| 5,000 | 4,950 | 50 | No |

As in Table 13.1, the first column of the table shows alternative possible values of output $Y$, and the second column shows the levels of planned aggregate expenditure $PAE$ implied by each value of output in the first column. Notice that 4,800, the value of short-run equilibrium output found in Table 13.1, is no longer an equilibrium; when output is 4,800, planned spending is 4,790, so output and planned spending are not equal. As the table shows, following the decline in planned aggregate expenditure, short-run equilibrium output is 4,750, the only value of output for which $Y = PAE$. Thus a drop of 10 units in autonomous expenditure has led to a 50-unit decline in short-run equilibrium output. If full-employment output is 4,800, then the recessionary gap shown in Figure 13.4 is 4,800 − 4,750 = 50 units.

**EXERCISE 13.3**

**In the economy described in Example 13.4, we found a recessionary gap of 50, relative to potential output of 4,800. Suppose that in this economy, the natural rate of unemployment $u^*$ is 5 percent. What will the actual unemployment rate be after the recessionary gap appears? (*Hint*: Recall Okun's law from Chapter 12.)**

Example 13.4 showed that a decline in autonomous expenditure, arising from a decreased willingness of consumers to spend, causes short-run equilibrium output to fall and opens up a recessionary gap. The same conclusion applies to declines in autonomous expenditure arising from other sources. Suppose, for example, that firms become disillusioned with new technologies and cut back their planned investment in new equipment. In terms of the model, this reluctance of firms to invest can be interpreted as a decline in planned investment spending $I^p$. Under our assumption that planned investment spending is given and does not depend on output, planned investment is part of autonomous expenditure. So a decline in planned investment spending depresses autonomous expenditure and output, in precisely the same way that a decline in the autonomous part of consumption spending does. Similar conclusions apply to declines in other components of autonomous expenditure, such as government purchases and net exports, as we will see in later applications.

**EXERCISE 13.4**

**Repeat the analysis of Example 13.4, except assume that consumers become *more* rather than less confident about the future. As a result, $\overline{C}$ rises by 10 units, which in turn raises autonomous expenditure by 10 units. Show graphically that this increase in consumers' willingness to spend leads to an expansionary output gap. Find the numerical value of the expansionary output gap.**

*"These are hard times for retailers, so we should show them our support in every way we can."*

ECONOMIC NATURALIST 13.3

### What caused the 1990–1991 recession?

As we saw in Chapter 12, the 1990–1991 recession came at the wrong time for the first President Bush, ultimately costing him the election. What caused the output of the U.S. economy to fall below its potential during that period?

Two factors have received a substantial part of the blame for the 1990–1991 recession, one being a *decline in consumer confidence.* Organizations such as the Conference Board and the Survey Research Center of the University of Michigan perform regular consumer surveys, in which people are asked their views about the future of the economy in general and their own fortunes in particular. Consumer responses are then summarized in measures of "consumer confidence." A high level of confidence implies that people are optimistic about both their own economic futures and the future of the economy in general. Economists have found that when consumers are optimistic, they are more likely to spend, particularly on big-ticket items such as cars and vacations. Hence, when consumer confidence took its sharpest-ever plunge following Iraq's invasion of Kuwait and the associated spike in oil prices in August 1990, economists and policymakers winced. As Americans became increasingly concerned both about U.S. energy security and the possibility of a ground war in the Middle East, planned aggregate expenditure and hence output fell, as analyzed by Example 13.4.

The second factor in the 1990–1991 recession, a *credit crunch,* arose from problems in the U.S. banking system. During the 1980s, many U.S. banks had made large real estate loans, taking undeveloped land or commercial real estate as collateral. When land and other real estate prices fell sharply in the late 1980s, banks suffered serious losses. Some regions of the country, such as New England, were hit especially hard. Many financially distressed banks either had no new funds to lend or were not permitted to lend by government regulators. This decline in the supply of credit from banks made credit costlier and more difficult to obtain for many borrowers, especially small and medium-sized firms. Without access to credit, these firms could not make capital investments, further reducing planned spending and output. In terms of the model presented in this chapter, a decline in planned investment spending brought about by a credit crunch can be thought of as a fall in planned investment spending $I^p$. Like the decline in $\overline{C}$ illustrated in Example 13.4, a fall in $I^p$ reduces planned expenditure and hence short-run equilibrium output.

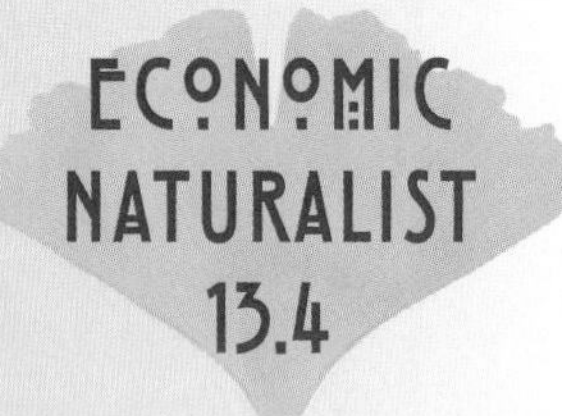


### Why was the deep Japanese recession of the 1990s bad news for the rest of East Asia?

During the 1990s Japan suffered a prolonged economic slump, still ongoing as of this writing. Japan's economic problems were a major concern not only of the Japanese but of policymakers in other East Asian countries, such as Thailand and Singapore. Why did East Asian policymakers worry about the effects of the Japanese slump on their own economies?

Although the economies of Japan and its East Asian neighbors are intertwined in many ways, one of the most important links is through trade. Much of the economic success of East Asia has been based on the development of export industries, and over the years Japan has been the most important customer for East Asian goods. When the economy slumped in the 1990s, Japanese households and firms reduced their purchases of imported goods sharply. This fall in demand dealt a major blow to the export industries of other East Asian countries.

Not just the owners and workers of export industries were affected, though; as wages and profits in export industries fell, so did domestic spending in the East Asian nations. The declines in domestic spending reduced sales at home as well as abroad, further weakening the East Asian economies. In terms of the model, the decline in exports to Japan reduced net exports $NX$, and thus autonomous expenditure, in East Asian countries. The fall in autonomous expenditure led to a recessionary gap, much like that shown in Figure 13.4.

Japan is not the only country whose economic ups and downs have had a major impact on its trading partners. Because the United States is the most important trading partner of both Canada and Mexico, the U.S. recession that began in 2001 (see Economic Naturalist 13.5) led to declining exports and recessions in Canada and Mexico as well. East Asia, which exports high-tech goods to the United States, was also hurt by the U.S recession, with GDP in countries such as Singapore dropping sharply.

**What caused the 2001 recession in the United States?**

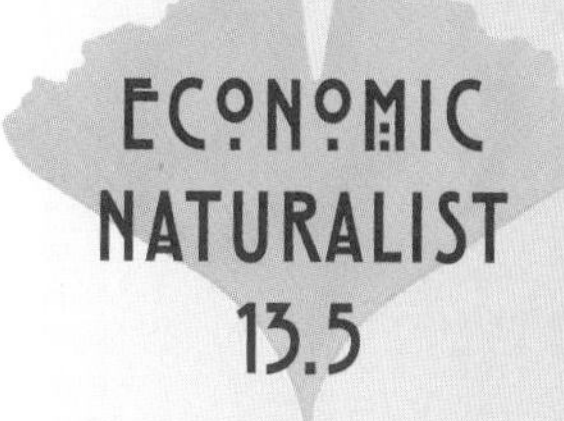


According to the National Bureau of Economic Research, a recession began in the United States in March 2001—the first U.S. recession in 10 years (see Economic Naturalist 12.1 for a discussion of the NBER's "call"). What caused the 2001 recession in the United States?

Consumer spending is nearly two-thirds of aggregate expenditure, so it should not be surprising that most recessions involve significant reductions in spending by households. In this respect, the 2001 recession in the United States was quite unusual, as consumer spending remained fairly strong throughout most of the downturn (see Economic Naturalist 13.2). Instead, this recession can be attributed primarily to a steep drop in investment spending by firms.

Why did investment expenditures fall? The period between 1995 and 2000 had been one of high rates of investment and rapid growth in the U.S. economy, fueled in large part by optimism about new technologies such as the Internet, fiber optics, and genetic engineering. However, by mid-2000 it was becoming apparent that some of the new technologies would not be as profitable as had been hoped. The prices of shares in high-tech companies fell sharply during the year and corporations cut back their investments in computers, software, telecommunications equipment, and the like. Although at first the decline was concentrated in the high-tech sector, the slowdown spread to other parts of the economy. Total employment peaked and began to decline in March 2001.

An additional shock occurred on September 11, 2001, when terrorist attacks destroyed the World Trade Center in New York City and inflicted heavy damage on the Pentagon in Washington, D.C. People became afraid to travel, and the demand for air travel, hotel rooms, and tourist services plummeted, worsening the downturn. However, consumers regained confidence surprisingly quickly—for example, spurred by generous rebates, they purchased a record number of automobiles during October. Increased government spending on security and defense also raised planned aggregate expenditure by the end of the year. The U.S. economy rebounded sharply in the first quarter of 2002.

## THE MULTIPLIER

In Example 13.4, earlier in the chapter, we analyzed a case in which the initial decline in consumer spending (as measured by the fall in $\overline{C}$) was only 10 units, and yet short-run equilibrium output fell by 50 units. Why did a relatively modest initial decline in consumer spending lead to a much larger fall in output?

The reason the impact on output was greater than the initial change in spending is the "vicious circle" effect suggested by Grandma's reminiscences about the Great Depression. Specifically, a fall in consumer spending not only reduces the sales of consumer goods directly; it also reduces the incomes of workers and owners in the industries that produce consumer goods. As their incomes fall, these workers and capital owners reduce their spending, which reduces the output and incomes of *other* producers in the economy. And these reductions in income lead to still further cuts in spending. Ultimately, these successive rounds of declines in spending and income may lead to a decrease in planned aggregate expenditure and output that is significantly greater than the change in spending that started the process.

***income-expenditure multiplier*** the effect of a 1-unit increase in autonomous expenditure on short-run equilibrium output; for example, a multiplier of 5 means that a 10-unit decrease in autonomous expenditure reduces short-run equilibrium output by 50 units

The effect on short-run equilibrium output of a 1-unit increase in autonomous expenditure is called the **income-expenditure multiplier,** or the *multiplier* for short. In the economy of Example 13.4 the multiplier is 5. That is, each 1-unit change in autonomous expenditure leads to a 5-unit change in short-run equilibrium output in the same direction (or, as we saw in Example 13.4, a 10-unit change in autonomous expenditure leads to a 50-unit change in short-run equilibrium output). The idea that a change in spending may lead to a significantly larger change in short-run equilibrium output is a key feature of the basic Keynesian model.

What determines how large the multiplier will be? An important factor is the marginal propensity to consume (MPC) out of disposable income $c$. If the MPC is large, then falls in income will cause people to reduce their spending sharply, and the multiplier effect will then also be large. If the marginal propensity to consume is small, then people will not reduce spending so much when income falls, and the multiplier will also be small. Appendix B to this chapter provides more details on the multiplier in the basic Keynesian model, including a formula that allows us to calculate the value of the multiplier under specific assumptions about the economy.

**RECAP** **FINDING SHORT-RUN EQUILIBRIUM OUTPUT**

Short-run equilibrium output is the level of output at which output equals planned aggregate expenditure, or, in symbols, $Y = PAE$. For a specific sample economy, short-run equilibrium output can be solved for numerically, as in Table 13.1, or graphically.

The graphical solution is based on a diagram called the Keynesian cross. The Keynesian cross diagram includes two lines: a 45° line that represents the condition $Y = PAE$, and the expenditure line, which shows the relationship of planned aggregate expenditure to output. Short-run equilibrium output is determined at the intersection of the two lines. If short-run equilibrium output differs from potential output, an output gap exists.

Increases in autonomous expenditure shift the expenditure line upward, increasing short-run equilibrium output; decreases in autonomous expenditure shift the expenditure line downward, leading to declines in short-run equilibrium output. Decreases in autonomous expenditure that drive actual output below potential output are a source of recessions.

Generally, a one-unit change in autonomous expenditure leads to a larger change in short-run equilibrium output, reflecting the working of the income-expenditure multiplier. The multiplier arises because a given initial increase in spending raises the incomes of producers, which leads them to spend more, raising the incomes and spending of other producers, and so on.

## STABILIZING PLANNED SPENDING: THE ROLE OF FISCAL POLICY

***stabilization policies*** government policies that are used to affect planned aggregate expenditure, with the objective of eliminating output gaps

***expansionary policies*** government policy actions intended to increase planned spending and output

According to the basic Keynesian model, inadequate spending is an important cause of recessions. To fight recessions—at least, those caused by insufficient demand rather than slow growth of potential output—policymakers must find ways to stimulate planned spending. Policies that are used to affect planned aggregate expenditure, with the objective of eliminating output gaps, are called **stabilization policies.** Policy actions intended to increase planned spending and output are called **expansionary policies;** expansionary policy actions are normally taken when the economy is in recession. It is also possible, as we have seen, for the economy to be "overheated," with output greater than potential output (an expansionary gap). The risk of an expansionary gap, as we will see in more detail later, is that it may lead to an increase in inflation. To offset an expansionary gap, policymakers will try to reduce spending and output. **Contractionary policies** are policy actions intended to reduced planned spending and output.

**contractionary policies** government policy actions designed to reduce planned spending and output

The two major tools of stabilization policy are *monetary policy* and *fiscal policy.* Recall that monetary policy refers to decisions about the size of the money supply, whereas fiscal policy refers to decisions about the government's budget—how much the government spends and how much tax revenue it collects. In the remainder of this chapter we will focus on how fiscal policy can be used to influence spending in the basic Keynesian model, as well as on some practical issues that arise in the use of fiscal policy in the real world. Monetary policy will be discussed in Chapters 14 and 15.

## GOVERNMENT PURCHASES AND PLANNED SPENDING

Decisions about government spending represent one of the two main components of fiscal policy, the other being decisions about taxes and transfer payments. As was mentioned earlier (see Box 13.1), Keynes himself felt that changes in government purchases were probably the most effective tool for reducing or eliminating output gaps. His basic argument was straightforward: Government purchases of goods and services, being a component of planned aggregate expenditure, directly affect total spending. If output gaps are caused by too much or too little total spending, then the government can help to guide the economy toward full employment by changing its own level of spending. Keynes's views seemed to be vindicated by the events of the 1930s, notably the fact that the Depression did not finally end until governments greatly increased their military spending in the latter part of the decade.

Example 13.5 shows how increased government purchases of goods and services can help to eliminate a recessionary gap. (The effects of government spending on transfer programs, such as unemployment benefits, are a bit different. We will return to that case shortly.)

### An increase in the government's purchases eliminates a recessionary gap

**EXAMPLE 13.5**

In Example 13.4, we found that a drop of 10 units in consumer spending creates a recessionary gap of 50 units. How can the government eliminate the output gap and restore full employment by changing its purchases of goods and services *G*?

In Example 13.4, we found that planned aggregate expenditure was given by the equation $PAE = 960 + 0.8Y$, so that autonomous expenditure equaled 960. The 10-unit drop in $\overline{C}$ implied a 10-unit drop in autonomous expenditure, to 950. Because the multiplier in that sample economy equaled 5, this 10-unit decline in autonomous expenditure resulted in turn in a 50-unit decline in short-run equilibrium output.

To offset the effects of the consumption decline, the government would have to restore autonomous expenditure to its original value, 960. Under our assumption that government purchases are simply given and do not depend on output, government purchases are part of autonomous expenditure, and changes in government purchases change autonomous expenditure one-for-one. Thus, to increase autonomous expenditure from 950 to 960, the government should simply increase its purchases by 10 units (for example, by increasing spending on military defense or road construction). According to the basic Keynesian model, this increase in government purchases should return autonomous expenditure and hence output to their original levels.

The effect of the increase in government purchases is shown graphically in Figure 13.5. After the 10-unit decline in the autonomous component of consumption

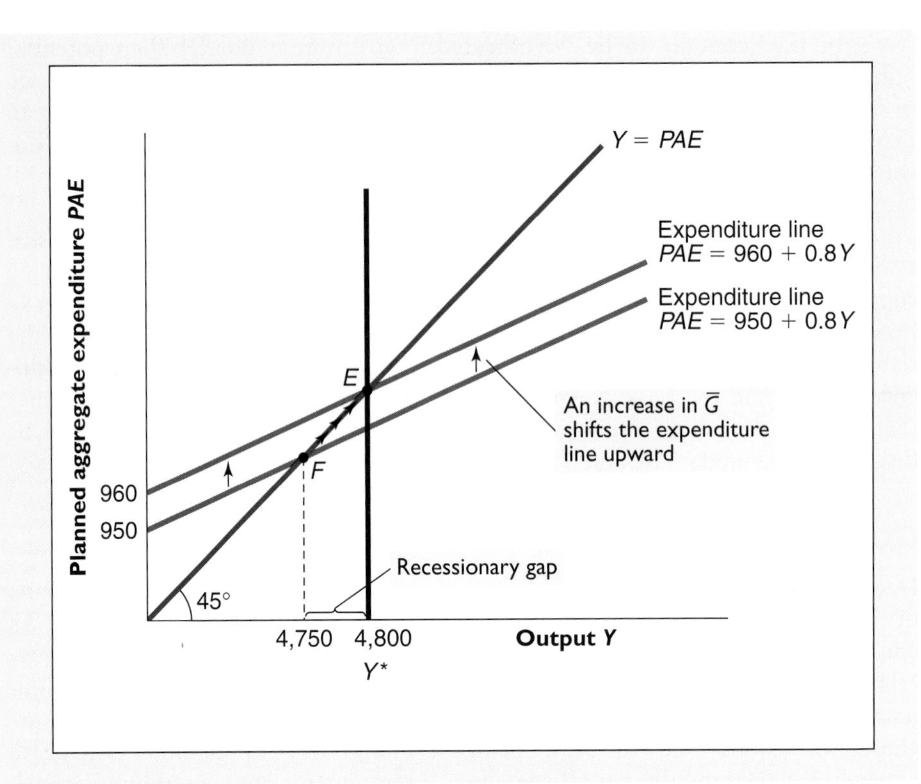


**FIGURE 13.5**
**An Increase in Government Purchases Eliminates a Recessionary Gap.**
After a 10-unit decline in the autonomous part of consumer spending $\overline{C}$, the economy is at point *F*, with a recessionary gap of 50 (see Figure 13.4). A 10-unit increase in government purchases raises autonomous expenditure by 10 units, shifting the expenditure line back to its original position and raising the equilibrium point from *F* to *E*. At point *E*, where output equals potential output ($Y = Y^* = 4{,}800$), the output gap has been eliminated.

spending $\overline{C}$, the economy is at point *F*, with a 50-unit recessionary gap. A 10-unit increase in government purchases raises autonomous expenditure by 10 units, raising the intercept of the expenditure line by 10 units and causing the expenditure line to shift upward in parallel fashion. The economy returns to point *E*, where short-run equilibrium output equals potential output ($Y = Y^* = 4{,}800$) and the output gap has been eliminated.

### EXERCISE 13.5

**In Exercise 13.4, you considered the case in which consumers become more rather than less confident, leading to an expansionary output gap. Discuss how a change in government purchases could be used to eliminate this output gap. Show your analysis graphically.**

To this point we have been considering the effect of fiscal policy on a hypothetical economy. Economic Naturalists 13.6 and 13.7 illustrate the application of fiscal policy in real economies.

ECONOMIC NATURALIST 13.6

**Why is Japan building roads nobody wants to use?**

A few years ago Japanese officials decided to build a 160-mile-long toll road on the northern island of Hokkaido. Thus far the costs of the road have been about \$60 million per mile. Very few drivers use the road, largely because an existing highway that runs parallel to the new toll road is free. Officials tried to attract drivers by offering prizes and running promotional contests. Though the campaign succeeded in increasing the average number of cars on the road to 862 per day, the route is still the least used highway in Japan (*The New York Times*, November 25, 1999, p. A1). Why is Japan building roads nobody wants to use?

As noted in Economic Naturalist 13.4, Japan spent the 1990s in a deep slump. In response, the Japanese government periodically initiated large spending programs to try to stimulate the economy. Indeed, during the 1990s the Japanese government spent more than \$1 trillion on public works projects. More than \$10 billion was spent on the Tokyo subway system alone, an amount so far over budget that subway tokens will have to cost an estimated $9.50 each if the investment is ever to be recouped. (Even more frustrating to passengers, the subway does not run in a complete circle, requiring them to make inconvenient transfers to traverse the city.) Other examples of government spending programs include the construction of multimillion-dollar concert halls in small towns, elaborate tunnels where simple roads would have been adequate, and the digging up and relaying of cobblestone sidewalks. Despite all this spending, the Japanese slump has dragged on.

The basic Keynesian model implies that increases in government spending such as those undertaken in Japan should help to increase output and employment. Japanese public works projects do appear to have stimulated the economy, though not enough to pull Japan out of recession. Why has Japan's fiscal policy proved inadequate to the task? Some critics have argued that the Japanese government was unconscionably slow in initiating the fiscal expansion, and that when spending was finally increased, it was simply not enough, relative to the size of the Japanese economy and the depth of the recession. Another possibility, which lies outside the basic Keynesian model, is that the wasteful nature of much of the government spending demoralized Japanese consumers, who realized that as taxpayers they would at some point be responsible for the costs incurred in building roads nobody wants to use. Reduced consumer confidence implies reduced consumption spending, which may to some extent have offset the stimulus from government spending. Very possibly, more productive investments of Japanese public funds would have had a greater impact on aggregate expenditure (by avoiding the fall in consumer confidence); certainly, they would have had a greater long-term benefit in terms of increasing the potential output of the economy.

### Does military spending stimulate the economy?

ECONOMIC NATURALIST 13.7

An antiwar poster from the 1960s bore the message "War is good business. Invest your son." War itself poses too many economic and human costs to be good business, but military spending could be a different matter. According to the basic Keynesian model, increases in planned aggregate expenditure resulting from stepped-up government purchases may help bring an economy out of a recession or depression. Does military spending stimulate aggregate demand? Figure 13.6 shows U.S. military spending as a share of GDP from 1940 to 2001. The shaded areas in the figure correspond to periods of recession as shown in Table 12.1.

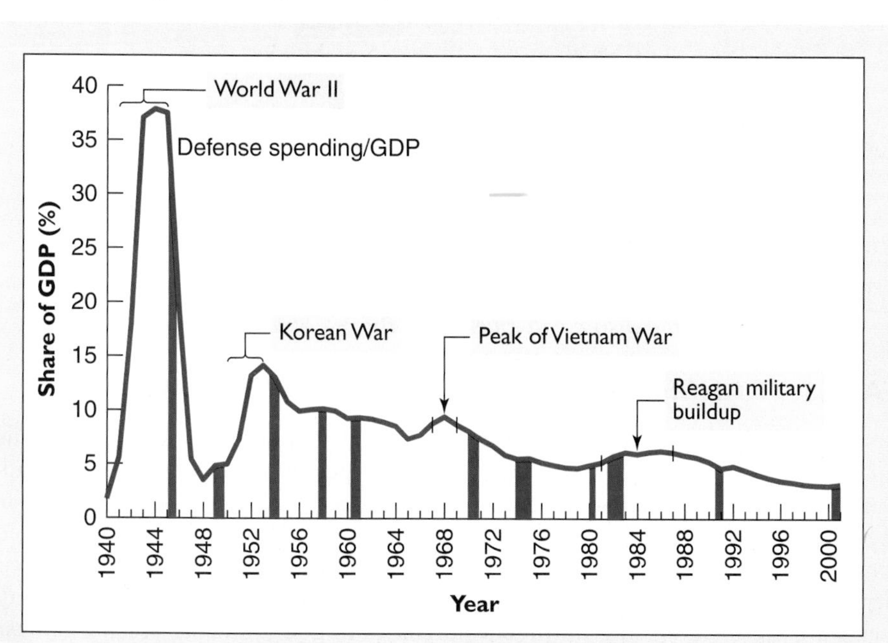


**FIGURE 13.6**
**U.S. Military Expenditures as a Share of GDP, 1940–2001.**
Military expenditures as a share of GDP rose during World War II, the Korean War, the Vietnam War, and the Reagan military buildup of the early 1980s. Increased military spending is generally associated with an expanding economy and declining unemployment. The shaded areas indicate periods of recession.

Note the spike that occurred during World War II (1941–1945), when military spending reached nearly 38 percent of U.S. GDP, as well as the surge during the Korean War (1950–1953). Smaller increases in military spending relative to GDP occurred at the peak of the Vietnam War in 1967–1969 and during the Reagan military buildup of the 1980s.

Figure 13.6 provides some support for the idea that expanded military spending tends to promote growth in aggregate demand. The clearest case is the World War II era, during which massive military spending helped the U.S. economy to recover from the Great Depression. The U.S. unemployment rate fell from 17.2 percent of the workforce in 1939 (when defense spending was less than 2 percent of GDP) to 1.2 percent in 1944 (when defense spending was greater than 37 percent of GDP). Two brief recessions, in 1945 and 1948–1949, followed the end of the war and the sharp decline in military spending. At the time, though, many people feared that the war's end would bring a resumption of the Great Depression, so the relative mildness of the two postwar recessions was something of a relief.

Increases in defense spending during the post–World War II period were also associated with economic expansions. The Korean War of 1950–1953 occurred simultaneously with a strong expansion, during which the unemployment rate dropped from 5.9 percent in 1949 to 2.9 percent in 1953. A recession began in 1954, the year after the armistice was signed, though military spending had not yet declined much. Economic expansions also occurred during the Vietnam-era military buildup in the 1960s and the Reagan buildup of the 1980s. Finally, on a smaller scale, increased government spending for military purposes and homeland security probably contributed to the relative mildness of the U.S. recession that began in 2001. These episodes support the idea that increases in government purchases—in this case, of weapons, other military supplies, and the services of military personnel—can help to stimulate the economy.

*"Your majesty, my voyage will not only forge a new route to the spices of the East but also create over three thousand new jobs."*

## TAXES, TRANSFERS, AND AGGREGATE SPENDING

Besides making decisions about government purchases of goods and services, fiscal policymakers also determine the level and types of taxes to be collected and transfer payments to be made. (Transfer payments, recall, are payments made by the government to the public, for which no current goods or services are received. Examples of transfer payments are unemployment insurance benefits, Social Security benefits, and income support payments to farmers. Once again, transfer payments are *not* included in government purchases of goods and services.) The basic Keynesian model implies that, like changes in government purchases, changes in the level of taxes or transfers can be used to affect planned aggregate expenditure and thus eliminate output gaps.

Unlike changes in government purchases, however, changes in taxes or transfers do not affect planned spending directly. Instead they work indirectly, by changing disposable income in the private sector. For example, either a tax cut or an increase in government transfer payments increases disposable income, equal to $Y - T$. According to the consumption function, when disposable income rises, households should spend more. Thus a tax cut or increase in transfers should increase planned aggregate expenditure. Likewise, an increase in taxes or a cut in transfers, by lowering households' disposable income, will tend to lower planned spending. Example 13.6 illustrates the effects of a tax cut on spending and output.

**EXAMPLE 13.6**

**Using a tax cut to close a recessionary gap**

In Example 13.4, we found that in our hypothetical economy, an initial drop in consumer spending of 10 units creates a recessionary gap of 50 units. Example 13.5 showed that this recessionary gap could be eliminated by a 10-unit increase in government purchases. Suppose that, instead of increasing government purchases, fiscal policymakers decided to stimulate consumer spending by changing the level of tax collections. By how much should they change taxes to eliminate the output gap?

A common first guess at the answer to this problem is that policymakers should cut taxes by 10, but that guess is not correct. Let's see why.

The source of the recessionary gap in Example 13.4 is the reduction that households made in their consumption spending by 10 units at each level of output $Y$—that is, the constant term $\overline{C}$ in the consumption function is assumed to have fallen 10 units. To eliminate this recessionary gap, the change in taxes must induce households to increase their consumption spending by 10 units at each output level. However, if taxes $T$ are cut by 10 units, raising disposable income $Y - T$ by 10 units, consumption at each level of output $Y$ will increase by only 8 units.

Why? The reason is that the marginal propensity to consume out of disposable income in our example is 0.8, so that consumption spending increases by only 0.8 times the amount of the tax cut. (The rest of the tax cut is saved.) An increase in autonomous expenditure of 8 units is not enough to return output to its full-employment level, in this example.

To raise consumption spending by 10 units, fiscal policymakers must instead cut taxes by 12.5 units. Because $0.8(12.5) = 10$, a tax cut of 12.5 will spur households to increase their consumption by 10 units at each level of output. That increase will just offset the 10-unit decrease in $\overline{C}$ in this example, restoring the economy to full employment.

Note that since $T$ refers to *net taxes,* or taxes less transfers, the same result could be obtained by increasing transfer payments by 12.5 units. Because households spend 0.8 times any increase in transfer payments they receive, this policy would also raise consumption spending by 10 units at any level of output.

Graphically, the effect of the tax cut is identical to the effect of the increase in government purchases, shown in Figure 13.5. Because it leads to a 10-unit increase in consumption at any level of output, the tax cut shifts the expenditure line up by 10 units. Equilibrium is attained at point $E$ in Figure 13.5, where output again equals potential output.

**EXERCISE 13.6**

**In a particular economy, a 20-unit increase in planned investment moved the economy from an initial situation with no output gap to a situation with an expansionary gap. Describe two ways in which fiscal policy could be used to offset this expansionary gap. Assume the marginal propensity to consume equals 0.5.**

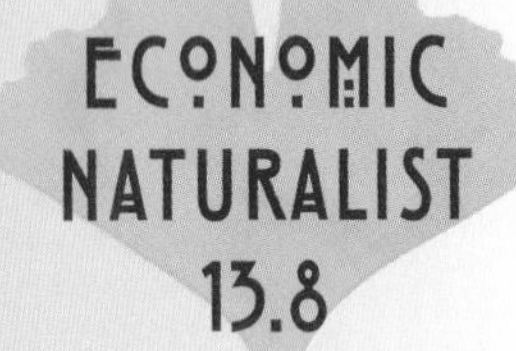


**Why did the federal government send out millions of $300 and $600 checks to households in 2001?**

On May 25, 2001, Congress passed the Economic Growth and Tax Relief Reconciliation Act (EGTRRA) of 2001, which President George W. Bush signed on June 7. The EGTRRA made significant cuts in income tax rates and also provided for one-time tax rebate checks of up to $300 for individual taxpayers and up to $600 for married taxpayers filing a joint return. Millions of families received these checks in August and September of 2001, with payments totaling about $38 billion. Why did the federal government send out these checks?

Although the 2001 recession was not officially "declared" until November 2001 (when the National Bureau of Economic Research announced that the recession had begun in March), there was clear evidence by the spring of 2001 that the economy was slowing. Congress and the president hoped that by sending tax rebate checks to households, they could stimulate spending and perhaps avoid recession. In retrospect, the timing of the tax rebate was quite good, since the economy and consumer confidence were further buffeted by the terrorist attacks on New York City and Washington on September 11, 2001.

Did the tax rebates have their intended effect of stimulating consumer spending? It is difficult to know with any certainty, since we do not know how much households would have spent if they had not received these extra funds. A University of Michigan survey conducted in August, September, and October 2001 found that only 22 percent of households anticipated spending most of their rebates.[4] If this number is accurate, it would suggest the rebates had a relatively small effect on planned spending. On the other hand, consumer spending held up remarkably well during the last quarter of 2001 and into 2002, assisting the economy's recovery substantially. Most economists would agree that fiscal policy generally—including not only the tax rebates, but also significantly increased spending for the military and for domestic security following September 11—was an important reason that the 2001 recession was relatively short and mild.

**RECAP FISCAL POLICY AND PLANNED SPENDING**

Fiscal policy includes two general tools for affecting total spending and eliminating output gaps: (1) changes in government purchases and (2) changes in taxes or transfer payments. An increase in government purchases increases autonomous expenditure by an equal amount. A reduction in taxes or an

[4]Matthew D. Shapiro and Joel Slemrod, *Consumer Response to Tax Rebates,* National Bureau of Economic Research Working Paper 8672, December 2001.

increase in transfer payments increases autonomous expenditure by an amount equal to the marginal propensity to consume times the reduction in taxes or increase in transfers. The ultimate effect of a fiscal policy change on short-run equilibrium output equals the change in autonomous expenditure times the multiplier. Accordingly, if the economy is in recession, an increase in government purchases, a cut in taxes, or an increase in transfers can be used to stimulate spending and eliminate the recessionary gap.

## FISCAL POLICY AS A STABILIZATION TOOL: THREE QUALIFICATIONS

The basic Keynesian model might lead you to think that precise use of fiscal policy can eliminate output gaps. But as is often the case, the real world is more complicated than economic models suggest. We close the chapter with three qualifications about the use of fiscal policy as a stabilization tool.

### FISCAL POLICY AND THE SUPPLY SIDE

We have focused so far on the use of fiscal policy to affect planned aggregate expenditure. However, most economists would agree that *fiscal policy may affect potential output as well as planned aggregate expenditure.* On the spending side, for example, investments in public capital, such as roads, airports, and schools, can play a major role in the growth of potential output, as we discussed in Chapter 7. On the other side of the ledger, tax and transfer programs may well affect the incentives, and thus the economic behavior, of households and firms. For example, a high tax rate on interest income may reduce the willingness of people to save for the future, while a tax break on new investment may encourage firms to increase their rate of capital formation. Such changes in saving or investment will in turn affect potential output. Many other examples could be given of how taxes and transfers affect economic behavior and thus possibly affect potential output as well.

Some critics of the Keynesian theory have gone so far as to argue that the *only* effects of fiscal policy that matter are effects on potential output. This was essentially the view of the so-called *supply-siders,* a group of economists and journalists whose influence reached a high point during the first Reagan term (1981–1985). Supply-siders focused on the need for tax cuts, arguing that lower tax rates would lead people to work harder (because they would be allowed to keep a larger share of their earnings), to save more, and to be more willing to innovate and take risks. Through their arguments that lower taxes would substantially increase potential output, with no significant effect on planned spending, the supply-siders provided crucial support for the large tax cuts that took place under the Reagan administration. Supply-sider ideas also were used to support the long-term income tax cut passed under President George W. Bush in 2001.

A more balanced view is that fiscal policy affects *both* planned spending *and* potential output. Thus, in making fiscal policy, government officials should take into account not only the need to stabilize planned aggregate expenditure but also the likely effects of government spending, taxes, and transfers on the economy's productive capacity.

### THE PROBLEM OF DEFICITS

A second consideration for fiscal policymakers thinking about stabilization policies is *the need to avoid large and persistent budget deficits.* Recall from Chapter 9 that the government's budget deficit is the excess of government spending over tax collections. Sustained government deficits can be harmful because they reduce national saving, which in turn reduces investment in new capital goods—an important source of long-run economic growth. The need to keep deficits under control may make increasing spending or cutting taxes to fight a slowdown a less attractive option, both economically and politically. For example, Japan has substantially increased government spending over the past decade in its attempts to stimulate its lagging economy (see Economic Naturalist 13.6). The Japanese government's budget deficit has in the process become so large that the Japanese prime minister ruled out additional fiscal stimulus until the deficit can be brought under better control.

## THE RELATIVE INFLEXIBILITY OF FISCAL POLICY

The third qualification about the use of fiscal policy is that *fiscal policy is not always flexible enough to be useful for stabilization.* Our examples have implicitly assumed that the government can change spending or taxes relatively quickly in order to eliminate output gaps. In reality, changes in government spending or taxes must usually go through a lengthy legislative process, which reduces the ability of fiscal policy to respond in a timely way to economic conditions. For example, budget and tax changes proposed by the president must typically be submitted to Congress 18 months or more before they go into effect. Another factor that limits the flexibility of fiscal policy is that fiscal policymakers have many other objectives besides stabilizing aggregate spending, from ensuring an adequate national defense to providing income support to the poor. What happens if, say, the need to strengthen the national defense requires an increase in government spending, but the need to contain planned aggregate expenditure requires a decrease in government spending? Such conflicts can be difficult to resolve through the political process.

This lack of flexibility means that fiscal policy is less useful for stabilizing spending than the basic Keynesian model suggests. Nevertheless, most economists view fiscal policy as an important stabilizing force, for two reasons. The first is the presence of **automatic stabilizers,** provisions in the law that imply *automatic* increases in government spending or decreases in taxes when real output declines. For example, some government spending is earmarked as "recession aid"; it flows to communities automatically when the unemployment rate reaches a certain level. Taxes and transfer payments also respond automatically to output gaps: When GDP declines, income tax collections fall (because households' taxable incomes fall) while unemployment insurance payments and welfare benefits rise—all without any explicit action by Congress. These automatic changes in government spending and tax collections help to increase planned spending during recessions and reduce it during expansions, without the delays inherent in the legislative process.

***automatic stabilizers*** provisions in the law that imply *automatic* increases in government spending or decreases in taxes when real output declines

The second reason that fiscal policy is an important stabilizing force is that although fiscal policy may be difficult to change quickly, it may still be useful for dealing with prolonged episodes of recession. The Great Depression of the 1930s and the Japanese slump of the 1990s are two cases in point. However, because of the relative lack of flexibility of fiscal policy, in modern economies aggregate spending is more usually stabilized through monetary policy. The stabilizing role of monetary policy is the subject of the next chapter.

## SUMMARY

- The basic Keynesian model shows how fluctuations in planned aggregate expenditure, or total planned spending, can cause actual output to differ from potential output. Too little spending leads to a recessionary output gap; too much spending creates an expansionary output gap. This model relies on the crucial assumption that firms do not respond to every change in demand by changing prices. Instead, they typically set a price for some period, then meet the demand forthcoming at that price. Firms do not change prices continually because changing prices entails costs, called *menu costs.*

- *Planned aggregate expenditure* is total planned spending on final goods and services. The four components of total spending are consumption, investment, government purchases, and net exports. Planned and actual consumption, government purchases, and net exports are generally assumed to be the same. Actual investment may differ from planned investment, because firms may sell a greater or lesser amount of their production than they expected. If firms sell less than they expected, for example, they are forced to add more goods to inventory than anticipated. And because additions to inventory are counted as part of investment, in this case actual investment (including inventory investment) is greater than planned investment.

- Consumption is related to disposable, or after-tax, income by a relationship called the *consumption function.* The amount by which desired consumption rises when disposable income rises by one dollar is called the *marginal propensity to consume* (*MPC*, or $c$). The marginal propensity to consume is always greater than zero but less than one (that is, $0 < c < 1$).

- An increase in real output raises planned aggregate expenditure, since higher output (and, equivalently, higher income) encourages households to consume more. Planned aggregate expenditure can be broken down into two components, autonomous expenditure and induced expenditure. *Autonomous expenditure* is the portion of planned spending that is independent of output; *induced expenditure* is the portion of spending that depends on output.

- In the period in which prices are fixed, *short-run equilibrium output* is the level of output that just equals planned aggregate expenditure. Short-run equilibrium can be determined numerically by a table which compares alternative values of output and the planned spending implied by each level of output. Short-run equilibrium output can also be determined graphically in a Keynesian cross diagram, drawn with planned aggregate expenditure on the vertical axis and output on the horizontal axis. The Keynesian cross contains two lines: an expenditure line, which relates planned aggregate expenditure to output, and a 45° line, which represents the condition that short-run equilibrium output equals planned aggregate expenditure. Short-run equilibrium output is determined at the point at which these two lines intersect.

- Changes in autonomous expenditure will lead to changes in short-run equilibrium output. In particular, if the economy is initially at full employment, a fall in autonomous expenditure will create a recessionary gap and a rise in autonomous expenditure will create an expansionary gap. The amount by which a one-unit increase in autonomous expenditure raises short-run equilibrium output is called the *multiplier.* An increase in autonomous expenditure not only raises spending directly; it also raises the incomes of producers, who in turn increase their spending, and so on. Hence the multiplier is greater than 1; that is, a one-dollar increase in autonomous expenditure tends to raise short-run equilibrium output by more than one dollar.

- To eliminate output gaps and restore full employment, the government employs *stabilization policies.* The two major types of stabilization policy are monetary policy and fiscal policy. Stabilization policies work by changing planned aggregate expenditure and hence short-run equilibrium output. For example, an increase in government purchases raises autonomous expenditure directly, so it can be used to reduce or eliminate a recessionary gap. Similarly, a cut in taxes or an increase in transfer payments increases the public's disposable income, raising consumption spending at each level of output by an amount equal to the marginal propensity to consume times the cut in taxes or increase in transfers. Higher consumer spending, in turn, raises short-run equilibrium output.

- Three qualifications must be made to the use of fiscal policy as a stabilization tool. First, fiscal policy may affect potential output as well as aggregate spending. Second, large and persistent government budget deficits reduce national saving and growth; the need to keep deficits under control may limit the use of expansionary fiscal policies. Finally, because changes in fiscal policy must go through a lengthy legislative process, fiscal policy is not always flexible enough to be useful for short-run stabilization. However, *automatic stabilizers*—provisions in the law that imply automatic increases in government spending or reductions in taxes when output declines—can overcome the problem of legislative delays to some extent and contribute to economic stability.

## KEY TERMS

automatic stabilizers (354)
autonomous expenditure (337)
consumption function (334)
contractionary policies (347)
expansionary policies (346)
induced expenditure (337)
income-expenditure multiplier (346)
marginal propensity to consume (MPC), or $c$ (335)
menu costs (330)
planned aggregate expenditure (*PAE*) (331)
short-run equilibrium output (338)
stabilization policies (346)
wealth effect (334)

## REVIEW QUESTIONS

1. What is the key assumption of the basic Keynesian model? Explain why this assumption is needed if one is to accept the view that aggregate spending is a driving force behind short-term economic fluctuations.
2. Give an example of a good or service whose price changes very frequently and one whose price changes relatively infrequently. What accounts for the difference?
3. Define *planned aggregate expenditure* and list its components. Why does planned spending change when output changes?
4. Explain how planned spending and actual spending can differ. Illustrate with an example.
5. Sketch a graph of the consumption function, labeling the axes of the graph. Discuss the economic meaning of (a) a movement from left to right along the graph of the consumption function; and (b) a parallel upward shift of the consumption function. Give an example of a factor that could lead to a parallel upward shift of the consumption function.
6. Sketch the Keynesian cross diagram. Explain in words the economic significance of the two lines graphed in the diagram. Given only this diagram, how could you determine autonomous expenditure, induced expenditure, the marginal propensity to consume, and short-run equilibrium output?
7. Using the Keynesian cross diagram, illustrate the two causes of the 1990–1991 recession discussed in Economic Naturalist 13.3.
8. Define the *multiplier*. In economic terms, why is the multiplier greater than 1?
9. The government is considering two alternative policies, one involving increased government purchases of 50 units, the other involving a tax cut of 50 units. Which policy will stimulate planned aggregate expenditure by more? Why?
10. Discuss three reasons why the use of fiscal policy to stabilize the economy is more complicated than suggested by the basic Keynesian model.

## PROBLEMS

1. Acme Manufacturing is producing $4,000,000 worth of goods this year and expects to sell its entire production. It is also planning to purchase $1,500,000 in new equipment during the year. At the beginning of the year the company has $500,000 in inventory in its warehouse. Find actual investment and planned investment if
   a. Acme actually sells $3,850,000 worth of goods.
   b. Acme actually sells $4,000,000 worth of goods.
   c. Acme actually sells $4,200,00 worth of goods.

   Assuming that Acme's situation is similar to that of other firms, in which of these three cases is output equal to short-run equilibrium output?
2. Data on before-tax income, taxes paid, and consumption spending for the Simpson family in various years are given below.

| Before-tax income ($) | Taxes paid ($) | Consumption spending ($) |
|---|---|---|
| 25,000 | 3,000 | 20,000 |
| 27,000 | 3,500 | 21,350 |
| 28,000 | 3,700 | 22,070 |
| 30,000 | 4,000 | 23,600 |

   a. Graph the Simpsons' consumption function and find their household's marginal propensity to consume.
   b. How much would you expect the Simpsons to consume if their income was $32,000 and they paid taxes of $5,000?
   c. Homer Simpson wins a lottery prize. As a result, the Simpson family increases its consumption by $1,000 at each level of after-tax income. ("Income" does not include the prize money.) How does this change affect the graph of their consumption function? How does it affect their marginal propensity to consume?

3. An economy is described by the following equations:

$$C = 1800 + 0.6(Y - T)$$
$$I^p = 900$$
$$G = 1{,}500$$
$$NX = 100$$
$$T = 1{,}500$$
$$Y^* = 9{,}000$$

   a. Find a numerical equation linking planned aggregate expenditure to output.
   b. Find autonomous expenditure and induced expenditure in this economy.

4. For the economy described in Problem 3:
   a. Construct a table like Table 13.1 to find short-run equilibrium output. Consider possible values for short-run equilibrium output ranging from 8,200 to 9,000.
   b. Show the determination of short-run equilibrium output for this economy using the Keynesian cross diagram.
   c. What is the output gap for this economy? If the natural rate of unemployment is 4 percent, what is the actual unemployment rate for this economy (use Okun's law)?

5. For the economy described in Problem 3, find the effect on short-run equilibrium output of
   a. An increase in government purchases from 1,500 to 1,600.
   b. A decrease in tax collections from 1,500 to 1,400 (leaving government purchases at their original value).
   c. A decrease in planned investment spending from 900 to 800.

   Assume each of these changes is a change in autonomous expenditure. Take as given that the multiplier for this economy is 2.5. If you have studied Appendix B in this chapter, show why this is so.

6. An economy is initially at full employment, but a decrease in planned investment spending (a component of autonomous expenditure) pushes the economy into recession. Assume that the MPC of this economy is 0.75 and that the multiplier is 4.
   a. How large is the recessionary gap after the fall in planned investment?
   b. By how much would the government have to change its purchases to restore the economy to full employment?
   c. Alternatively, by how much would the government have to change taxes?
   d.* Suppose that the government's budget is initially in balance, with government spending equal to taxes collected. A balanced-budget law forbids the government from running a deficit. Is there anything that fiscal policymakers could do to restore full employment in this economy, assuming they do not want to violate the balanced-budget law?

7. An economy is described by the following equations:

$$C = 40 + 0.8(Y - T)$$
$$I^p = 70$$
$$G = 120$$
$$NX = 10$$
$$T = 150$$
$$Y^* = 580$$

   The multiplier in this economy is 5.

   a. Find a numerical equation relating planned aggregate expenditure to output.
   b. Construct a table to find the value of short-run equilibrium output. (*Hint:* The economy is fairly close to full employment.)
   c. By how much would government purchases have to change in order to eliminate any output gap? By how much would taxes have to change? Show the effects of these fiscal policy changes in a Keynesian cross diagram.
   d. Repeat part c assuming that $Y^* = 630$.

*Problems marked by an asterisk (*) are more difficult.

8.* For the following economy, find autonomous expenditure, the multiplier, short-run equilibrium output, and the output gap. By how much would autonomous expenditure have to change to eliminate the output gap?

$$C = 3{,}000 + 0.5(Y - T)$$
$$I^p = 1{,}500$$
$$G = 2{,}500$$
$$NX = 200$$
$$T = 2{,}000$$
$$Y^* = 12{,}000$$

9.* An economy has zero net exports. Otherwise, it is identical to the economy described in Problem 7.

a. Find short-run equilibrium output.
b. Economic recovery abroad increases the demand for the country's exports; as a result, $NX$ rises to 100. What happens to short-run equilibrium output?
c. Repeat part b, but this time assume that foreign economies are slowing, reducing the demand for the country's exports, so that $NX = -100$. (A negative value of net exports means that exports are less than imports.)
d. How do your results help to explain the tendency of recessions and expansions to spread across countries?

10.* This problem illustrates the workings of automatic stabilizers. Suppose that the components of planned spending in an economy are as described in Appendix A: $C = \overline{C} + c(Y - T)$, $I^p = \overline{I}$, $G = \overline{G}$, and $NX = \overline{NX}$. However, suppose that, realistically, taxes are not fixed but depend on income. Specifically, we assume

$$T = tY,$$

where $t$ is the fraction of income paid in taxes (the tax rate). As we will see in this problem, a tax system of this sort serves as an automatic stabilizer, because taxes collected automatically fall when incomes fall.

a. Find an algebraic expression for short-run equilibrium output in this economy.
b. Find an algebraic expression for the multiplier, that is, the amount that output changes when autonomous expenditure changes by 1 unit. Compare the expression you found to the formula for the multiplier when taxes are fixed. Show that making taxes proportional to income reduces the multiplier.
c. Explain how reducing the size of the multiplier helps to stabilize the economy, holding constant the typical size of fluctuations in the components of autonomous expenditure.
d. Suppose $\overline{C} = 500$, $\overline{I} = 1{,}500$, $\overline{G} = 2{,}000$, $\overline{NX} = 0$, $c = 0.8$, and $t = 0.25$. Calculate numerical values for short-run equilibrium output and the multiplier.

## ANSWERS TO IN-CHAPTER EXERCISES

13.1 First we need to find an equation that relates planned aggregate expenditure $PAE$ to output $Y$. We start with the definition of planned aggregate expenditure and then substitute the numerical values given in the problem:

$$\begin{aligned} PAE &= C + I^p + G + NX \\ &= [\overline{C} + c(Y - T)] + I^p + G + NX \\ &= [820 + 0.7(Y - 600)] + 600 + 600 + 200 \\ &= 1{,}800 + 0.7Y. \end{aligned}$$

*Problems marked by an asterisk (*) are more difficult.

Using this relationship we construct a table analogous to Table 13.1. Some trial and error is necessary to find an appropriate range of guesses for output (column 1).

**Determination of short-run equilbrium output**

| (1) Output $Y$ | (2) Planned aggregate expenditure $PAE = 1{,}800 + 0.7Y$ | (3) $Y - PAE$ | (4) $Y = PAE$? |
|---|---|---|---|
| 5,000 | 5,300 | −300 | No |
| 5,200 | 5,440 | −240 | No |
| 5,400 | 5,580 | −180 | No |
| 5,600 | 5,720 | −120 | No |
| 5,800 | 5,860 | −60 | No |
| 6,000 | 6,000 | 0 | **Yes** |
| 6,200 | 6,140 | 60 | No |
| 6,400 | 6,280 | 120 | No |
| 6,600 | 6,420 | 180 | No |

Short-run equilibrium output equals 6,000, as that is the only level of output that satisfies the condition $Y = PAE$.

13.2 The graph shows the determination of short-run equilibrium output, $Y = 6{,}000$. The intercept of the expenditure line is 1,800 and its slope is 0.7. Notice that the intercept equals autonomous expenditure and the slope equals the marginal propensity to consume.

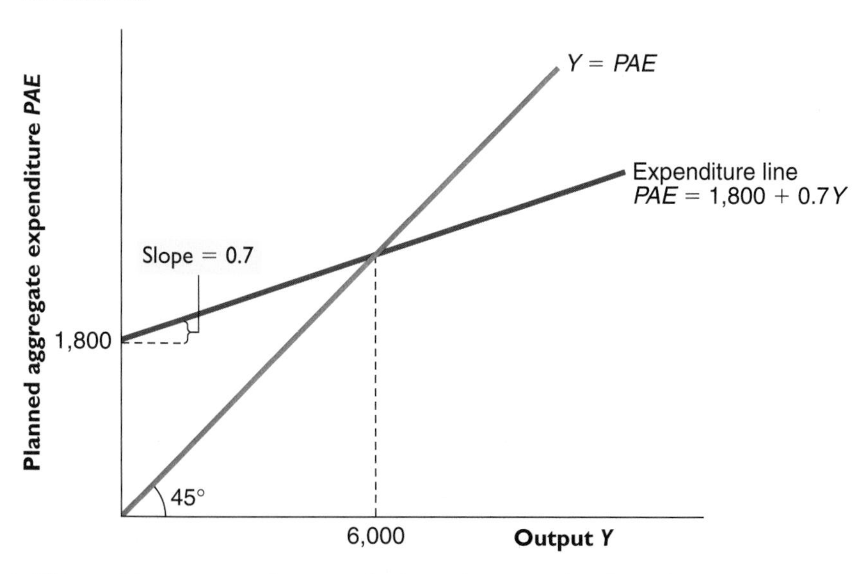


13.3 This problem is an application of Okun's law, introduced in Chapter 12. The recessionary gap in this example is 50/4,800, or about 1.04 percent, of potential output. By Okun's law, cyclical unemployment is one-half the percentage size of the output gap, or 0.52 percent. As the natural rate of unemployment is 5 percent, total unemployment rate after the recessionary gap appears will be approximately 5.52 percent.

13.4 This exercise is just the reverse of Example 13.4. An increase in $\overline{C}$ of 10 units raises autonomous expenditure and hence the intercept of the expenditure line by 10 units. The expenditure line shifts up, in parallel fashion, by 10 units, leading to an increase in output and an expansionary output gap. As output falls by 50 units in Example 13.4, it rises by 50 units, to 4,850, in the case analyzed here. To verify that short-run equilibrium output equals 4,850, note that an increase of 10 units in autonomous expenditure implies that $PAE$ rises from $960 + 0.8Y$ to $970 + 0.8Y$. When $Y = 4{,}850$, then $PAE = 970 + 0.8(4{,}850) = 4{,}850$, so that we have $Y = PAE$.

13.5 In Exercise 13.4 we saw that a 10-unit increase in $\overline{C}$ increases autonomous expenditure and hence the intercept of the expenditure line by 10 units. The expenditure line shifts upward, in parallel fashion, by 10 units, leading to an expansionary output gap. To offset this gap, the government should reduce its purchases by 10 units, returning autonomous expenditure to its original level. The expenditure line shifts back down to its original position, restoring output to its initial full-employment level. The graph is just the reverse of Figure 13.5, with the expenditure line being shifted up by the increase in consumption and down by the offsetting reduction in government purchases.

13.6 The 20-unit increase in planned investment is a 20-unit increase in autonomous expenditure, which will lead to an even greater increase in short-run equilibrium output. To offset the 20-unit increase in autonomous expenditure by means of fiscal policy, the government can reduce its purchases by 20 units. Alternatively, it could raise taxes (or cut transfers) to reduce consumption spending. Since the $MPC = 0.5$, to reduce consumption spending by 20 units at each level of output, the government will need to increase taxes (or reduce transfers) by 40 units. At each level of output, a 40-unit tax increase will reduce disposable income by 40 units and cause consumers to reduce their spending by $0.5 \times 40 = 20$ units, as needed to eliminate the expansionary output gap.

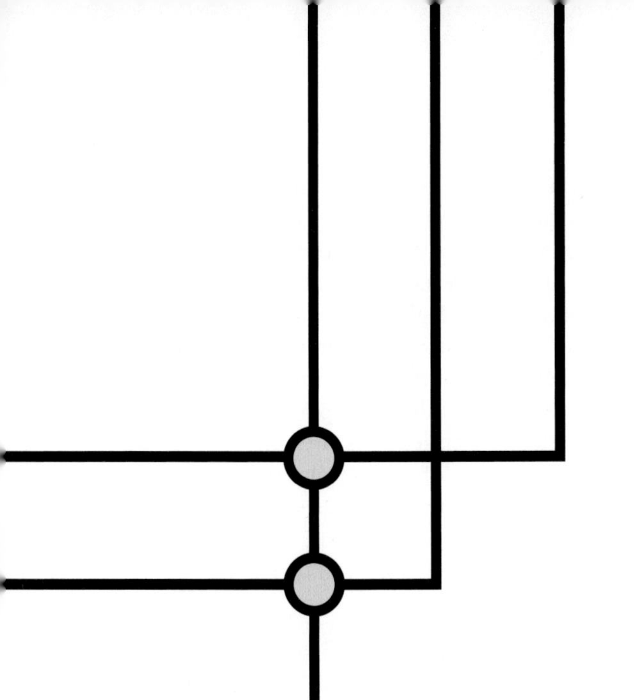

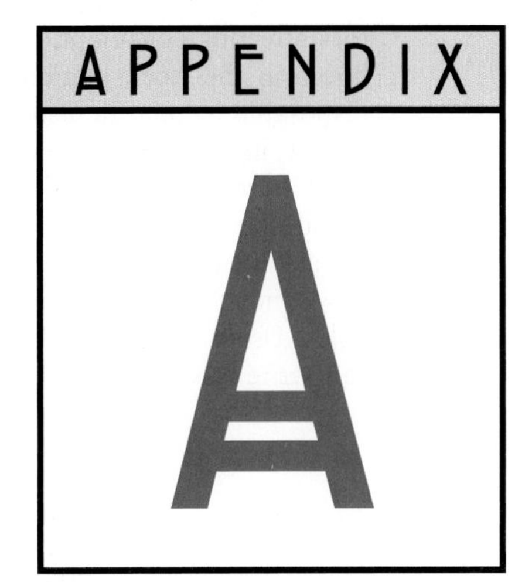


# AN ALGEBRAIC SOLUTION OF THE BASIC KEYNESIAN MODEL

This chapter has shown how to solve the basic Keynesian model numerically and graphically, using the Keynesian cross diagram. In this appendix we will show how to find a more general algebraic solution for short-run equilibrium output in the basic Keynesian model. This solution has the advantage of showing clearly the links between short-run equilibrium output, the multiplier, and autonomous expenditure. The general method can also be applied when we make changes to the basic Keynesian model, as we will see in following chapters.

The model we will work with is the same one presented in the main part of the chapter. Start with the definition of planned aggregate expenditure, Equation 13.1:

$$PAE = C + I^p + G + NX. \tag{13.1}$$

Equation 13.1 says that planned aggregate expenditure is the sum of the four types of planned spending: consumption spending by households $C$; planned investment spending by firms $I^p$; government purchases $G$; and net exports purchased by foreigners $NX$.

The first component of planned aggregate expenditure, consumption spending, is determined by the *consumption function,* Equation 13.2:

$$C = \overline{C} + c(Y - T). \tag{13.2}$$

The consumption function says that consumption spending increases when disposable (after-tax) income $Y - T$ increases. Each dollar increase in disposable income raises consumption spending by $c$ dollars, where $c$, known as the *marginal propensity to consume,* is a number between 0 and 1. Other factors affecting consumption spending are captured by the term $\overline{C}$. For example, a boom in the stock market that leads consumers to spend more at each level of disposable income (a *wealth effect*) would be represented as an increase in $\overline{C}$.

As in the body of the chapter, we assume that planned investment, government purchases, net exports, and net tax collections are simply given numbers. A variable whose value is fixed and given from outside the model is called an *exogenous* variable; so, in other words, we are assuming that planned investment, government purchases, net exports, and net tax collections are exogenous variables. Using an overbar to denote the given value of an exogenous variable, we can write this assumption as

$I^p = \overline{I}$ Planned investment,

$G = \overline{G}$ Government purchases,

$NX = \overline{NX}$ Net exports,

$T = \overline{T}$ Net taxes (taxes less transfers).

So, for example, $\overline{I}$ is the given value of planned investment spending, as determined outside the model. In our examples we will set $\overline{I}$ and the other exogenous variables equal to some particular number.

Our goal is to solve algebraically for *short-run equilibrium output,* the level of output that prevails during the period in which prices are predetermined. The first step is to relate planned aggregate expenditure *PAE* to output *Y*. Starting with the definition of planned aggregate expenditure (Equation 13.1), use the consumption function (Equation 13.2) to substitute for consumption spending *C* and replace $I^p$, *G*, *NX*, and *T* with their exogenous values. With these substitutions, planned aggregate expenditure can be written as

$$PAE = [\overline{C} - c(Y - \overline{T})] + \overline{I} + \overline{G} + \overline{NX}.$$

Rearranging this equation to separate the terms that do and do not depend on output *Y*, we get

$$PAE = [\overline{C} - c\overline{T} + \overline{I} + \overline{G} + \overline{NX}] + cY. \qquad (13A.1)$$

Equation 13A.1 is an important equation, because it shows the relationship between planned aggregate expenditure *PAE* and output *Y*. The bracketed term on the right side of the equation represents *autonomous expenditure,* the part of planned spending that does not depend on output. The term $cY$ represents *induced expenditure,* the part of planned spending that does depend on output. Equation 13A.1 is also the equation that describes the *expenditure line* in the Keynesian cross diagram; it shows that the intercept of the expenditure line equals autonomous expenditure and the slope of the expenditure line equals the marginal propensity to consume.

We can illustrate how Equation 13A.1 works numerically by using Example 13.2 in the text. That example assumed the following numerical values: $\overline{C} = 620$, $\overline{I} = 220$, $\overline{G} = 300$, $\overline{NX} = 20$, $\overline{T} = 250$, and $c = 0.8$. Plugging these values into Equation 13A.1 and simplifying, we get

$$PAE = 960 + 0.8Y,$$

which is the same answer we found in Example 13.2. Autonomous expenditure in this example equals 960, and induced expenditure equals $0.8Y$.

The second step in solving for short-run equilibrium output begins with the definition of short-run equilibrium output (Equation 13.3)

$$Y = PAE.$$

Remember that short-run equilibrium output is the value of output at which output equals planned aggregate expenditure. Using Equation 13A.1 above to substitute for *PAE* in the definition of short-run equilibrium output, we get

$$Y = [\overline{C} - c\overline{T} + \overline{I} + \overline{G} + \overline{NX}] + cY.$$

The value of *Y* that solves this equation is the value of short-run equilibrium output. To solve for *Y*, group all terms involving *Y* on the left side of the equation:

$$Y - cY = [\overline{C} - c\overline{T} + \overline{I} + \overline{G} + \overline{NX}]$$

or

$$Y(1 - c) = [\overline{C} - c\overline{T} + \overline{I} + \overline{G} + \overline{NX}].$$

Dividing both sides of the equation by $(1 - c)$ gives

$$Y = \left(\frac{1}{1-c}\right)[\overline{C} - c\overline{T} + \overline{I} + \overline{G} + \overline{NX}]. \qquad (13A.2)$$

Equation 13A.2 gives short-run equilibrium output for our model economy in terms of the exogenous values $\overline{C}$, $\overline{I}$, $\overline{G}$, $\overline{NX}$, and $\overline{T}$ and the marginal propensity to consume, $c$. We can use this formula to solve for short-run equilibrium output in specific numerical examples. For example, suppose that we once again plug in the numerical values assumed in Example 13.2: $\overline{C} = 620$, $\overline{I} = 220$, $\overline{G} = 300$, $\overline{NX} = 20$, $\overline{T} = 250$, and $c = 0.8$. We get

$$Y = \left(\frac{1}{1-0.8}\right)[620 - 0.8(250) + 220 + 300 + 20] = \frac{1}{0.2}(960) = 5(960) = 4{,}800,$$

which is the same answer we found more laboriously using Table 13.1.

**EXERCISE 13A.1**

**Use Equation 13A.2 to find short-run equilibrium output for the economy described in Exercise 13.1 in the text. What are the intercept and the slope of the expenditure line?**

Equation 13A.2 shows clearly the relationship between autonomous expenditure and short-run equilibrium output. Autonomous expenditure is the first term on the right side of Equation 13A.1, equal to $\overline{C} - c\overline{T} + \overline{I} + \overline{G} + \overline{NX}$. The equation shows that a one-unit increase in autonomous expenditure increases short-run equilibrium output by $1/(1 - c)$ units. In other words, we can see from Equation 13A.2 that the *multiplier* for this model equals $1/(1 - c)$. Further discussion of the multiplier is given in the second appendix to the chapter.

## ANSWERS TO IN-APPENDIX EXERCISE

13A.1 The equation describing short-run equilibrium output is

$$Y = \left(\frac{1}{1-c}\right)(\overline{C} - c\overline{T} + \overline{I} + \overline{G} + \overline{NX}). \qquad (13A.2)$$

Using data from Exercise 13.1, set $\overline{C} = 820$, $c = 0.7$, $\overline{I} = 600$, $\overline{G} = 600$, $\overline{NX} = 200$, and $\overline{T} = 600$. Plugging these values into Equation (13.A.2) we get

$$Y = \left(\frac{1}{1 - 0.7}\right) [820 - 0.7(600 + 600 + 600 + 200)] = 3.33 \times 1{,}800 = 6{,}000,$$

which is the same result obtained in Exercise 13.1.

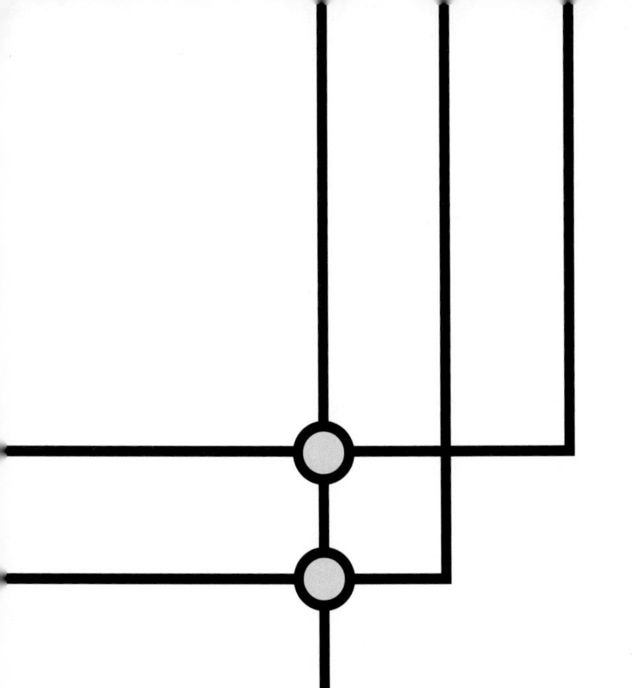

APPENDIX

B

# THE MULTIPLIER IN THE BASIC KEYNESIAN MODEL

This appendix builds on Example 13.4 in the text to give a more complete explanation of the *income-expenditure multiplier* in the basic Keynesian model. In Example 13.4, we saw that a drop in autonomous expenditure of 10 units caused a decline in short-run equilibrium output of 50 units, five times as great as the initial change in spending. Hence the multiplier in this example is 5.

To see why this multiplier effect occurs, note that the initial decrease of 10 in consumer spending (more precisely, in the constant term of the consumption function, $\overline{C}$) in Example 13.4 has two effects. First, the fall in consumer spending directly reduces planned aggregate expenditure by 10 units. Second, the fall in spending also reduces by 10 units the incomes of producers (workers and firm owners) of consumer goods. Under the assumption of Example 13.4 that the marginal propensity to consume is 0.8, the producers of consumer goods will therefore reduce *their* consumption spending by 8, or 0.8 times their income loss of 10. This reduction in spending cuts the income of *other* producers by 8 units, leading them to reduce their spending by 6.4, or 0.8 times their income loss of 8. These income reductions of 6.4 lead still other producers to cut their spending by 5.12, or 0.8 times 6.4, and so on. In principle this process continues indefinitely, although after many rounds of spending and income reductions the effects become quite small.

When all these "rounds" of income and spending reductions are added, the *total* effect on planned spending of the initial reduction of 10 in consumer spending is

$$10 + 8 + 6.4 + 5.12 + \cdots.$$

The three dots indicate that the series of reductions continues indefinitely. The total effect of the initial decrease in consumption can also be written as

$$10[1 + 0.8 + (0.8)^2 + (0.8)^3 + \cdots].$$

This expression highlights the fact that the spending that takes place in each round is 0.8 times the spending in the previous round (0.8), because that is the marginal propensity to consume out of the income generated by the previous round of spending.

A useful algebraic relationship, which applies to any number $x$ greater than 0 but less than 1, is

$$1 + x + x^2 + x^3 + \cdots = \frac{1}{1 - x}.$$

If we set $x = 0.8$, this formula implies that the total effect of the decline in consumption spending on aggregate demand and output is

$$10\left(\frac{1}{1 - 0.8}\right) = 10\left(\frac{1}{0.2}\right) = 10 \times 5 = 50.$$

This answer is consistent with our earlier calculation, which showed that short-run equilibrium output fell by 50 units, from 4,800 to 4,750.

By a similar analysis we can also find a general algebraic expression for the multiplier in the basic Keynesian model. Recalling that $c$ is the marginal propensity to consume out of disposable income, we know that a 1-unit increase in autonomous expenditure raises spending and income by 1 unit in the first round; by $c \times 1 = c$ units in the second round; by $c \times c = c^2$ units in the second round; by $c \times c^2 = c^3$ units in the third round; and so on. Thus the total effect on short-run equilibrium output of a 1-unit increase in autonomous expenditure is given by

$$1 + c + c^2 + c^3 + \cdots.$$

Applying the algebraic formula given above, and recalling that $0 < c < 1$, we can rewrite this expression as $1/(1 - c)$. Thus, in a basic Keynesian model with a marginal propensity to consume of $c$, the multiplier equals $1/(1 - c)$, the same result found in Appendix A to this chapter. Note that if $c = 0.8$, then $1/(1 - c) = 1/(1 - 0.8) = 5$, which is the same value of the multiplier we found numerically above.

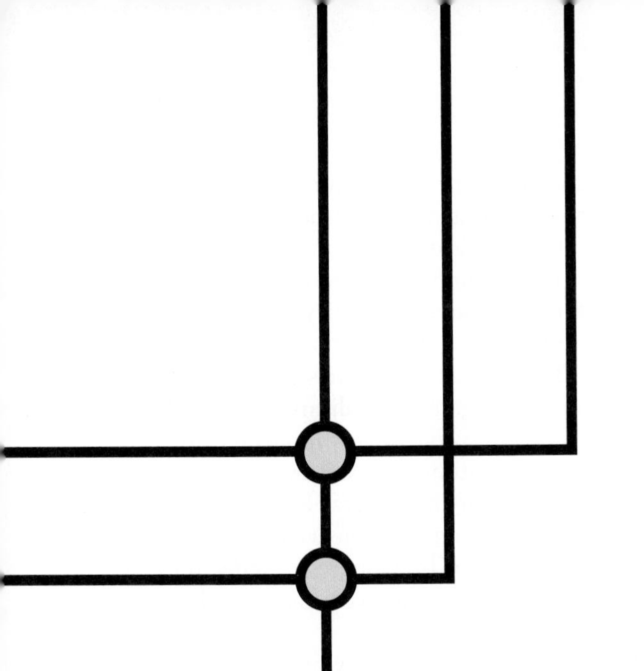

# CHAPTER 14

# STABILIZING THE ECONOMY: THE ROLE OF THE FED

Financial market participants and commentators go to remarkable lengths to try to predict the actions of the Federal Reserve. For a while, the CNBC financial news program *Squawk Box* reported regularly on what the commentators called the Greenspan Briefcase Indicator. The idea was to spot Fed chairman Alan Greenspan on his way to meet with the Federal Open Market Committee, the group that determines U.S. monetary policy. If Greenspan's briefcase was packed full, presumably with macroeconomic data and analyses, the guess was that the Fed planned to change interest rates. A slim briefcase meant no change in rates was likely.

"It was right 17 out of the first 20 times," the program's anchor Mark Haines noted, "but it has a built-in self-destruct mechanism, because Greenspan packs his [own] briefcase. He can make it wrong or right. He has never publicly acknowledged the indicator, but we have reason to believe that he knows about it. We have to consider the fact that he wants us to stop doing it because the last two times the briefcase has been wrong, and that's disturbing."[1]

The Briefcase Indicator is but one example of the close public scrutiny that the chairman of the Federal Reserve and other monetary policymakers face. Every speech, every congressional testimony, every interview from a member of the Board of Governors is closely analyzed for clues about the future course of monetary policy. In self-defense, Greenspan and other policymakers have become masters of the carefully worded but often ambiguous public statement

[1]Robert H. Frank, "Safety in Numbers," *New York Times Magazine,* November 28, 1999, p. 35.

calculated to leave the "Fed-watchers" guessing. The reason for the intense public interest in the Federal Reserve's decisions about monetary policy—and especially the level of interest rates—is that those decisions have important implications both for financial markets and for the economy in general.

In this chapter we examine the workings of monetary policy, one of the two major types of *stabilization policy.* (The other type, fiscal policy, was discussed in Chapter 13.) As we saw in Chapter 13, stabilization policies are government policies that are meant to influence planned aggregate expenditure, with the goal of eliminating output gaps. Both types of stabilization policy, monetary and fiscal, are important and have been useful at various times. However, monetary policy, which can be changed quickly by a decision of the Federal Reserve's Federal Open Market Committee (FOMC), is more flexible and responsive than fiscal policy, which can be changed only by legislative action by Congress. Under normal circumstances, therefore, monetary policy is used more actively than fiscal policy to help stabilize the economy.

We will begin this chapter by discussing how the Fed uses its ability to control the money supply to influence the level of interest rates. We then turn to the economic effects of changes in interest rates. Building on our analysis of the basic Keynesian model in Chapter 13, we will see that, in the short run, monetary policy works by affecting planned spending and thus short-run equilibrium output. We will defer discussion of the other major effect of monetary policy actions, changes in the rate of inflation. The effects of monetary policy on inflation are addressed in Chapter 15.

## THE FEDERAL RESERVE AND INTEREST RATES

When we introduced the Federal Reserve System in Chapter 10, we focused on the Fed's tools for controlling the *money supply,* that is, the quantity of currency and checking accounts held by the public. Determining the nation's money supply is the primary task of monetary policymakers. But if you follow the economic news regularly, you may find the idea that the Fed's job is to control the money supply a bit foreign, because the news media nearly always focus on the Fed's decisions about *interest rates*. Indeed, the announcement the Fed makes after each meeting of the Federal Open Market Committee nearly always concerns its plan for a particular short-term interest rate, called the *federal funds rate* (more on the federal funds rate later).

Actually, there is no contradiction between the two ways of looking at monetary policy—as control of the money supply or as the setting of interest rates. As we will see in this section, controlling the money supply and controlling the nominal interest rate are two sides of the same coin: Any value of the money supply chosen by the Fed implies a specific setting for the nominal interest rate, and vice versa. The reason for this close connection is that the nominal interest rate is effectively the "price" of money (or, more accurately, its opportunity cost). So, by controlling the quantity of money supplied to the economy, the Fed also controls the "price" of money (the nominal interest rate).

To better understand how the Fed determines interest rates, we will look first at the market for money, beginning with the demand side of that market. We will see that given the demand for money by the public, the Fed can control interest rates by changing the amount of money it supplies. Later we will show how the Fed uses control of interest rates to influence planned spending and the state of the economy.

### THE DEMAND FOR MONEY

Recall from Chapter 10 that *money* refers to the set of assets, such as cash and checking accounts, that are usable in transactions. Money is also a store of value, like stocks, bonds, or real estate—in other words, a type of financial asset. As a financial asset, money is a way of holding wealth.

Anyone who has some wealth must determine the *form* in which he or she wishes to hold that wealth. For example, if Louis has wealth of $10,000, he could if he wished hold all $10,000 in cash. Or he could hold $5,000 of his wealth in the form of cash and $5,000 in government bonds. Or he could hold $1,000 in cash, $2,000 in a checking account, $2,000 in government bonds, and $5,000 in rare stamps. Indeed, there are thousands of different real and financial assets to choose from, all of which can be held in different amounts and combinations, so Louis's choices are virtually infinite. The decision about the forms in which to hold one's wealth is called the **portfolio allocation decision.**

***portfolio allocation decision*** the decision about the forms in which to hold one's wealth

What determines the particular mix of assets that Louis or another wealth holder will choose? All else being equal, people generally prefer to hold assets that they expect to pay a high *return* and do not carry too much *risk*. They may also try to reduce the overall risk they face through *diversification*—that is, by owning a variety of different assets.[2] Many people own some real assets, such as a car or a home, because they provide services (transportation or shelter) and often a financial return (an increase in value, as when the price of a home rises in a strong real estate market).

Here we do not need to analyze the entire portfolio allocation decision, but only one part of it—namely, the decision about how much of one's wealth to hold in the form of *money* (cash and checking accounts). The amount of wealth an individual chooses to hold in the form of money is that individual's **demand for money.** So if Louis decided to hold his entire $10,000 in the form of cash, his demand for money would be $10,000. But if he were to hold $1,000 in cash, $2,000 in a checking account, $2,000 in government bonds, and $5,000 in rare stamps, his demand for money would be only $3,000—that is, $1,000 in cash plus the $2,000 in his checking account.

***demand for money*** the amount of wealth an individual chooses to hold in the form of money

**EXAMPLE 14.1**

### Consuelo's demand for money

Example 9.1 presented the balance sheet of an individual named Consuelo (see Table 9.1). What is her demand for money? If Consuelo wanted to increase her money holdings by $100, how could she do so? What if she wanted to reduce her money holdings by $100?

Looking back at Table 9.1, p. 233, we see that Consuelo's balance sheet shows five different asset types: cash, a checking account, shares of stock, a car, and furniture. Of these assets, the first two (the cash and the checking account) are forms of money. As shown in Table 9.1, Consuelo's money holdings consist of $80 in cash and $1,200 in her checking account. Thus Consuelo's demand for money—the amount of wealth she chooses to hold in the form of money—is $1,280.

There are many different ways in which Consuelo could increase her money holdings, or demand for money, by $100. She could sell $100 worth of stock and deposit the proceeds in the bank. That action would leave the total value of her assets and her wealth unchanged (because the decrease in her stockholdings would be offset by the increase in her checking account) but would increase her money holdings by $100. Another possibility would be to take a $100 cash advance on her credit card. That action would increase both her money holdings and her assets by $100 but would also increase her liabilities—specifically, her credit card balance—by $100. Once again, her total wealth would not change, though her money holdings would increase.

To reduce her money holdings, Consuelo need only use some of her cash or checking account balance to acquire a nonmoney asset or pay down a liability.

[2]Chapter 11 discusses risk, return, and diversification in more detail.

For example, if she were to buy an additional $100 of stock by writing a check against her bank account, her money holdings would decline by $100. Similarly, writing a check to reduce her credit card balance by $100 would reduce her money holdings by $100. You can confirm that though her money holdings decline, in neither case does Consuelo's total wealth change.

How much money should an individual (or household) choose to hold? Application of the *cost-benefit principle* tells us that an individual should increase his or her money holdings only so long as the benefit of doing so exceeds the cost. As we saw in Chapter 10, the principal *benefit* of holding money is its usefulness in carrying out transactions. Consuelo's shares of stock, her car, and her furniture are all valuable assets, but she cannot use them to buy groceries or pay her rent. She can make routine payments using cash or her checking account, however. Because of its usefulness in daily transactions, Consuelo will almost certainly want to hold some of her wealth in the form of money. Furthermore, if Consuelo is a high-income individual, she will probably choose to hold more money than someone with a lower income would, because she is likely to spend more and carry out more transactions that the low-income person.

© Aaron Haupt/Stock Boston

Innovations such as ATM machines have reduced the amount of money that people need to hold for routine transactions.

Consuelo's benefit from holding money is also affected by the technological and financial sophistication of the society she lives in. For example, in the United States, developments such as credit cards, debit cards, and ATM machines have generally reduced the amount of money people need to carry out routine transactions, decreasing the public's demand for money at given levels of income. In the United States in 1960, for example, money holdings in the form of cash and checking account balances (the monetary aggregate M1) were about 28 percent of GDP. By 2002 that ratio had fallen to 11 percent of GDP.

Although money is an extremely useful asset, there is also a cost to holding money—more precisely, an opportunity cost—which arises from the fact that most forms of money pay little or no interest. Cash pays zero interest, and most checking accounts pay either no interest or very low rates. For the sake of simplicity, we will just assume that *the nominal interest rate on money is zero*. In contrast, most alternative assets, such as bonds or stocks, pay a positive nominal return. A bond, for example, pays a fixed amount of interest each period to the holder, while stocks pay dividends and may also increase in value (capital gains).

The cost of holding money arises because, in order to hold an extra dollar of wealth in the form of money, a person must reduce by one dollar the amount of wealth held in the form of higher-yielding assets, such as bonds or stocks. The *opportunity cost* of holding money is measured by the interest rate that could have been earned if the person had chosen to hold interest-bearing assets instead of money. All else being equal, the higher the nominal interest rate, the higher the opportunity cost of holding money, and hence the less money people will choose to hold.

We have been talking about the demand for money by individuals, but businesses also hold money to carry out transactions with customers and to pay workers and suppliers. The same general factors that determine individuals' money demand also affect the demand for money by businesses. That is, in choosing how much money to hold, a business, like an individual, will compare the benefits of holding money for use in transactions with the opportunity cost of holding a non-interest-bearing asset. Although we will not differentiate between the money held by individuals and the money held by businesses in discussing money demand, you should be aware that in the U.S. economy, businesses hold a significant portion—more than half—of the total money stock. Example 14.2 illustrates the determination of money demand by a businessowner.

EXAMPLE 14.2

**How much money should Kim's restaurants hold?**

Kim owns several successful restaurants. Her accountant informs her that on the typical day her restaurants are holding a total of $50,000 in cash on the premises. The accountant points out that if Kim's restaurants reduced their cash holdings, Kim could use the extra cash to purchase interest-bearing government bonds.

The accountant proposes two methods of reducing the amount of cash Kim's restaurants hold. First, she could increase the frequency of cash pickups by her armored car service. The extra service would cost $500 annually but would allow Kim's restaurants to reduce their average cash holding to $40,000. Second, in addition to the extra pickups, Kim could employ a computerized cash management service to help her keep closer tabs on the inflows and outflows of cash at her restaurants. The service costs $700 a year, but the accountant estimates that, together with more frequent pickups, the more efficient cash management provided by the service could help Kim reduce average cash holdings at her restaurants to $30,000.

The interest rate on government bonds is 6 percent. How much money should Kim's restaurants hold? What if the interest rate on government bonds is 8 percent?

Kim's restaurants need to hold cash to carry out their normal business, but holding cash also has an opportunity cost, which is the interest those funds could be earning if they were held in the form of government bonds instead of zero-interest cash. As the interest rate on government bonds is 6 percent, each $10,000 by which Kim can reduce her restaurants' money holdings yields an annual benefit of $600 (6 percent of $10,000).

If Kim increases the frequency of pickups by her armored car service, reducing the restaurants' average money holdings from $50,000 to $40,000, the benefit will be the additional $600 in interest income that Kim will earn. The cost is the $500 charged by the armored car company. Since the benefit exceeds the cost, Kim should purchase the extra service and reduce the average cash holdings at her restaurants to $40,000.

Should Kim go a step further and employ the cash management service as well? Doing so would reduce average cash holdings at the restaurants from $40,000 to $30,000, which has a benefit in terms of extra interest income of $600 per year. However, this benefit is less than the cost of the cash management service, which is $700 per year. So Kim should *not* employ the cash management service and instead should maintain average cash holdings in her restaurants of $40,000.

If the interest rate on government bonds rises to 8 percent, then the benefit of each $10,000 reduction in average money holdings is $800 per year (8 percent of $10,000) in extra interest income. In this case the benefit of employing the cash management service, $800, exceeds the cost of doing so, which is $700. So Kim should employ the service, reducing the average cash holdings of her business to $30,000. The example shows that a higher nominal interest rate on alternative assets reduces the quantity of money demanded.

EXERCISE 14.1

**The interest rate on government bonds falls from 6 percent to 4 percent. How much cash should Kim's restaurants hold now?**

## MACROECONOMIC FACTORS THAT AFFECT THE DEMAND FOR MONEY

In any household or business the demand for money will depend on a variety of individual circumstances. For example, a high-volume retail business that serves thousands of customers each day will probably choose to have more money on hand than a legal firm that bills clients and pays employees monthly. But while individuals and businesses vary considerably in the amount of money they choose to hold, three macroeconomic factors affect the demand for money quite broadly: the nominal interest rate, real output, and the price level. As we see next, the nominal interest rate affects the cost of holding money throughout the economy, while real output and the price level affect the benefits of money.

- *The nominal interest rate (i).* We have seen that the interest rate paid on alternatives to money, such as government bonds, determines the opportunity cost of holding money. The higher the prevailing nominal interest rate, the greater the opportunity cost of holding money, and hence the less money individuals and businesses will demand.

What do we mean by *the* nominal interest rate? As we have discussed, there are thousands of different assets, each with its own interest rate (rate of return). So can we really talk about *the* nominal interest rate? The answer is that, while there are many different assets, each with its own corresponding interest rate, the rates on those assets tend to rise and fall together. This is to be expected, because if the interest rates on some assets were to rise sharply while the rates on other assets declined, financial investors would flock to the assets paying high rates and refuse to buy the assets paying low rates. So, although there are many different interest rates in practice, speaking of the general level of interest rates usually does make sense. In this book, when we talk about *the* nominal interest rate, what we have in mind is some average measure of interest rates. This simplification is one more application of the macroeconomic concept of *aggregation,* introduced in Chapter 4.

The nominal interest rate is a macroeconomic factor that affects the cost of holding money. A macroeconomic factor that affects the *benefit* of holding money is

- *Real income or output (Y).* An increase in aggregate real income or output—as measured, for example, by real GDP—raises the quantity of goods and services that people and businesses want to buy and sell. When the economy enters a boom, for example, people do more shopping and stores have more customers. To accommodate the increase in transactions, both individuals and businesses need to hold more money. Thus higher real output raises the demand for money.

A second macroeconomic factor affecting the benefit of holding money is

- *The price level (P).* The higher the prices of goods and services, the more dollars (or yen, or euros) are needed to make a given set of transactions. Thus a higher price level is associated with a higher demand for money.

Today, when a couple of teenagers go out for a movie and snacks on Saturday night, they need probably five times as much cash as their parents did 25 years ago. Because the prices of movie tickets and popcorn have risen steeply over 25 years, more money (that is, more dollars) is needed to pay for a Saturday night date than in the past. By the way, the fact that prices are higher today does *not* imply that people are worse off today than in the past, because nominal wages and salaries have also risen substantially. In general, however, higher prices do imply that people need to keep a greater number of dollars available, in cash or in a checking account.

## THE MONEY DEMAND CURVE

For the purposes of monetary policymaking, economists are most interested in the aggregate, or economywide, demand for money. The interaction of the aggregate demand for money, determined by the public, and the supply of money,

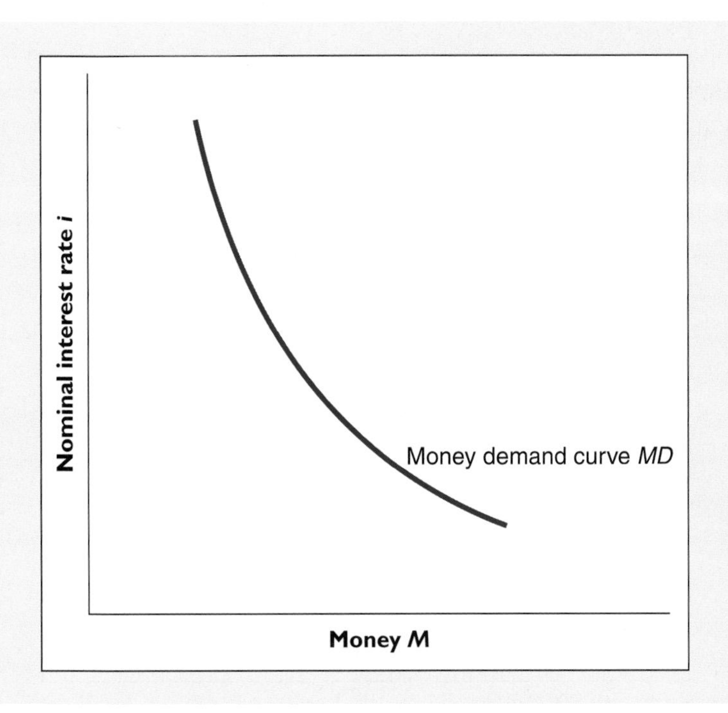


**FIGURE 14.1**
**The Money Demand Curve.**
The money demand curve relates the economywide demand for money to the nominal interest rate. Because an increase in the nominal interest rate raises the opportunity cost of holding money, the money demand curve slopes down.

which is set by the Fed, determines the nominal interest rate that prevails in the economy.

The economywide demand for money can be represented graphically by the *money demand curve* (see Figure 14.1). The **money demand curve** relates the aggregate quantity of money demanded $M$ to the nominal interest rate $i$. The quantity of money demanded $M$ is a nominal quantity, measured in dollars (or yen, or euros, depending on the country). Because an increase in the nominal interest rate increases the opportunity cost of holding money, which reduces the quantity of money demanded, the money demand curve slopes down.

***money demand curve*** Shows the relationship between the aggregate quantity of money demanded $M$ and the nominal interest rate $i$; because an increase in the nominal interest rate increases the opportunity cost of holding money, which reduces the quantity of money demanded, the money demand curve slopes down

If we think of the nominal interest rate as the "price" (more precisely, the opportunity cost) of money and the amount of money people want to hold as the "quantity," the money demand curve is analogous to the demand curve for a good or service. As with a standard demand curve, the fact that a higher price of money leads people to demand less of it is captured in the downward slope of the demand curve. Furthermore, as in a standard demand curve, changes in factors other than the price of money (the nominal interest rate) can cause the demand curve for money to shift. For a given nominal interest rate, any change that makes people want to hold more money will shift the money demand curve to the right, and any change that makes people want to hold less money will shift the money demand curve to the left. We have already identified two macroeconomic factors other than the nominal interest rate that affect the economywide demand for money: real output and the price level. Because an increase in either of these variables increases the demand for money, it shifts the money demand curve rightward, as shown in Figure 14.2. Similarly, a fall in real output or the general price level reduces money demand, shifting the money demand curve leftward.

The money demand curve may also shift in response to other changes that affect the cost or benefit of holding money, such as the technological and financial advances we mentioned earlier. For example, the introduction of ATM machines reduced the amount of money people choose to hold and thus shifted the economywide money demand curve to the left. Economic Naturalist 14.1 describes another potential source of shifts in the demand for money, holdings of U.S. dollars by foreigners.

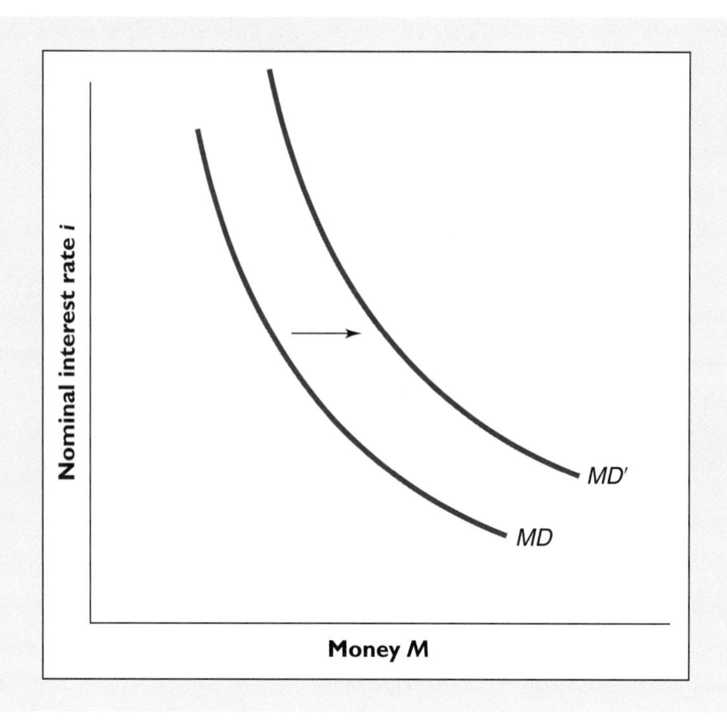


**FIGURE 14.2**
**A Shift in the Money Demand Curve.**
At a given nominal interest rate, any change that makes people want to hold more money—such as an increase in the general price level or in real GDP—will shift the money demand curve to the right.

ECONOMIC NATURALIST 14.1

**Why does the average Argentine hold more U.S. dollars than the average U.S. citizen?**

Estimates are that the value of U.S. dollars circulating in Argentina exceeds $1,000 per person, which is higher than the per capita dollar holdings in the United States. A number of other countries, including those that once belonged to the former Soviet Union, also hold large quantities of dollars. In all, as much as $300 billion in U.S. currency—more than half the total amount issued—may be circulating outside the borders of the United States. Why do Argentines and other non-U.S. residents hold so many dollars?

U.S. residents and businesses hold dollars primarily for transactions purposes, rather than as a store of value. As a store of value, interest-bearing bonds and dividend-paying stocks are a better choice for Americans than zero-interest money. But this is not necessarily the case for the citizens of other countries, particularly nations that are economically or politically unstable. Argentina, for example, endured many years of high and erratic inflation in the 1970s and 1980s, which sharply eroded the value of financial investments denominated in Argentine pesos. Lacking better alternatives, many Argentines began saving in the form of U.S. currency—dollar bills hidden in the mattress or plastered into the wall—which they correctly believed to be more stable in value than peso-denominated assets.

Argentina's use of dollars became officially recognized in 1990. In that year the country instituted a new monetary system, called a currency board, under which U.S. dollars and Argentine pesos by law traded freely one for one. Under the currency board system, Argentines became accustomed to carrying U.S. dollars in their wallets for transactions purposes, along with pesos. However, in 2001 Argentina's monetary problems returned with a vengeance, as the currency board system broke down, the peso plummeted in value relative to the dollar, and inflation returned. The Argentine demand for dollars is thus likely only to increase in the next few years.

The countries formed as a result of the breakup of the Soviet Union have endured not only high inflation, but political instability and uncertainty as well. In a politically volatile environment, citizens face the risk that their savings, including their bank deposits, will be confiscated or heavily taxed by the government.

Often they conclude that a hidden cache of U.S. dollars—an estimated $1 million in 100-dollar bills can be stored in a suitcase—is the safest way to hold wealth.

In practice, changes in the foreign demand for U.S. dollars are an important source of fluctuation in the U.S. money demand curve. For example, large quantities of U.S. dollars flowed abroad during the Gulf War of 1990–1991, reflecting concerns among residents of the Middle East about regional instability. This increase in the demand for dollars shifted the U.S. money demand curve substantially to the right, as in Figure 14.2. Because policymakers at the Federal Reserve are concerned primarily with the number of dollars circulating in the U.S. economy, rather than in the world as a whole, they pay close attention to these international flows of greenbacks.

**RECAP MONEY DEMAND**

For the economy as a whole, the demand for money is the amount of wealth that individuals, households, and businesses choose to hold in the form of money. The opportunity cost of holding money is measured by the nominal interest rate $i$, which is the return that could be earned on alternative assets such as bonds. The benefit of holding money is its usefulness in transactions.

Increases in real GDP ($Y$) or the price level ($P$) raise the nominal volume of transactions and thus the economywide demand for money. The demand for money is also affected by technological and financial innovations, such as the introduction of ATM machines, that affect the costs or benefits of holding money.

The money demand curve relates the economywide demand for money to the nominal interest rate. Because an increase in the nominal interest rate raises the opportunity cost of holding money, the money demand curve slopes downward.

Changes in factors other than the nominal interest rate that affect the demand for money can shift the money demand curve. For example, increases in real GDP or the price level raise the demand for money, shifting the money demand curve to the right, whereas decreases shift the money demand curve to the left.

## THE SUPPLY OF MONEY AND MONEY MARKET EQUILIBRIUM

Where there is demand, can supply be far behind? As we have seen, the *supply* of money is controlled by the central bank—in the United States, the Federal Reserve, or Fed. The Fed's primary tool for controlling the money supply is *open-market operations*. For example, to increase the money supply, the Fed can use newly created money to buy government bonds from the public (an open-market purchase), which puts the new money into circulation.

Figure 14.3 shows the demand for and the supply of money in a single diagram. The nominal interest rate is on the vertical axis, and the nominal quantity of money (in dollars) is on the horizontal axis. As we have seen, because a higher nominal interest rate increases the opportunity cost of holding money, the money demand curve slopes downward. And because the Fed fixes the supply of money, we have drawn the *money supply curve* as a vertical line that intercepts the horizontal axis at the quantity of money chosen by the Fed, denoted $M$.

As in standard supply and demand analysis, equilibrium in the market for money occurs at the intersection of the supply and demand curves, shown as point $E$ in Figure 14.3. The equilibrium amount of money in circulation, $M$, is simply the amount of money the Fed chooses to supply. The equilibrium nominal

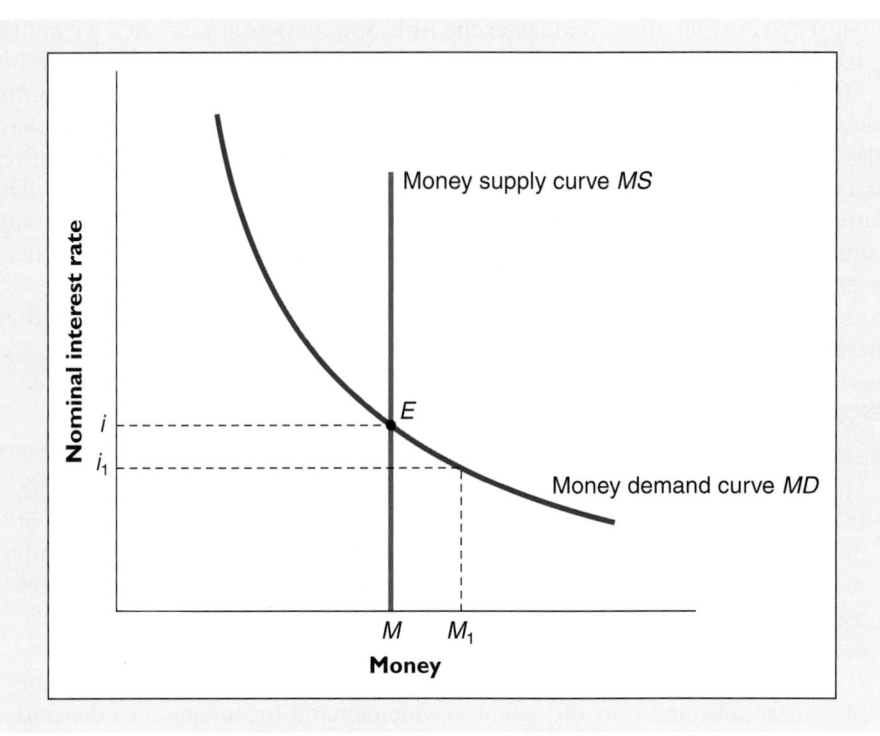


**FIGURE 14.3**
**Equilibrium in the Market for Money.**
Equilibrium in the market for money occurs at point *E*, where the demand for money by the public equals the amount of money supplied by the Federal Reserve. The equilibrium nominal interest rate, which equates the supply of and demand for money, is *i*.

interest rate *i* is the interest rate at which the quantity of money demanded by the public, as determined by the money demand curve, equals the fixed supply of money made available by the Fed.

To understand how the market for money reaches equilibrium, it may be helpful to recall the relationship between interest rates and the market price of bonds that was introduced in Chapter 11 (see Example 11.1). As we saw in the earlier chapter, the prices of existing bonds are *inversely related* to the current interest rate. Higher interest rates imply lower bond prices, and lower interest rates imply higher bond prices. With this relationship between interest rates and bond prices in mind, let's ask what happens if, say, the nominal interest rate is initially below the equilibrium level in the market for money—for example, at a value such as $i_1$ in Figure 14.3. At that interest rate the public's demand for money is $M_1$, which is greater than the actual amount of money in circulation, equal to $M$. How will the public—households and firms—react if the amount of money they hold is less than they would like? To increase their holdings of money, people will try to sell some of the interest-bearing assets they hold, such as bonds. But if everyone is trying to sell bonds and there are no willing buyers, then all the attempt to reduce bond holdings will achieve is to drive down the price of bonds, in the same way that a glut of apples will drive down the price of apples.

A fall in the price of bonds, however, is equivalent to an increase in interest rates. Thus the public's collective attempt to increase its money holdings by selling bonds and other interest-bearing assets, which has the effect of lowering bond prices, also implies higher market interest rates. As interest rates rise, the quantity of money demanded by the public will decline (represented by a right-to-left movement along the money demand curve), as will the desire to sell bonds. Only when the interest rate reaches its equilibrium value, *i* in Figure 14.3, will people be content to hold the quantities of money and other assets that are actually available in the economy.

**EXERCISE 14.2**

**Describe the adjustment process in the market for money if the nominal interest rate is initially above rather than below its equilibrium value. What happens to the price of bonds as the money market adjusts toward equilibrium?**

## HOW THE FED CONTROLS THE NOMINAL INTEREST RATE

We began this section by noting that the public and the press usually talk about Fed policy in terms of decisions about the nominal interest rate rather than the money supply. Indeed, Fed policymakers themselves usually describe their plans in terms of a target value for the interest rate. We now have the necessary background to understand how the Fed translates the ability to determine the economy's money supply into control of the nominal interest rate.

Figure 14.3 showed that the nominal interest rate is determined by equilibrium in the market for money. Let's suppose that for some reason the Fed decides to lower the interest rate. As we will see, to lower the interest rate the Fed must increase the supply of money, which can be accomplished by using newly created money to purchase government bonds from the public (an open-market purchase).

Figure 14.4 shows the effects of such an increase in the money supply by the Fed. If the initial money supply is $M$, then equilibrium in the money market occurs at point $E$ in the figure, and the equilibrium nominal interest rate is $i$. Now suppose the Fed, by means of open-market purchases of bonds, increases the money supply to $M'$. This increase in the money supply shifts the vertical money supply curve to the right, which shifts the equilibrium in the money market from point $E$ to point $F$ (see Figure 14.4). Note that at point $F$ the equilibrium nominal interest rate has declined, from $i$ to $i'$. The nominal interest rate must decline if the public is to be persuaded to hold the extra money that has been injected into the economy.

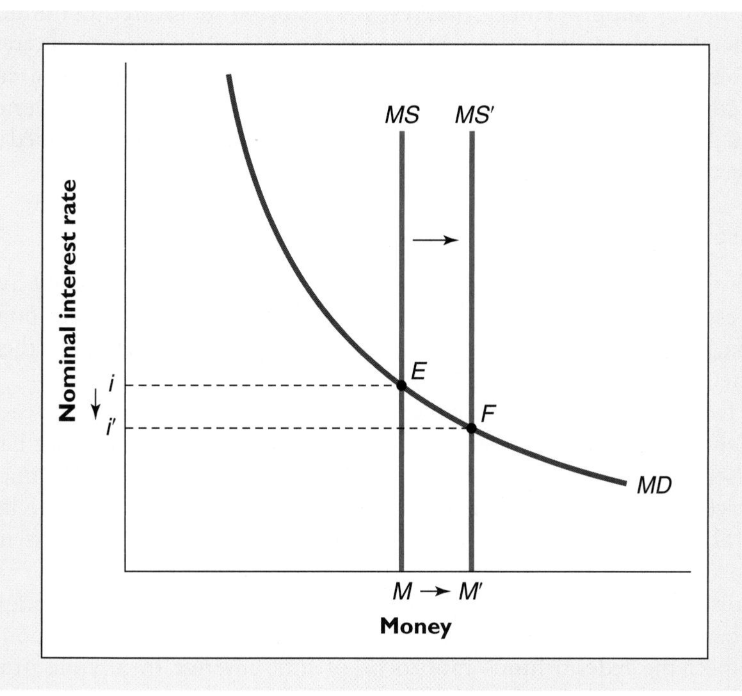


**FIGURE 14.4**
**The Fed Lowers the Nominal Interest Rate.**
The Fed can lower the equilibrium nominal interest rate by increasing the supply of money. For the given money demand curve, an increase in the money supply from $M$ to $M'$ shifts the equilibrium point in the money market from $E$ to $F$, lowering the equilibrium nominal interest rate from $i$ to $i'$.

To understand what happens in financial markets when the Fed expands the money supply, recall once again the inverse relationship between interest rates and the price of bonds. To increase the money supply, the Fed must buy government bonds from the public. However, if households and firms are initially satisfied with their asset holdings, they will be willing to sell bonds only at a price that is higher than the initial price. That is, the Fed's bond purchases will drive up the price of bonds in the open market. But we know that higher bond prices imply lower interest rates. Thus the Fed's bond purchases lower the prevailing nominal interest rate.

A similar scenario unfolds if the Fed decides to raise interest rates. To raise interest rates, the Fed must *reduce* the money supply. Reduction of the money supply is accomplished by an open-market sale—the sale of government bonds to the public in exchange for money. (The Fed keeps a large inventory of government bonds, acquired through previous open-market purchases, for use in open-market operations.) But in the attempt to sell bonds on the open market, the Fed will drive down the price of bonds. Given the inverse relationship between the price of bonds and the interest rate, the fall in bond prices is equivalent to a rise in the interest rate. In terms of money demand and money supply, the higher interest rate is necessary to persuade the public to hold less money.

As Figures 14.3 and 14.4 illustrate, control of the interest rate is not separate from control of the money supply. If Fed officials choose to set the nominal interest rate at a particular level, they can do so only by setting the money supply at a level consistent with the target interest rate. The Fed *cannot* set the interest rate and the money supply independently, since for any given money demand curve, a particular interest rate implies a particular size of the money supply, and vice versa.

Since monetary policy actions can be expressed in terms of either the interest rate or the money supply, why does the Fed (and almost every other central bank) choose to communicate its policy decisions to the public in terms of a target nominal interest rate rather than a target money supply? One reason, as we will see shortly, is that the main effects of monetary policy on both the economy and financial markets are exerted through interest rates. Consequently, the interest rate is often the best summary of the overall impact of the Fed's actions. Another reason for focusing on interest rates is that they are more familiar to the public than the money supply. Finally, interest rates can be monitored continuously in the financial markets, which makes the effects of Fed policies on interest rates easy to observe. By contrast, measuring the amount of money in the economy requires collecting data on bank deposits, with the consequence that several weeks may pass before policymakers and the public know precisely how Fed actions have affected the money supply.

### What's so important about the federal funds rate?

Although thousands of interest rates and other financial data are easily available, the interest rate that is perhaps most closely watched by the public, politicians, the media, and the financial markets is the *federal funds rate.* What is the federal funds rate, and why is it so important?

***federal funds rate*** the interest rate that commercial banks charge each other for very short-term (usually overnight) loans; because the Fed frequently sets its policy in the form of a target for the federal funds rate, this rate is closely watched in financial markets

The **federal funds rate** is the interest rate commercial banks charge each other for very short-term (usually overnight) loans. For example, a bank that has insufficient reserves to meet its legal reserve requirements (see Chapter 10) might borrow reserves for a few days from a bank that has extra reserves. Despite its name, the federal funds rate is not an official government interest rate and is not connected to the federal government.

Because the market for loans between commercial banks is tiny compared to some other financial markets, such as the market for government bonds, one might expect the federal funds rate to be of little interest to anyone other than the managers of commercial banks. But enormous attention is paid to this interest rate, because over most of the past 40 years, the Fed has expressed its policies in terms of a target value for it. Indeed, at the close of every meeting of the Federal Open Market Committee, the Fed announces whether the federal funds rate will be increased, decreased, or left unchanged. The Fed may also indicate the likely direction of future changes in the federal funds rate. Thus more than any other financial variable, changes in the federal funds rate indicate the Fed's plans for monetary policy.

Why does the Fed choose to focus on this particular nominal interest rate over all others? As we saw in Chapter 10, in practice the Fed affects the money supply through its control of bank reserves. Because open-market operations directly affect the supply of bank reserves, the Fed's control over the federal funds rate is particularly tight. However, if Fed officials chose to do so, they could probably signal their intended policies just as effectively in terms of another short-term nominal interest rate, such as the rate on short-term government debt.

Figure 14.5 shows the behavior of the federal funds rate since 1970. As you can see, the Fed has allowed this interest rate to vary considerably in response to economic conditions. Later in the chapter we will consider two specific episodes in which the Fed changed the federal funds rate in response to an economic slowdown.

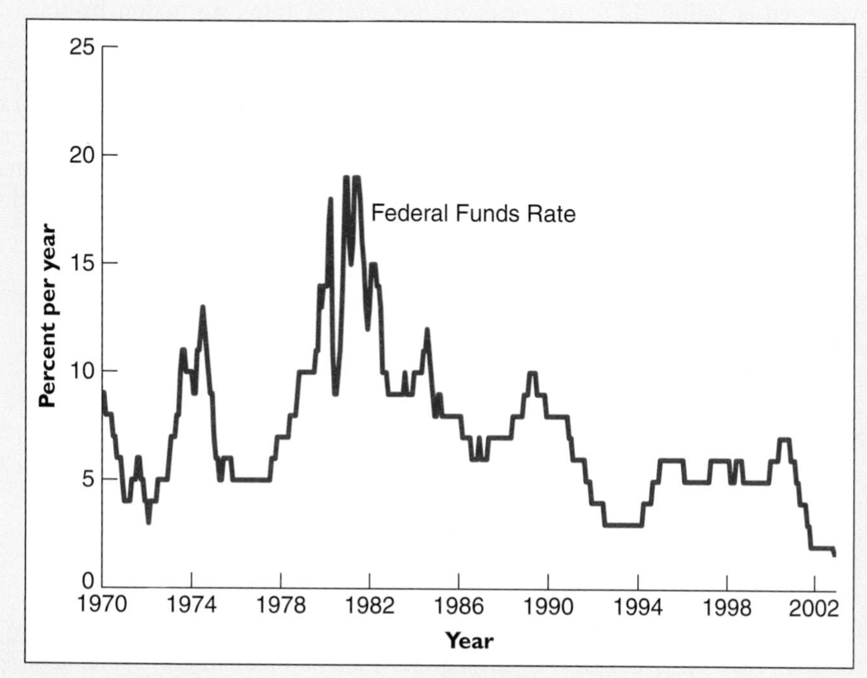


SOURCE: Federal Reserve Bank of St. Louis (http://www.stls.frb.org/fred/).

**FIGURE 14.5**
**The Federal Funds Rate, 1970–2002.**
The federal funds rate is the interest rate commercial banks charge each other for short-term loans. It is closely watched because the Fed expresses its policies in terms of a target value for the federal funds rate. The Fed has allowed the federal funds rate to vary considerably in response to economic conditions.

## CAN THE FED CONTROL THE REAL INTEREST RATE?

Through its control of the money supply the Fed can control the economy's *nominal* interest rate. But many important economic decisions, such as the decisions to save and invest, depend on the *real* interest rate. To affect those decisions, the Fed must exert some control over the real interest rate.

Most economists believe that the Fed can control the real interest rate, at least for some period. To see why, recall the definition of the real interest rate from Chapter 6:

$$r = i - \pi.$$

The real interest rate $r$ equals the nominal interest rate $i$ minus the rate of inflation $\pi$. As we have seen, the Fed can control the nominal interest rate quite precisely through its ability to determine the money supply. Furthermore, inflation appears to change relatively slowly in response to changes in policy or economic conditions, for reasons we will discuss in the next chapter. Because inflation tends to adjust slowly, actions by the Fed to change the nominal interest rate generally lead the real interest rate to change by about the same amount.

The idea that the Fed can set the real interest rate appears to contradict the analysis in Chapter 9, which concluded that the real interest rate is determined by the condition that national saving must equal investment in new capital goods.

This apparent contradiction is rooted in a difference in the time frame being considered. Because inflation does not adjust quickly, the Fed can control the real interest rate over the short run. In the long run, however—that is, over periods of several years or more—the inflation rate and other economic variables will adjust, and the balance of saving and investment will determine the real interest rate. Thus the Fed's ability to influence consumption and investment spending through its control of the real interest rate is strongest in the short run.

In discussing the Fed's control over interest rates, we should also return to a point mentioned earlier in this chapter: In reality, not just one but many thousands of interest rates are seen in the economy. Because interest rates tend to move together (allowing us to speak of *the* interest rate), an action by the Fed to change the federal funds rate generally causes other interest rates to change in the same direction. However, the tendency of other interest rates (such as the long-term government bond rate or the rate on bonds issued by corporations) to move in the same direction as the federal funds rate is only a tendency, not an exact relationship. In practice, then, the Fed's control of other interest rates may be somewhat less precise than its control of the federal funds rate—a fact that complicates the Fed's policymaking.

**RECAP THE FEDERAL RESERVE AND INTEREST RATES**

In the market for money, the money demand curve slopes downward, reflecting the fact that a higher nominal interest rate increases the opportunity cost of holding money and thus reduces the amount of money people want to hold. The money supply curve is vertical at the quantity of money that the Fed chooses to supply. The equilibrium nominal interest rate $i$ is the interest rate at which the quantity of money demanded by the public equals the fixed supply of money made available by the Fed.

The Federal Reserve controls the nominal interest rate by changing the supply of money. An open-market purchase of government bonds increases the money supply and lowers the equilibrium nominal interest rate. Conversely, an open-market sale of bonds reduces the money supply and increases the nominal interest rate. The Fed can prevent changes in the demand for money from affecting the nominal interest rate by adjusting the quantity of money supplied appropriately. The Fed typically expresses its policy intentions in terms of a target for a specific nominal interest rate, the federal funds rate.

Because inflation is slow to adjust, in the short run the Fed can control the real interest rate (equal to the nominal interest rate minus the inflation rate) as well as the nominal interest rate. In the long run, however, the real interest rate is determined by the balance of saving and investment.

## THE EFFECTS OF FEDERAL RESERVE ACTIONS ON THE ECONOMY

Now that we have seen how the Fed can influence interest rates (both nominal and real), we can consider how monetary policy can be used to eliminate output gaps and stabilize the economy. The basic idea is relatively straightforward. As we will see in this section, planned aggregate expenditure is affected by the level of real interest rate prevailing in the economy. Specifically, a lower real interest rate encourages higher planned spending by households and firms, while a higher real interest rate reduces spending. By adjusting the real interest rate, the Fed can move planned spending in the desired direction. Under the assumption of the basic Keynesian model that firms produce just enough goods and services to meet the demand for their output, the Fed's stabilization of planned spending leads to stabilization of aggregate output and employment as well. In this section we will first explain how planned aggregate expenditure is related to the real interest rate. Then we will show how the Fed can use changes in the real interest rate to fight a recession or inflation.

## PLANNED AGGREGATE EXPENDITURE AND THE REAL INTEREST RATE

In Chapter 13 we saw how planned spending is affected by changes in real output $Y$. Changes in output affect the private sector's disposable income $(Y - T)$, which in turn influences consumption spending—a relationship captured by the consumption function.

A second variable that has potentially important effects on aggregate expenditure is the real interest rate $r$. In Chapter 11, in our discussion of saving and investment, we saw that the real interest rate influences the behavior of both households and firms.

For households, the effect of a higher real interest rate is to increase the reward for saving, which leads households to save more.[3] At a given level of income, households can save more only if they consume less. Thus, saying that a higher real interest rate *increases* saving is the same as saying that a higher real interest rate *reduces* consumption spending at each level of income. The idea that higher real interest rates reduce household spending makes intuitive sense. Think, for example, about people's willingness to buy consumer durables, such as automobiles or furniture. Purchases of consumer durables, which are part of consumption spending, are often financed by borrowing from a bank, credit union, or finance company. When the real interest rate rises, the monthly finance charges associated with the purchase of a car or a piano are higher, and people become less willing or able to make the purchase. Thus a higher real interest rate reduces people's willingness to spend on consumer goods, holding constant disposable income and other factors that affect consumption.

When the real interest rate rises, financing a new car becomes more expensive and fewer cars are purchased.

Besides reducing consumption spending, a higher real interest rate also discourages firms from making capital investments. As in the case of a consumer thinking of buying a car or a piano, when a rise in the real interest rate increases financing costs, firms may reconsider their plans to invest. For example, upgrading a computer system may be profitable for a manufacturing firm when the cost of the system can be financed by borrowing at a real interest rate of 3 percent. However, if the real interest rate rises to 6 percent, doubling the cost of funds to the firm, the same upgrade may not be profitable and the firm may choose not to invest. We should also remember that residential investment—the building of houses and apartment buildings—is also part of investment spending. Higher interest rates, in the form of higher mortgage rates, certainly discourage this kind of investment spending as well.

The conclusion is that, at any given level of output, *both consumption spending and planned investment spending decline when the real interest rate increases.* Conversely, a fall in the real interest rate tends to stimulate consumption and investment spending by reducing financing costs. Example 14.3 is a numerical illustration of how planned aggregate expenditure can be related to the real interest rate and output.

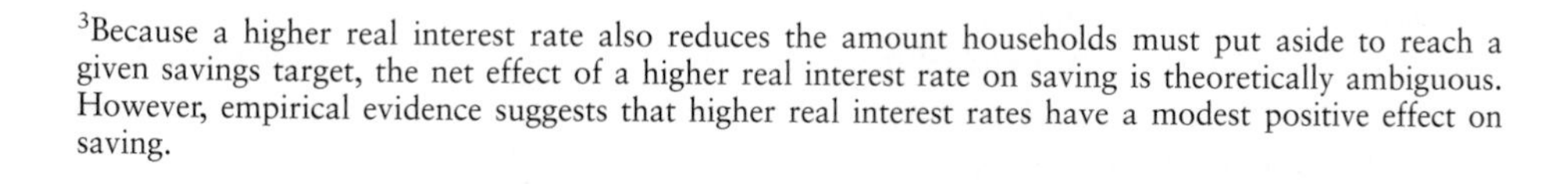

[3]Because a higher real interest rate also reduces the amount households must put aside to reach a given savings target, the net effect of a higher real interest rate on saving is theoretically ambiguous. However, empirical evidence suggests that higher real interest rates have a modest positive effect on saving.

**EXAMPLE 14.3**

**Planned aggregate expenditure and the real interest rate**

In a certain economy, the components of planned spending are given by

$$C = 640 + 0.8(Y - T) - 400r,$$
$$I^P = 250 - 600r,$$
$$G = 300,$$
$$NX = 20,$$
$$T = 250.$$

Find the relationship of planned aggregate expenditure to the real interest rate $r$ and output $Y$ in this economy. Find autonomous expenditure and induced expenditure.

This example is similar to Example 13.2, except that now the real interest rate $r$ is allowed to affect both consumption and planned investment. For example, the final term in the equation describing consumption, $-400r$, implies that a 1 percent (0.01) increase in the real interest rate, from 4 percent to 5 percent, for example, reduces consumption spending by $400(0.01) = 4$ units. Similarly, the final term in the equation for planned investment tells us that in this example, a 1 percent increase in the real interest rate lowers planned investment by $600(0.01) = 6$ units. Thus the overall effect of a 1 percent increase in the real interest rate is to lower planned aggregate expenditure by 10 units, the sum of the effects on consumption and investment. As in the earlier examples, disposable income $(Y - T)$ is assumed to affect consumption spending through a marginal propensity to consume of 0.8 (see the first equation), and government purchases $G$, net exports $NX$, and taxes $T$ are assumed to be fixed numbers.

To find a numerical equation that describes the relationship of planned aggregate expenditure ($PAE$) to output, we can begin as in Chapter 13 with the general definition of planned aggregate expenditure:

$$PAE = C + I^P + G + NX.$$

Substituting for the four components of expenditure, using the equations describing each type of spending above, we get

$$PAE = [640 + 0.8(Y - 250) - 400r] + [250 - 600r] + 300 + 20.$$

The first term in brackets on the right side of this equation is the expression for consumption, using the fact that taxes $T = 250$; the second bracketed term is planned investment; and the last two terms correspond to the given numerical values of government purchases and net exports. If we simplify this equation and group together the terms that do not depend on output $Y$ and the terms that do depend on output, we get

$$PAE = [(640 - 0.8 \times 250 - 400r) + (250 - 600r) + 300 + 20] + 0.8Y,$$

or, simplifying further,

$$PAE = [1{,}010 - 1{,}000r] + 0.8Y. \tag{14.1}$$

In Equation 14.1, the term in brackets is *autonomous expenditure,* the portion of planned aggregate expenditure that does not depend on output. *Notice that in this example autonomous expenditure depends on the real interest rate r.* Induced expenditure, the portion of planned aggregate expenditure that does depend on output, equals $0.8Y$ in this example.

EXAMPLE 14.4

### The real interest rate and short-run equilibrium output

In the economy described in Example 14.3, the real interest rate $r$ is set by the Fed to equal 0.05 (5 percent). Find short-run equilibrium output.

We found in Example 14.3 that, in this economy, planned aggregate expenditure is given by Equation 14.1. We are given that the Fed sets the real interest rate at 5 percent. Setting $r = 0.05$ in Equation 14.1 gives

$$PAE = [1{,}010 - 1{,}000 \times (0.05)] + 0.8Y.$$

Simplifying, we get

$$PAE = 960 + 0.8Y.$$

So, when the real interest rate is 5 percent, autonomous expenditure is 960 and induced expenditure is $0.8Y$. Short-run equilibrium output is the level of output that equals planned aggregate spending. To find short-run equilibrium output, we could now apply the tabular method used in Chapter 13, comparing alternative values of output with the planned aggregate expenditure at that level of output. Short-run equilibrium output would be determined as the value of output such that output just equals spending, or

$$Y = PAE.$$

However, conveniently, when we compare this example with Example 13.2 in the last chapter, we see that the equation for planned aggregate expenditure, $PAE = 960 + 0.8Y$, is identical to what we found there. Thus Table 13.1, which we used to solve Example 13.2, applies to this example as well, and we get the same answer for short-run equilibrium output, which is $Y = 4{,}800$.

Short-run equilibrium output can also be found graphically, using the Keynesian cross diagram from Chapter 13. Again, since the equation for planned aggregate output is the same as in Example 13.2, Figure 13.3 applies equally well here.

EXERCISE 14.3

**For the economy described in Example 14.4, suppose the Fed sets the real interest rate at 3 percent rather than at 5 percent. Find short-run equilibrium output. (*Hint:* Consider values between 4,500 and 5,500.)**

## THE FED FIGHTS A RECESSION

We have seen that the Fed can control the real interest rate, and that the real interest rate in turn affects planned spending and short-run equilibrium output. Putting these two results together, we can see how Fed actions may help to stabilize the economy.

Suppose the economy faces a recessionary gap—a situation in which real output is below potential output, and planned spending is "too low." To fight a recessionary gap, the Fed should reduce the real interest rate, stimulating consumption and investment spending. According to the theory we have developed, this increase in planned spending will cause output to rise, restoring the economy to full employment. Example 14.5 illustrates this point by extending Example 14.4.

EXAMPLE 14.5

### The Fed fights a recession

For the economy described in Example 14.4, suppose potential output $Y^*$ equals 5,000. As before, the Fed has set the real interest rate equal to 5 percent. At that real interest rate, what is the output gap? What should the Fed do to eliminate the output gap and restore full employment? You are given that the multiplier in this economy is 5.

In Example 14.4 we showed that with the real interest rate at 5 percent, short-run equilibrium output for this economy is 4,800. Potential output is 5,000, so the output gap $(Y - Y^*)$ equals $5{,}000 - 4{,}800 = 200$. Because actual output is below potential, this economy faces a recessionary gap.

To fight the recession, the Fed should lower the real interest rate, raising aggregate expenditure until output reaches 5,000, the full-employment level. That is, the Fed's objective is to increase output by 200. Because the multiplier equals 5, to increase output by 200 the Fed must increase autonomous expenditure by $200/5 = 40$ units. By how much should the Fed reduce the real interest rate to increase autonomous expenditure by 40 units? Autonomous expenditure in this economy is $[1{,}010 - 1{,}000r]$, as you can see from Equation 14.1, so that each percentage point reduction in $r$ increases autonomous expenditure by $1{,}000 \times (0.01) = 10$ units. To increase autonomous expenditure by 40, then, the Fed should lower the real interest rate by 4 percentage points, from 5 percent to 1 percent.

In summary, to eliminate the recessionary gap of 200, the Fed should lower the real interest rate from 5 percent to 1 percent. Notice that the Fed's decrease in the real interest rate increases short-run equilibrium output, as economic logic suggests.

The Fed's recession-fighting policy is shown graphically in Figure 14.6. The reduction in the real interest rate raises planned spending at each level of output, shifting the expenditure line upward. When the real interest rate equals 1 percent, the expenditure line intersects the $Y = PAE$ line at $Y = 5{,}000$, so that output and potential output are equal. A reduction in interest rates by the Fed, made with the intention of reducing a recessionary gap in this way, is an example of an *expansionary* monetary policy—or, less formally, a *monetary easing*.

**FIGURE 14.6**
**The Fed Fights a Recession.**
When the real interest rate is 5 percent, the expenditure line intersects the $Y = PAE$ line at point *E*. At that point output is 4,800, below the economy's potential output of 5,000 (a recessionary gap of 200). If the Fed reduces the real interest rate to 1 percent, stimulating consumption and investment spending, the expenditure line will shift upward. At the new point of intersection *F*, output will equal potential output at 5,000.

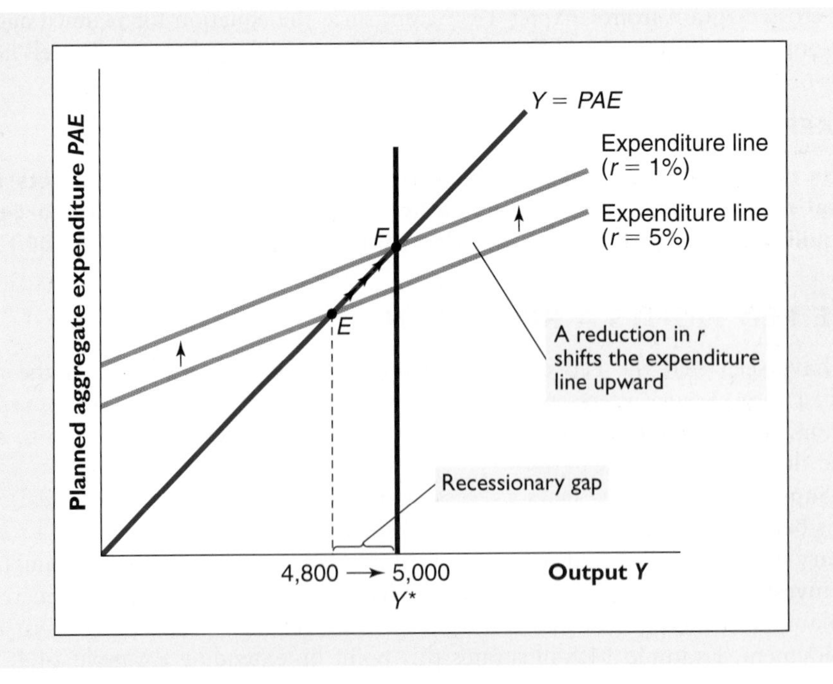


### EXERCISE 14.4

**Continuing Example 14.5, suppose that potential output is 4,850 rather than 5,000. By how much should the Fed cut the real interest rate to restore full employment? You may take as given that the multiplier is 5.**

**Why did the Fed cut the federal funds rate 23 times between 1989 and 1992?**

ECONOMIC NATURALIST 14.3

The federal funds rate fell sharply between 1989 and 1992 (see Figure 14.5). At its peak in March 1989, the federal funds rate was 9.9 percent; by December 1992 it had fallen to 2.9 percent. During this period the Federal Reserve announced reductions in the target federal funds rate on 23 separate occasions. Why did the Fed cut the federal funds rate so much in this period?

The U.S. economy went into recession in the summer of 1991, about the time Iraq invaded Kuwait. In an attempt to stimulate the economy, the Fed cut the federal funds rate numerous times. By March 1991, the trough of the recession, this key interest rate stood at 6.1 percent—almost 4 percentage points below its value two years earlier. Despite the Fed's actions, however, the economy recovered only slowly. Although the recession ended in March 1991, the unemployment rate continued to climb. The media dubbed the weak expansion a "jobless recovery."

The unusually slow recovery from the 1990–1991 recession puzzled economists as well as the media. Alan Greenspan, the chairman of the Fed, suggested that financial problems such as the credit crunch in the banking sector (see Economic Naturalist 13.3) were creating "financial headwinds" that held back economic growth. The Fed reacted to the sluggish performance of the economy by continuing to cut the federal funds rate, which reached 2.9 percent in December 1992 and remained around 3 percent until the spring of 1994. The Fed's policy appears ultimately to have been successful. Economic growth eventually picked up and the unemployment rate began to fall, reaching record lows by the end of the 1990s. Indeed, the expansion that began in March 1991 ultimately became the longest in U.S. history.

**How did the Fed respond to recession and the terror attacks in 2001?**

ECONOMIC NATURALIST 14.4

The U.S. economy began slowing in the fall of 2000, with investment in high-tech equipment falling particularly sharply (see Economic Naturalist 13.5). According to the National Bureau of Economic Research, a recession began in March 2001. To make matters worse, on September 11, 2001, terrorist attacks on New York City and Washington shocked the nation and led to serious problems in the travel and financial industries, among others. How did the Federal Reserve react to these events?

The Fed first began to respond to growing evidence of an economic slowdown at the end of the year 2000. At the time the federal funds rate stood at about 6.5 percent (see Figure 14.5). The Fed's most dramatic move was a surprise cut of 0.5 percentage points in the funds rate in January 2001, between regularly scheduled meetings of the Federal Open Market Committee. Further rate cuts followed, and by July the funds rate was below 4 percent. By summer's end, however, there was still considerable uncertainty about the likely severity of the economic slowdown.

The picture changed suddenly on September 11, 2001, when the terror attacks on the World Trade Center and the Pentagon killed more than 3,000 people. The terrorist attacks imposed great economic as well as human costs. The physical damage in lower Manhattan was in the billions of dollars, and many offices and businesses in the area had to close. The Fed, in its role as supervisor of the financial system, worked hard to assist in the restoration of normal operations in the financial district of New York City. (The Federal Reserve Bank of New York, which actually conducts open-market operations, is only a block from the site of the World Trade Center.) The Fed also tried to ease financial conditions by temporarily lowering the federal funds rate to as low as 1.25 percent, in the week following the attack.

In the weeks and months following September 11, the Fed turned its attention from the direct impact of the attack to the possible indirect effects on the

U.S. economy. The Fed was worried that consumers, nervous about the future, would severely cut back their spending; together with the ongoing weakness in investment, a fall in consumption spending could sharply worsen the recession. To stimulate spending, the Fed continued to cut the federal funds rate. By January 2002, the funds rate was at 1.75 percent, nearly 5 percentage points lower than a year earlier. The Fed kept the interest rate at that low level until November 2002, when it lowered the federal funds rate another .5 percentage point, to 1.25 percent.

As of this writing, a (weak) recovery from the recession that began in 2001 appears to be under way. A variety of factors have helped the economy, including expansionary fiscal policy (see Economic Naturalist 13.8). Most economists agree that expansionary actions by the Fed also played a constructive role in reducing the economic impact of the recession and the September 11 attacks.

## THE FED FIGHTS INFLATION

To this point we have focused on the problem of stabilizing output, without considering inflation. In the next chapter we will see how ongoing inflation can be incorporated into our analysis. For now we will simply note that one important cause of inflation is an expansionary output gap—a situation in which planned spending, and hence actual output, exceeds potential output. When an expansionary gap exists, firms find that the demand for their output exceeds their normal rate of production. Although firms may be content to meet this excess demand at previously determined prices for some time, if the high demand persists, they will ultimately raise their prices, spurring inflation.

Because an expansionary gap tends to lead to inflation, the Fed moves to eliminate expansionary gaps as well as recessionary gaps. The procedure for getting rid of an expansionary gap—a situation in which output is "too high" relative to potential output—is the reverse of that for fighting a recessionary gap, a situation in which output is "too low." As we have seen, the cure for a recessionary gap is to reduce the real interest rate, an action that stimulates planned spending and increases output. The cure for an expansionary gap is to *raise* the real interest rate, which reduces consumption and planned investment by raising the cost of borrowing. The resulting fall in planned spending leads in turn to a decline in output and to a reduction in inflationary pressures.

**EXAMPLE 14.6**

### The Fed fights inflation

For the economy studied in Examples 14.4 and 14.5, assume that potential output is 4,600 rather than 5,000. At the initial real interest rate of 5 percent, short-run equilibrium output is 4,800, so this economy has an expansionary gap of 200. How should the Fed change the real interest rate to eliminate this gap?

In Example 14.5 we were told that the multiplier in this economy is 5. Hence, to reduce total output by 200, the Fed needs to reduce autonomous expenditure by $200/5 = 40$ units. From Equation 14.1, we know that autonomous expenditure in this economy is $[1{,}010 - 1{,}000r]$, so that each percentage point (0.01) increase in the real interest rate lowers autonomous expenditure by 10 units ($1{,}000 \times 0.01$). We conclude that to eliminate the inflationary gap, the Fed should raise the real interest rate by 4 percentage points (0.04), from 5 percent to 9 percent. The higher real interest rate will reduce planned aggregate expenditure and output to the level of potential output, 4,600, eliminating inflationary pressures.

The effects of the Fed's inflation-fighting policy are shown in Figure 14.7. With the real interest rate at 5 percent, the expenditure line intersects the

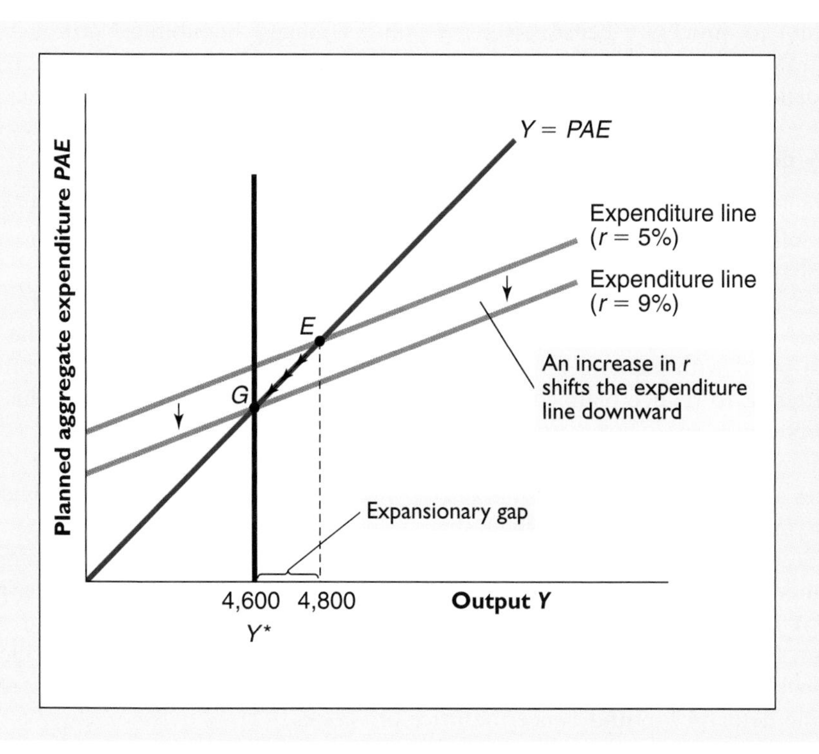


**FIGURE 14.7**
**The Fed Fights Inflation.**
When the real interest rate is 5 percent, the expenditure line intersects the $Y = PAE$, or 45°, line at point $E$, where short-run equilibrium output equals 4,800. If potential output is 4,600, an expansionary output gap of 200 exists. If the Fed raises the real interest rate to 9 percent, reducing planned aggregate expenditure, the expenditure line shifts downward. At the new intersection point $G$, actual output equals potential output at 4,600, and the expansionary gap is eliminated.

$Y = PAE$ line at point $E$ in the figure, where output equals 4,800. To reduce planned spending and output, the Fed raises the real interest rate to 9 percent. The higher real interest rate slows consumption and investment spending, moving the expenditure line downward. At the new equilibrium point $G$, actual output equals potential output at 4,600. The Fed's raising the real interest rate—a contractionary policy action—has thus eliminated the expansionary output gap, and with it, the threat of inflation.

*"Personally, I liked this roller coaster a lot better before the Federal Reserve Board got hold of it."*

The Fed's interest rate policies affect the economy as a whole, but they have a particularly important effect on financial markets. The introduction to this chapter noted the tremendous lengths financial market participants will go to in an attempt to anticipate Federal Reserve policy changes. Economic Naturalist 14.5 illustrates the type of information financial investors look for, and why it is so important to them.

**Why does news of inflation hurt the stock market?**

Financial market participants watch data on inflation extremely closely. A report that inflation is increasing or is higher than expected often causes stock prices to fall sharply. Why does bad news about inflation hurt the stock market?

Investors in the financial markets worry about inflation because of its likely impact on Federal Reserve policy. Financial investors understand that the Fed, when faced with signs of an expansionary gap, is likely to raise interest rates in an attempt to reduce planned spending and "cool down" the economy. This type of contractionary policy action hurts stock prices in two ways. First, it slows down economic activity, reducing the expected sales and profits of companies whose shares are traded in the stock market. Lower profits, in turn, reduce the dividends those firms are likely to pay their shareholders.

Second, higher real interest rates reduce the value of stocks by increasing the required return for holding stocks. We saw in Chapter 11 that an increase in the return financial investors require in order to hold stocks lowers current stock prices. Intuitively, if interest rates rise, interest-bearing alternatives to stocks such as newly issued government bonds will become more attractive to investors, reducing the demand for, and hence the price of, stocks.

*"Interest rates gyrated wildly today, on rumors that the Federal Reserve Board would be replaced by the cast of 'Saturday Night Live.'"*

ECONOMIC NATURALIST 14.6

**Should the Federal Reserve respond to changes in stock prices?**

Many credit the Federal Reserve and its chairman, Alan Greenspan, for effective monetary policymaking that set the stage for sustained economic growth and rising asset prices throughout the 1990s, in particular the second half of the decade. Between January 1995 and March 2000 the S&P 500 stock market index rose from a value of 459 to 1,527, a phenomenal 233 percent increase in just over five years, as the U.S. economy enjoyed a record-long business cycle expansion. Indeed, the stock market's strong, sustained rise helped to fuel additional consumer spending, which in turn promoted further economic expansion.

However, as stock prices fell sharply in the two years after their March 2000 peak, some people questioned whether the Federal Reserve should have preemptively raised interest rates to constrain investors' "irrational exuberance."[4] Overly optimistic investor sentiment led to a speculative run-up in stock prices that eventually burst in 2000 as investors began to realize that firms' earnings could not support the stock prices that were being paid. Earlier intervention by the Federal Reserve, critics argued, would have slowed down the dramatic increase in stock prices and therefore could have prevented the resulting stock market "crash" and the resulting loss of consumer wealth. As this chapter makes clear, the Federal Reserve's primary focus is on reducing output gaps and keeping inflation low. Should the Fed also respond to changing stock prices when it makes decisions about monetary policy?

At a symposium in August 2002, Alan Greenspan defended the Fed's monetary policymaking performance in the late 1990s, pointing out that it is very difficult to identify asset bubbles—surges in prices of assets to unsustainable levels—"until after the fact—that is, when its bursting confirm(s) its existence."[5] Even if such a speculative bubble could be identified, Greenspan noted, the Federal Reserve could have done little—short of "inducing a substantial contraction in economic activity"—to prevent investors' speculation from driving up stock prices. Indeed, Greenspan claimed, "the notion that a well-timed incremental tightening could have been calibrated to prevent the late 1990s bubble is almost surely an illusion." Rather, the Federal Reserve was focusing as early as 1999 on policies that would "mitigate the fallout when it occurs and, hopefully, ease the transition to the next expansion."[6]

Greenspan's remarks strongly suggest that in the absence of macroeconomic models that provide a comprehensive understanding of how dramatic movements in stock prices develop in the economy, the Federal Reserve should maintain its policymaking focus on reducing output gaps and keeping inflation low. Current research on the appropriate role of stock prices in the formulation of monetary policy is mixed: some researchers have come to the conclusion that changes in stock prices should affect monetary policy only to the extent that those changes affect future inflationary expectations, while others find strong support for including stock prices in the Fed's monetary policy decisions. Until economic research provides a definitive answer to this important question, the Fed will continue to monitor stock market conditions, but focus its monetary policymaking attention on the economic effects of changes in stock prices on inflation, spending, and output, rather than the stock prices themselves.

## THE FED'S POLICY REACTION FUNCTION

The Fed attempts to stabilize the economy by manipulating the real interest rate. When the economy faces a recessionary gap, the Fed reduces the real interest rate in order to stimulate spending. When an expansionary gap exists, so that inflation threatens to become a problem, the Fed restrains spending by raising the real interest rate. Economists sometimes find it convenient to summarize the behavior of the Fed in terms of a *policy reaction function*. In general, a **policy reaction function** describes how the action a policymaker takes depends on the state of the economy. Here, the policymaker's action is the Fed's choice of the real interest rate, and the state of the economy is given by factors such as the output gap or the inflation rate. Economic Naturalist 14.7 describes one attempt to quantify the Fed's policy reaction function.

***policy reaction function*** describes how the action a policymaker takes depends on the state of the economy

[4] Fed Chairman Alan Greenspan mentioned the possibility of "irrational exuberance" driving investor behavior in a December 5, 1996 speech, which is available online at http://www.federalreserve.gov/boarddocs/speeches/1996/19961205.htm.

[5] The text of Greenspan's speech is available online at http://www.federalreserve.gov/boarddocs/speeches/2002/20020830/default.htm.

[6] *The Federal Reserve's semiannual report on monetary policy,* testimony of Chairman Alan Greenspan before the Committee on Banking and Financial Services, U.S. House of Representatives July 22, 1999. Available online at: http://www.federalreserve.gov/boarddocs/hh/1999/July/Testimony.htm.

**What is the Taylor rule?**

In 1993 economist John Taylor proposed a "rule," now known as the Taylor rule, to describe the behavior of the Fed.[7] What is the Taylor rule? Does the Fed always follow it?

The rule Taylor proposed is not a rule in any legal sense but is instead an attempt to describe the Fed's behavior in terms of a quantitative policy reaction function. Taylor's "rule" can be written as

$$r = 0.01 - 0.5\left(\frac{Y^* - Y}{Y^*}\right) + 0.5\pi$$

where $r$ is the real interest rate set by the Fed, expressed as a decimal (for example, $5\% = 0.05$); $Y^* - Y$ is the current output gap (the difference between potential and actual output); $(Y^* - Y)/Y^*$ is the output gap relative to potential output; and $\pi$ is the inflation rate, expressed as a decimal (for example, a 2 percent inflation rate is expressed as 0.02). According to the Taylor rule, the Fed responds to both output gaps and the rate of inflation. For example, the formula implies that if a recessionary gap equal to a fraction 0.01 of potential output develops, the Fed will reduce the real interest rate by 0.5 percentage points (that is, 0.005). Similarly, if inflation rises by 1 percentage point (0.01), according to the Taylor rule the Fed will increase the real interest rate by 0.5 percentage points (0.005). Taylor has shown that his rule does in fact describe the behavior of the Fed under Chairman Alan Greenspan reasonably accurately. Thus the Taylor rule is a real-world example of a policy reaction function.

Although the Taylor rule has worked well as a description of the Fed's behavior, we reiterate that it is not a rule in any legal sense. The Fed is perfectly free to deviate from it and does so when circumstances warrant. Still, the Taylor rule provides a useful benchmark for assessing, and predicting, the Fed's actions.

**EXERCISE 14.5**

**This exercise asks you to apply the Taylor rule. Suppose inflation is 3 percent and the output gap is zero. According to the Taylor rule, at what value should the Fed set the real interest rate? The nominal interest rate? Suppose the Fed were to receive new information showing that there is a 1 percent recessionary gap (inflation is still 3 percent). According to the Taylor rule, how should the Fed change the real interest rate, if at all?**

Notice that according to the Taylor rule, the Fed responds to two variables—the output gap and inflation. In principle, any number of economic variables, from stock prices to the value of the dollar in terms of the Japanese yen, could affect Fed policy and thus appear in the policy reaction function. For the sake of simplicity, in applying the policy reaction function idea in the next chapter, we will assume that the Fed's choice of the real interest rate depends on only one variable—the rate of inflation. This simplification will not change our main results in any significant way. Furthermore, as we will see, having the Fed react only to inflation captures the most important aspect of Fed behavior—namely, its tendency to raise the real interest rate when the economy is "overheating" (experiencing an expansionary gap) and to reduce it when the economy is sluggish (experiencing a recessionary gap).

Table 14.1 describes an example of a policy reaction function according to which the Fed reacts only to inflation. According to the policy reaction function given in the table, the higher the rate of inflation, the higher the real interest rate

[7]John Taylor, "Discretion versus Policy Rules in Practice," *Carnegie-Rochester Conference Series on Public Policy,* 1963, pp. 195–214.

**TABLE 14.1**
**A Policy Reaction Function for the Fed**

| Rate of inflation, $\pi$ | Real interest rate set by Fed, $r$ |
|---|---|
| 0.00 (= 0%) | 0.02 (= 2%) |
| 0.01 | 0.03 |
| 0.02 | 0.04 |
| 0.03 | 0.05 |
| 0.04 | 0.06 |

set by the Fed. This relationship is consistent with the idea that the Fed responds to an expansionary gap (which threatens to lead to increased inflation) by raising the real interest rate. Figure 14.8 is a graph of this policy reaction function. The vertical axis of the graph shows the real interest rate chosen by the Fed; the horizontal axis shows the rate of inflation. The upward slope of the policy reaction function captures the idea that the Fed reacts to increases in inflation by raising the real interest rate.

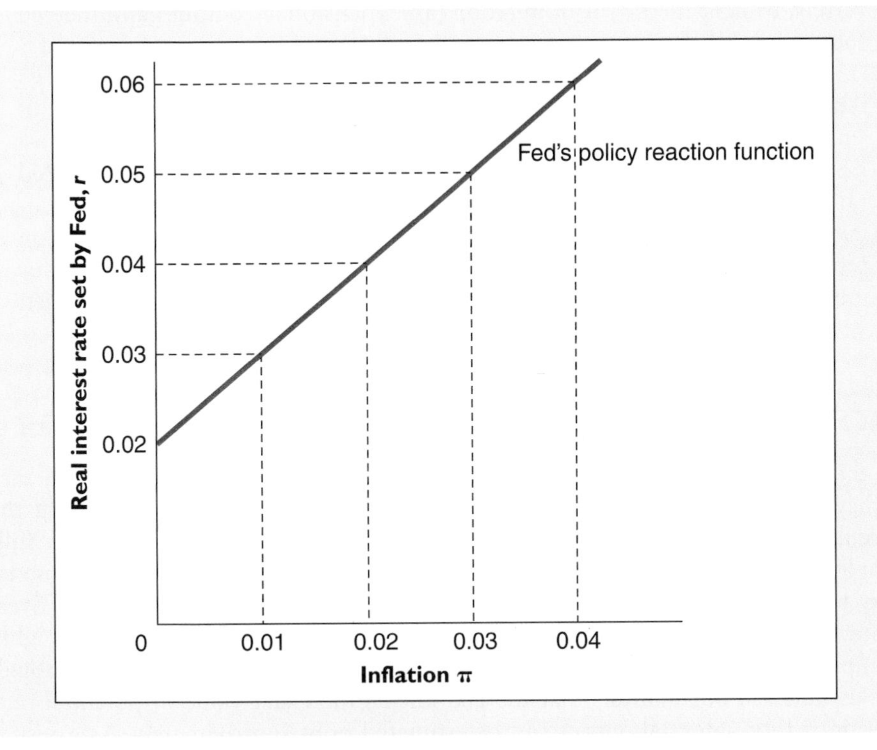


**FIGURE 14.8**
**An Example of a Fed Policy Reaction Function.** This hypothetical example of a policy reaction function for the Fed shows the real interest rate the Fed sets in response to any given value of the inflation rate. The upward slope captures the idea that the Fed raises the real interest rate when inflation rises. The numerical values in the figure are from Table 14.1.

How does the Fed determine its policy reaction function? In practice the process is a complex one, involving a combination of statistical analysis of the economy and human judgment. However, two useful insights into the process can be drawn even from the simplified policy reaction function shown in Table 14.1 and Figure 14.8. First, as we mentioned earlier in the chapter, though the Fed controls the real interest rate in the short run, in the long run the real interest rate is determined by the balance of saving and investment. To illustrate the implication of this fact for the Fed's choice of policy reaction function, suppose that the Fed estimates the long-run value of the real interest rate (as determined by the supply and demand for saving) to be 4 percent, or 0.04. By examining Table 14.1, we can see that the Fed's policy reaction function implies a long-run value of the real interest rate of 4 percent only if the inflation rate in the long run is 2 percent. Thus the Fed's choice of this policy reaction function makes sense only if the Fed's long-run target rate of inflation is 2 percent. We conclude that one important determinant of the Fed's policy reaction function is the policymakers' objective for inflation.

Second, the Fed's policy reaction function contains information not only about the central bank's long-run inflation target but also about how aggressively the Fed plans to pursue that target. To illustrate, suppose the Fed's policy reaction function was very flat, implying that the Fed changes the real interest rate rather modestly in response to increases or decreases in inflation. In this case we would conclude that the Fed does not intend to be very aggressive in its attempts to offset movements in inflation away from the target level. In contrast, if the reaction function slopes steeply upward, so that a given change in inflation elicits a large adjustment of the real interest rate by the Fed, we would say that the Fed plans to be quite aggressive in responding to changes in inflation.

**RECAP** **MONETARY POLICY AND THE ECONOMY**

An increase in the real interest rate reduces both consumption spending and planned investment spending. Through its control of the real interest rate, the Fed is thus able to influence planned spending and short-run equilibrium output. To fight a recession (a recessionary output gap), the Fed should lower the real interest rate, stimulating planned spending and output. Conversely, to fight the threat of inflation (an expansionary output gap), the Fed should raise the real interest rate, reducing planned spending and output.

The Fed's policy reaction function relates its policy action (specifically, its setting of the real interest rate) to the state of the economy. For the sake of simplicity, we consider a policy reaction function in which the real interest rate set by the Fed depends only on the rate of inflation. Because the Fed raises the real interest rate when inflation rises, in order to restrain spending, the Fed's policy reaction is upward-sloping. The Fed's policy reaction function contains information about the central bank's long-run target for inflation and the aggressiveness with which it intends to pursue that target.

## MONETARY POLICYMAKING: ART OR SCIENCE?

In this chapter we analyzed the basic economics underlying real-world monetary policy. As part of the analysis we worked through some examples showing the calculation of the real interest rate that is needed to restore output to its full-employment level. While those examples are useful in understanding how monetary policy works—as with our analysis of fiscal policy in Chapter 13—they overstate the precision of monetary policymaking. The real-world economy is highly complex, and our knowledge of its workings is imperfect. For example, though we assumed in our analysis that the Fed knows the exact value of potential output, in reality potential output can be estimated only approximately. As a result, at any given time the Fed has only a rough idea of the size of the output gap. Similarly, Fed policymakers have only an approximate idea of the effect of a given change in the real interest rate on planned spending, or the length of time before that effect will occur. Because of these uncertainties, the Fed tends to proceed cautiously. Fed policymakers avoid large changes in interest rates and rarely raise or lower the federal funds rate more than one-half of a percentage point (from 5.50 percent to 5.00 percent, for example) at any one time. Indeed, the typical change in the interest rate is one-quarter of a percentage point.

Is monetary policymaking an art or a science, then? In practice it appears to be both. Scientific analyses, such as the development of detailed statistical models of the economy, have proved useful in making monetary policy. But human judgment based on long experience—what has been called the "art" of monetary policy—plays a crucial role in successful policymaking and is likely to continue to do so.

## SUMMARY

- Monetary policy is one of two types of stabilization policy, the other being fiscal policy. Although the Federal Reserve operates by controlling the money supply, the media's attention nearly always focuses on the Fed's decisions about interest rates, not the money supply. There is no contradiction between these two ways of looking at monetary policy, however, as the Fed's ability to control the money supply is the source of its ability to control interest rates.

- The nominal interest rate is determined in the market for money, which has both a demand side and a supply side. For the economy as a whole, the *demand for money* is the amount of wealth households and businesses choose to hold in the form of money (such as cash or checking accounts). The demand for money is determined by a comparison of cost and benefits. The opportunity cost of holding money, which pays either zero interest or very low interest, is the interest that could have been earned by holding interest-bearing assets instead of money. Because the nominal interest rate measures the opportunity cost of holding a dollar in the form of money, an increase in the nominal interest rate reduces the quantity of money demanded. The benefit of money is its usefulness in carrying out transactions. All else being equal, an increase in the volume of transactions increases the demand for money. At the macroeconomic level, an increase in the price level or in real GDP increases the dollar volume of transactions, and thus the demand for money.

- The *money demand curve* relates the aggregate quantity of money demanded to the nominal interest rate. Because an increase in the nominal interest rate increases the opportunity cost of holding money, which reduces the quantity of money demanded, the money demand curve slopes down. Factors other than the nominal interest rate that affect the demand for money will shift the demand curve to the right or left. For example, an increase in the price level or real GDP increases the demand for money, shifting the money demand curve to the right.

- The Federal Reserve determines the supply of money through the use of open-market operations. The supply curve for money is vertical at the value of the money supply set by the Fed. Money market equilibrium occurs at the nominal interest rate at which money demand equals the money supply. The Fed can reduce the nominal interest rate by increasing the money supply (shifting the money supply curve to the right) or increase the nominal interest rate by reducing the money supply (shifting the money supply curve to the left). The nominal interest rate that the Fed targets most closely is the *federal funds rate,* which is the rate commercial banks charge each other for very short-term loans.

- In the short run, the Fed can control the real interest rate as well as the nominal interest rate. Recall that the real interest rate equals the nominal interest rate minus the inflation rate. Because the inflation rate adjusts relatively slowly, the Fed can change the real interest rate by changing the nominal interest rate. In the long run, the real interest rate is determined by the balance of saving and investment (Chapter 9).

- The Federal Reserve's actions affect the economy because changes in the real interest rate affect planned spending. For example, an increase in the real interest rate raises the cost of borrowing, reducing consumption and planned investment. Thus, by increasing the real interest rate, the Fed can reduce planned spending and short-run equilibrium output. Conversely, by reducing the real interest rate, the Fed can stimulate planned aggregate expenditure and thereby raise short-run equilibrium output. The Fed's ultimate objective is to eliminate output gaps. To eliminate a recessionary output gap, the Fed will lower the real interest rate. To eliminate an expansionary output gap the Fed will raise the real interest rate.

- A *policy reaction function* describes how the action a policymaker takes depends on the state of the economy. For example, a policy reaction function for the Fed could specify the real interest rate set by the Fed for each value of inflation.

- In practice, the Fed's information about the level of potential output and the size and speed of the effects of its actions is imprecise. Thus monetary policymaking is as much an art as a science.

## KEY TERMS

demand for money (369)
federal funds rate (378)
money demand curve (373)
policy reaction function (389)
portfolio allocation decision (369)

## REVIEW QUESTIONS

1. What is the *demand for money*? How does the demand for money depend on the nominal interest rate? On the price level? On income? Explain in terms of the costs and benefits of holding money.
2. Show graphically how the Fed controls the nominal interest rate. Can the Fed control the real interest rate?
3. What effect does an open-market purchase of bonds by the Fed have on nominal interest rates? Discuss in terms

of (a) the effect of the purchase on bond prices and (b) the effect of the purchase on the supply of money.

4. You hear a news report that employment growth is lower than expected. How do you expect that report to affect market interest rates? Explain. (*Hint:* Assume that Fed policymakers have access to the same data that you do.)

5. Why does the real interest rate affect planned aggregate expenditure? Give examples.

6. The Fed faces a recessionary gap. How would you expect it to respond? Explain step by step how its policy change is likely to affect the economy.

7. The Fed decides to take a *contractionary* policy action. What would you expect to happen to the nominal interest rate, the real interest rate, and the money supply? Under what circumstances would this type of policy action most likely be appropriate?

8. Define *policy reaction function.* Sketch a policy reaction function relating the Fed's setting of the real interest rate to inflation.

9. Discuss why the analysis of this chapter overstates the precision with which monetary policy can be used to eliminate output gaps.

## PROBLEMS

1. During the heavy Christmas shopping season, sales of retail stores, online sales firms, and other merchants rise significantly.
   a. What would you expect to happen to the money demand curve during the Christmas season? Show graphically.
   b. If the Fed took no action, what would happen to nominal interest rates around Christmas?
   c. In fact, nominal interest rates do not change significantly in the fourth quarter of the year, due to deliberate Fed policy. Explain and show graphically how the Fed can ensure that nominal interest rates remain stable around Christmas.

2. The following table shows Uma's estimated annual benefits of holding different amounts of money:

| Average money holdings ($) | Total benefit ($) |
|---|---|
| 500 | 35 |
| 600 | 47 |
| 700 | 57 |
| 800 | 65 |
| 900 | 71 |
| 1,000 | 75 |
| 1,100 | 77 |
| 1,200 | 77 |

   a. How much money will Uma hold on average if the nominal interest rate is 9 percent? 5 percent? 3 percent? Assume that she wants her money holding to be a multiple of $100. (*Hint:* Make a table comparing the extra benefit of each additional $100 in money holdings with the opportunity cost, in terms of forgone interest, of additional money holdings.)
   b. Graph Uma's money demand curve for interest rates between 1 percent and 12 percent.

3. How would you expect each of the following to affect the economywide demand for money? Explain.
   a. Competition among brokers forces down the commission charge for selling holdings of bonds or stocks.
   b. Grocery stores begin to accept credit cards in payment.
   c. Financial investors become concerned about increasing riskiness of stocks.
   d. Online banking allows customers to check balances and transfer funds between checking and mutual fund investments 24 hours a day.
   e. The economy enters a boom period.
   f. Political instability increases in developing nations.

4. Suppose the economywide demand for money is given by $P(0.2Y - 25{,}000i)$. The price level $P$ equals 3.0, and real output $Y$ equals 10,000. At what value should the Fed set the nominal money supply if
   a. It wants to set the nominal interest rate at 4 percent?
   b. It wants to set the nominal interest rate at 6 percent?

5. An economy is described by the following equations:

$$\begin{aligned} C &= 2{,}600 + 0.8(Y - T) - 10{,}000r, \\ I^P &= 2{,}000 - 10{,}000r, \\ G &= 1{,}800, \\ NX &= 0, \\ T &= 3{,}000. \end{aligned}$$

   The real interest rate, expressed as a decimal, is 0.10 (that is, 10 percent). Find a numerical equation relating planned aggregate expenditure to output. Using a table or other method, solve for short-run equilibrium output. Show your result graphically using the Keynesian cross diagram.

6. For the economy described in Problem 5 above:
   a. Potential output $Y^*$ equals 12,000. What real interest rate should the Fed set to bring the economy to full employment? You may take as given that the multiplier for this economy is 5.
   b. Repeat part a for the case in which potential output $Y^* = 9{,}000$.
   c.* Show that the real interest rate you found in part a sets national saving at potential output, defined as $Y^* - C - G$, equal to planned investment, $I^P$. This result shows that the real interest rate must be consistent with equilibrium in the market for saving when the economy is at full employment.

7.* Here is another set of equations describing an economy:

$$\begin{aligned} C &= 14{,}400 + 0.5(Y - T) - 40{,}000r, \\ I^P &= 8{,}000 - 20{,}000r, \\ G &= 7{,}000, \\ NX &= -1{,}800, \\ T &= 8{,}000, \\ Y^* &= 40{,}000. \end{aligned}$$

   a. Find a numerical equation relating planned aggregate expenditure to output and to the real interest rate.
   b. At what value should the Fed set the real interest rate to eliminate any output gap? (*Hint:* Set output $Y$ equal to the value of potential output given above in the equation you found in part a. Then solve for the real interest rate that also sets planned aggregate expenditure equal to potential output.)

8. Supposing that the Fed follows the Taylor rule (the Economic Naturalist 14.7), find the real interest rate and the nominal interest rate that the Fed will set in each of the following situations:
   a. Inflation of 4 percent and an expansionary gap equal to 1 percent of potential output.
   b. Inflation of 2 percent and a recessionary gap equal to 2 percent of potential output.
   c. Inflation of 6 percent and no output gap.
   d. Inflation of 2 percent and a recessionary gap of 5 percent. (Can the Fed set a negative real interest rate? If so, how?)

9. In mid-2002, with inflation at 2 percent, some economists estimated the size of the recessionary gap to be about 2 percent of potential output. At that time, the Fed was holding the (nominal) federal funds rate at 1.75 percent. How does the Fed's setting of the federal funds rate compare with what would be predicted by the Taylor rule?

*Problems marked with an asterisk (*) are more difficult.

10. By law, the Federal Reserve must report twice each year to Congress about monetary policy and the state of the economy. When the Monetary Policy Report is presented, it is customary for the Fed chairman to testify before Congress, to update legislators on the economic situation.

    Obtain a copy of the most recent Monetary Policy Report from the Fed's web page http://www.federalreserve.gov/ (click "Monetary Policy" and follow the links). In the period covered by the testimony, did monetary policy ease, tighten, or remain neutral? What principal developments in the economy led the Fed to take the actions that it did?

## ANSWERS TO IN-CHAPTER EXERCISES

14.1 At 4 percent interest, the benefit of each \$10,000 reduction in cash holdings is \$400 per year (4% × \$10,000). In this case the cost of the extra armored car service, \$500 a year, exceeds the benefit of reducing cash holdings by \$10,000. Kim's restaurants should therefore continue to hold \$50,000 in cash. Comparing this result with Example 14.2, you can see that the demand for money by Kim's restaurants is lower, the higher the nominal interest rate.

14.2 If the nominal interest rate is above its equilibrium value, then people are holding more money than they would like. To bring their money holdings down, they will use some of their money to buy interest-bearing assets such as bonds.

If everyone is trying to buy bonds, however, the price of bonds will be bid up. An increase in bond prices is equivalent to a fall in market interest rates. As interest rates fall, people will be willing to hold more money. Eventually interest rates will fall enough that people are content to hold the amount of money supplied by the Fed, and the money market will be in equilibrium.

14.3 If $r = 0.03$, then consumption is $C = 640 + 0.8(Y - 250) - 400(0.03) = 428 + 0.8Y$, and planned investment is $I^P = 250 - 600(0.03) = 232$. Planned aggregate expenditure is given by

$$\begin{aligned} PAE &= C + I^P + G + NX \\ &= (428 + 0.8Y) + 232 + 300 + 20 \\ &= 980 + 0.8Y. \end{aligned}$$

To find short-run equilibrium output, we can construct a table analogous to Table 14.1. As usual, some trial and error is necessary to find an appropriate range of guesses for output (column 1).

**Determination of short-run equilibrium output**

| (1) Output Y | (2) Planned aggregate expenditure PAE = 980 + 0.8Y | (3) Y − PAE | (4) Y = PAE? |
|---|---|---|---|
| 4,500 | 4,580 | −80 | No |
| 4,600 | 4,660 | −60 | No |
| 4,700 | 4,740 | −40 | No |
| 4,800 | 4,820 | −20 | No |
| 4,900 | 4,900 | 0 | **Yes** |
| 5,000 | 4,980 | 20 | No |
| 5,100 | 5,060 | 40 | No |
| 5,200 | 5,140 | 60 | No |
| 5,300 | 5,220 | 80 | No |
| 5,400 | 5,300 | 100 | No |
| 5,500 | 5,380 | 120 | No |

Short-run equilibrium output equals 4,900, as that is the only level of output that satisfies the condition $Y = PAE$.

The answer can be obtained more quickly by simply setting $Y = PAE$ and solving for short-run equilibrium output $Y$. Remembering that $PAE = 980 + 0.8Y$ and substitution for $PAE$, we get

$$Y = 980 + 0.8Y$$
$$Y(1 - 0.8) = 980$$
$$Y = 5 \times 980 = 4{,}900.$$

So lowering the real interest rate from 5 percent to 3 percent increases short-run equilibrium output from 4,800 (as found in Example 14.4) to 4,900.

If you have read Appendix 13B on the multiplier, there is yet another way to find the answer. Using Appendix 13B we can determine that the multiplier in this model is 5, since $1/(1 - c) = 1/(1 - 0.8) = 5$. Each percentage point reduction in the real interest rate increases consumption by 4 units and planned investment by 6 units, for a total impact on planned spending of 10 units per percentage point reduction. Reducing the real interest rate by 2 percentage points, from 5 percent to 3 percent, thus increases autonomous expenditure by 20 units. Because the multiplier is 5, an increase of 20 in autonomous expenditure raises short-run equilibrium output by $20 \times 5 = 100$ units, from the value of 4,800 we found in Example 14.4 to the new value of 4,900.

14.4 When the real interest rate is 5 percent, output is 4,800. Each percentage point reduction in the real interest rate increases autonomous expenditure by 10 units. Since the multiplier in this model is 5, to raise output by 50 units the real interest rate should be cut by 1 percentage point, from 5 percent to 4 percent. Increasing output by 50 units, to 4,850, eliminates the output gap.

14.5 If $\pi = 0.03$ and the output gap is zero, we can plug these values into the Taylor rule to obtain

$$r = 0.01 - 0.5(0) + 0.5(0.03) = 0.025 = 2.5\%.$$

So the real interest rate implied by the Taylor rule when inflation is 3 percent and the output gap is zero is 2.5 percent. The nominal interest rate equals the real rate plus the inflation rate, or $2.5\% + 3\% = 5.5\%$.

If there is a recessionary gap of 1 percent of potential output, the Taylor rule formula becomes

$$r = 0.01 - 0.5(0.01) + 0.5(0.03) = 0.02 = 2\%.$$

The nominal interest rate implied by the Taylor rule in this case is the 2 percent real rate plus the 3 percent inflation rate, or 5 percent. So the Taylor rule has the Fed lowering the interest rate when the economy goes into recession, which is both sensible and realistic.

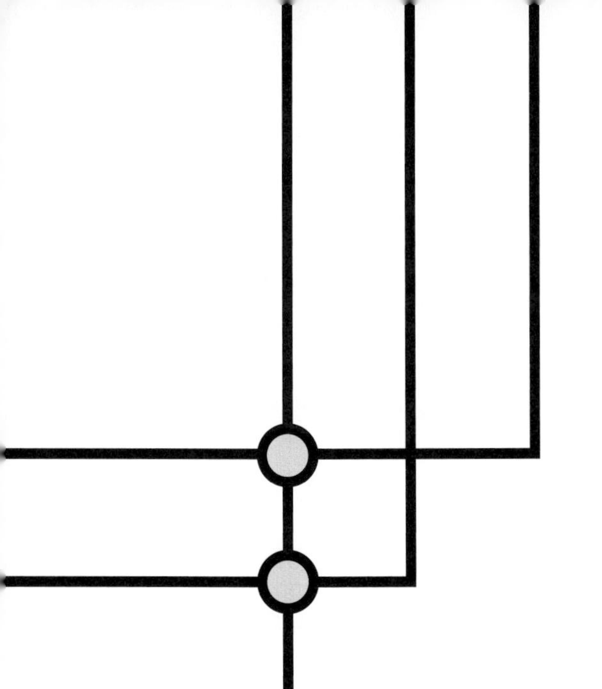

APPENDIX

# MONETARY POLICY IN THE BASIC KEYNESIAN MODEL

This appendix extends the algebraic analysis of the basic Keynesian model that was presented in Appendix 13A to include the role of monetary policy. The main difference from Appendix 13A is that in this analysis the real interest rate is allowed to affect planned spending. We will not describe the supply and demand for money algebraically but will simply assume that the Fed can set the real interest rate $r$ at any level it chooses.

The real interest rate affects consumption and planned investment. To capture these effects, we will modify the equations for those two components of spending as follows:

$$C = \overline{C} + c(Y - T) - ar,$$
$$I^P = \overline{I} - br.$$

The first equation is the consumption function with an additional term, equal to $-ar$. Think of $a$ as a fixed number, greater than zero, that measures the strength of the interest rate effect on consumption. Thus the term $-ar$ captures the idea that when the real interest rate $r$ rises, consumption declines by $a$ times the increase in the interest rate. Likewise, the second equation adds the term $-br$ to the equation for planned investment spending. The parameter $b$ is a fixed positive number that measures how strongly changes in the real interest rate affect planned investment; for example, if the real interest rate $r$ rises, planned investment is assumed to decline by $b$ times the increase in the real interest rate. We continue to assume that government purchases, taxes, and net exports are exogenous variables, so that $G = \overline{G}$, $T = \overline{T}$, and $NX = \overline{NX}$.

To solve for short-run equilibrium output, we start as usual by finding the relationship of planned aggregate expenditure to output. The definition of planned aggregate expenditure is

$$PAE = C + I^P + G + NX.$$

Substituting the modified equations for consumption and planned investment into this definition, along with the exogenous values of government spending, net exports, and taxes, we get

$$PAE = [\overline{C} + c(Y - \overline{T}) - ar] + [\overline{I} - br] + \overline{G} + \overline{NX}.$$

The first term in brackets on the right side describes the behavior of consumption, and the second bracketed term describes planned investment. Rearranging this equation in order to group together terms that depend on the real interest rate and terms that depend on output, we find

$$PAE = [\overline{C} - c\overline{T} + \overline{I} + \overline{G} + \overline{NX}] - (a + b)r + cY.$$

This equation is similar to Equation 13A.1, in the first appendix to Chapter 13, except that it has an extra term, $-(a + b)r$, on the right side. This extra term captures the idea that an increase in the real interest rate reduces consumption and planned investment, lowering planned spending. Notice that the term $-(a + b)r$ is part of autonomous expenditure, since it does not depend on output. Since autonomous expenditure determines the intercept of the expenditure line in the Keynesian cross diagram, changes in the real interest rate will shift the expenditure line up (if the real interest rate decreases) or down (if the real interest rate increases).

To find short-run equilibrium output, we uses the definition of short-run equilibrium output to set $Y = PAE$ and solve for $Y$:

$$\begin{aligned} Y &= PAE \\ &= [\overline{C} - c\overline{T} + \overline{I} + \overline{G} + \overline{NX}] - (a + b)r + cY \\ Y(1 - c) &= [\overline{C} - c\overline{T} + \overline{I} + \overline{G} + \overline{NX}] - (a + b)r \\ Y &= \left(\frac{1}{1 - c}\right)[(\overline{C} - c\overline{T} + \overline{I} + \overline{G} + \overline{NX}) - (a + b)r]. \qquad (14A.1) \end{aligned}$$

Equation 14A.1 shows that short-run equilibrium output once again equals the multiplier, $1/(1 - c)$, times autonomous expenditure, $\overline{C} - c\overline{T} + \overline{I} + \overline{G} + \overline{NX} - (a + b)r$. Autonomous expenditure in turn depends on the real interest rate $r$. The equation also shows that the impact of a change in the real interest rate on short-run equilibrium output depends on two factors: (1) the effect of a change in the real interest rate on consumption and planned investment, which depends on the magnitude of $(a + b)$; and (2) the size of the multiplier, $1/(1 - c)$, which relates changes in autonomous expenditure to changes in short-run equilibrium output. The larger the effect of the real interest rate on planned spending, and the larger the multiplier, the more powerful will be the effect of a given change in the real interest rate on short-run equilibrium output.

To check Equation 14A.1, we can use it to resolve Example 14.4 (see page 383). In that example we are given $\overline{C} = 640$, $\overline{I} = 250$, $\overline{G} = 300$, $\overline{NX} = 20$, $\overline{T} = 250$, $c = 0.8$, $a = 400$, and $b = 600$. The real interest rate set by the Fed is 5 percent, or 0.05. Substituting these values into Equation 14A.1 and solving, we obtain

$$\begin{aligned} Y &= \left(\frac{1}{1 - 0.8}\right)[640 - 0.8 \times 250 + 250 + 300 + 20 - (400 + 600) \times 0.05] \\ &= 5 \times 960 = 4{,}800 \end{aligned}$$

This is the same result we found in Example 14.4.

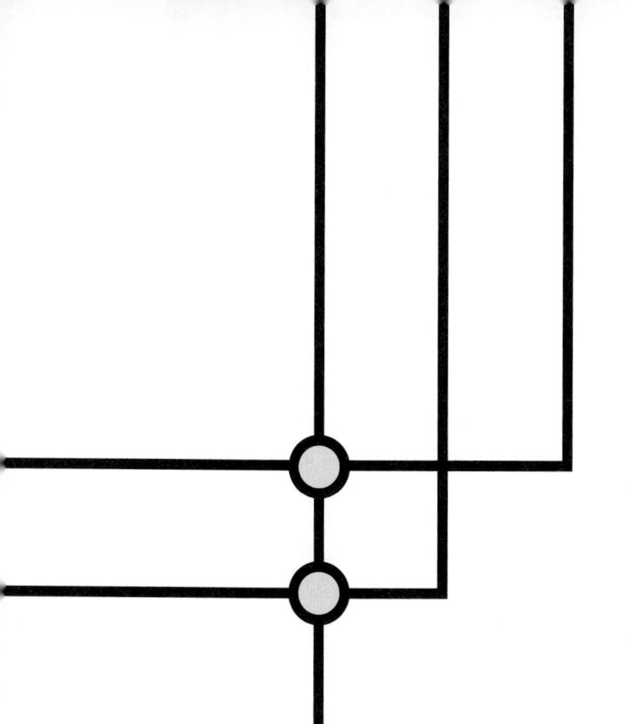

CHAPTER 17

# EXCHANGE RATES AND THE OPEN ECONOMY

Two Americans visiting London were commiserating over their problems understanding English currency. "Pounds, shillings, tuppence, thruppence, bob, and quid, it's driving me crazy," said the first American. "This morning it took me 20 minutes to figure out how much to pay the taxi driver."

The second American was more upbeat. "Actually," he said, "since I adopted my new system, I haven't had any problems at all."

The first American looked interested. "What's your new system?"

"Well," replied the second, "now, whenever I take a taxi, I just give the driver all the English money I have. And would you believe it, I have got the fare exactly right every time!"

Dealing with unfamiliar currencies—and translating the value of foreign money into dollars—is a problem every international traveler faces. The traveler's problem is complicated by the fact that *exchange rates*—the rates at which one country's money trades for another—may change unpredictably. Thus the number of British pounds, Russian rubles, Japanese yen, or Australian dollars that a U.S. dollar can buy may vary over time, sometimes quite a lot.

The economic consequences of variable exchange rates are much broader than their impact on travel and tourism, however. For example, the competitiveness of U.S. exports depends in part on the prices of U.S. goods in terms of foreign currencies, which in turn depend on the exchange rate between the U.S. dollar and those currencies. Likewise, the prices Americans pay for imported goods depend in part on the value of the dollar relative to the currencies of the countries that produce those goods. Exchange rates also affect

the value of financial investments made across national borders. For countries that are heavily dependent on trade and international capital flows—the majority of the world's nations—fluctuations in the exchange rate may have a significant economic impact.

This chapter discusses exchange rates and the role they play in open economies. We will start by distinguishing between the *nominal exchange rate*—the rate at which one national currency trades for another—and the *real exchange rate*—the rate at which one country's goods trade for another's. We will show how exchange rates affect the prices of exports and imports, and thus the pattern of trade.

Next we will turn to the question of how exchange rates are determined. Exchange rates may be divided into two broad categories, flexible and fixed. The value of a *flexible* exchange rate is determined freely in the market for national currencies, known as the *foreign exchange market*. Flexible exchange rates vary continually with changes in the supply of and demand for national currencies. In contrast, the value of a *fixed* exchange rate is set by the government at a constant level. Because most large industrial countries, including the United States, have a flexible exchange rate, we will focus on that case first. We will see that a country's monetary policy plays a particularly important role in determining the exchange rate. Furthermore, in an open economy with a flexible exchange rate, the exchange rate becomes a tool of monetary policy, in much the same way as the real interest rate.

Although most large industrial countries have a flexible exchange rate, many small and developing economies fix their exchange rates, so we will consider the case of fixed exchange rates as well. We will explain first how a country's government (usually, its central bank) goes about maintaining a fixed exchange rate at the officially determined level. Though fixing the exchange rate generally reduces day-to-day fluctuations in the value of a nation's currency, we will see that, at times, a fixed exchange rate can become severely unstable, with potentially serious economic consequences. We will close the chapter by discussing the relative merits of fixed and flexible exchange rates.

## EXCHANGE RATES

The economic benefits of trade between nations in goods, services, and assets are similar to the benefits of trade within a nation. In both cases, trade in goods and services permits greater specialization and efficiency, whereas trade in assets allows financial investors to earn higher returns while providing funds for worthwhile capital projects. However, there is a difference between the two cases, which is that trade in goods, services, and assets *within* a nation normally involves a single currency—dollars, yen, pesos, or whatever the country's official form of money happens to be—whereas trade *between* nations usually involves dealing in different currencies. So, for example, if an American resident wants to purchase an automobile manufactured in South Korea, she (or more likely, the automobile dealer) must first trade dollars for the Korean currency, called the won. The Korean car manufacturer is then paid in won. Similarly, an Argentine who wants to purchase shares in a U.S. company (a U.S. financial asset) must first trade his Argentine pesos for dollars and then use the dollars to purchase the shares.

### NOMINAL EXCHANGE RATES

Because international transactions generally require that one currency be traded for another, the relative values of different currencies are an important factor in international economic relations. The rate at which two currencies can be traded for each other is called the **nominal exchange rate,** or more simply the *exchange rate,* between the two currencies. For example, if one U.S. dollar can be exchanged for 110 Japanese yen, the nominal exchange rate between the U.S. and Japanese currencies is 110 yen per dollar. Each country has many nominal exchange rates, one corresponding to each currency against which its own currency is traded. Thus the dollar's value can be quoted in terms of English pounds, Swedish kroner, Israeli shekels, Russian rubles, or dozens of other currencies. Table 17.1 gives exchange rates between the dollar and six other important currencies as of the close of business in New York City on September 4, 2002.

***nominal exchange rate*** the rate at which two currencies can be traded for each other

**TABLE 17.1**
**Nominal Exchange Rates for the U.S. Dollar**

| Country | Foreign currency/dollar | Dollar/foreign currency |
|---|---|---|
| United Kingdom (pound) | 0.6393 | 1.5643 |
| Canada (Canadian dollar) | 1.5674 | 0.6380 |
| Mexico (peso) | 9.9850 | 0.1002 |
| Japan (yen) | 118.0000 | 0.00848 |
| Switzerland (Swiss franc) | 1.4760 | 0.6775 |
| South Korea (won) | 1,190.1000 | 0.00084 |

SOURCE: *The Wall Street Journal,* September 5, 2002.

As Table 17.1 shows, exchange rates can be expressed either as the amount of foreign currency needed to purchase one dollar (left column) or as the number of dollars needed to purchase one unit of the foreign currency (right column). These two ways of expressing the exchange rate are equivalent: Each is the reciprocal of the other. For example, on September 4, 2002, the U.S.–Canadian exchange rate could have been expressed either as 1.5674 Canadian dollars per U.S. dollar or as 0.6380 U.S. dollars per Canadian dollar, where $0.6380 = 1/1.5674$.

**EXAMPLE 17.1**

**Nominal exchange rates**

Based on Table 17.1, find the exchange rate between the British and Canadian currencies. Express the exchange rate in both Canadian dollars per pound and pounds per Canadian dollar.

From Table 17.1, we see that 0.6393 British pounds will buy a U.S. dollar, and that 1.5674 Canadian dollars will buy a U.S. dollar. Therefore 0.6393 British pounds and 1.5674 Canadian dollars are equal in value:

$$0.6393 \text{ pounds} = 1.5674 \text{ Canadian dollars.}$$

Dividing both sides of this equation by 1.5674, we get

$$0.4079 \text{ pounds} = 1 \text{ Canadian dollar.}$$

In other words, the British–Canadian exchange rate can be expressed as 0.4079 pounds per Canadian dollar. Alternatively, the exchange rate can be expressed as $1/0.4079 = 2.45$ Canadian dollars per pound.

**EXERCISE 17.1**

**From the business section of the newspaper or an online source (try the Federal Reserve Bank of St. Louis FRED database, http://research.**

**stlouisfed.org/fred/data/exchange.html), find recent quotations of the value of the U.S. dollar against the British pound, the Canadian dollar, and the Japanese yen. Based on these data find the exchange rate (a) between the pound and the Canadian dollar and (b) between the Canadian dollar and the yen. Express the exchange rates you derive in two ways (e.g., both as pounds per Canadian dollar and as Canadian dollars per pound).**

Figure 17.1 shows the nominal exchange rate for the U.S. dollar for 1973 to 2002. Rather than showing the value of the dollar relative to that of an individual foreign currency, such as the Japanese yen or the British pound, the figure expresses the value of the dollar as an average of its values against other major currencies. This average value of the dollar is measured relative to a base value of 100 in 1973. So, for example, a value of 120 for the dollar in a particular year implies that the dollar was 20 percent more valuable in that year, relative to other major currencies, than it was in 1973.

You can see from Figure 17.1 that the dollar's value has fluctuated over time, sometimes increasing (as in the period 1980–1985) and sometimes decreasing (as in 1985–1987). An increase in the value of a currency relative to other currencies is known as an **appreciation;** a decline in the value of a currency relative to other currencies is called a **depreciation.** So we can say that the dollar appreciated in 1980–1985 and depreciated in 1985–1987. We will discuss the reasons a currency may appreciate or depreciate later in this chapter.

***appreciation*** an increase in the value of a currency relative to other currencies

***depreciation*** a decrease in the value of a currency relative to other currencies

In this chapter we will use the symbol $e$ to stand for a country's nominal exchange rate. Although the exchange rate can be expressed either as foreign currency units per unit of domestic currency, or vice versa, as we saw in Table 17.1, let's agree to define $e$ as *the number of units of the foreign currency that the domestic currency will buy.* For example, if we treat the United States as the "home" or "domestic" country and Japan as the "foreign" country, $e$ will be defined as the number of Japanese yen that one dollar will buy. Defining the nominal exchange rate this way implies that an *increase* in $e$ corresponds to an *appreciation,* or a strengthening, of the home currency, while a *decrease* in $e$ implies a *depreciation,* or weakening, of the home currency.

**FIGURE 17.1**
**The U.S. Nominal Exchange Rate, 1973–2002.**
This figure expresses the value of the dollar from 1973 to 2002 as an average of its values against other major currencies, relative to a base value of 100 in 1973.

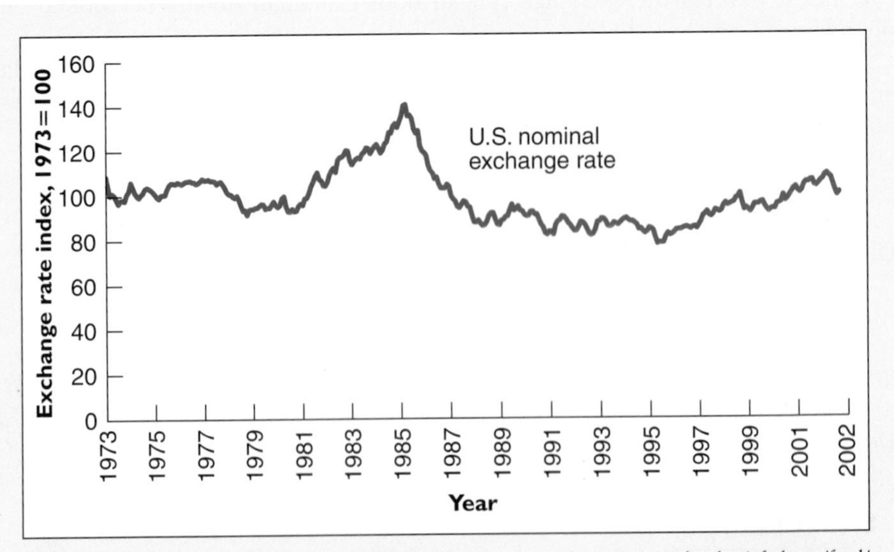


SOURCE: Federal Reserve Bank of St. Louis, FRED database (http://research.stlouisfed.org/fred/data/exchange.html).

*"On the foreign-exchange markets today, the dollar fell against all major currencies and the doughnut."*

## FLEXIBLE VERSUS FIXED EXCHANGE RATES

As we saw in Figure 17.1, the exchange rate between the U.S. dollar and other currencies isn't constant but varies continually. Indeed, changes in the value of the dollar occur daily, hourly, even minute by minute. Such fluctuations in the value of a currency are normal for countries like the United States, which have a *flexible* or *floating exchange rate.* The value of a **flexible exchange rate** is not officially fixed but varies according to the supply and demand for the currency in the **foreign exchange market**—the market on which currencies of various nations are traded for one another. We will discuss the factors that determine the supply and demand for currencies shortly.

***flexible exchange rate*** an exchange rate whose value is not officially fixed but varies according to the supply and demand for the currency in the foreign exchange market

***foreign exchange market*** the market on which currencies of various nations are traded for one another

Some countries do not allow their currency values to vary with market conditions but instead maintain a *fixed exchange rate.* The value of a **fixed exchange rate** is set by official government policy. (A government that establishes a fixed exchange rate typically determines the exchange rate's value independently, but sometimes exchange rates are set according to an agreement among a number of governments.) Some countries fix their exchange rates in terms of the U.S. dollar (Argentina, for example), but there are other possibilities. Some French-speaking African countries have traditionally fixed the value of their currencies in terms of the French franc. Under the gold standard, which many countries used until its collapse during the Great Depression, currency values were fixed in terms of ounces of gold. In the next part of the chapter we will focus on flexible exchange rates, but we will return later to the case of fixed rates. We will also discuss the costs and benefits of each type of exchange rate.

***fixed exchange rate*** an exchange rate whose value is set by official government policy

## THE REAL EXCHANGE RATE

The nominal exchange rate tells us the price of the domestic currency in terms of a foreign currency. As we will see in this section, the *real exchange rate* tells us the price of the average domestic *good or service* in terms of the average foreign *good or service.* We will also see that a country's real exchange rate has important implications for its ability to sell its exports abroad.

To provide background for discussing the real exchange rate, imagine you are in charge of purchasing for a U.S. corporation that is planning to acquire a large number of new computers. The company's computer specialist has identified two models, one Japanese-made and one U.S.-made, that meet the necessary specifications. Since the two models are essentially equivalent, the company will buy the one with the lower price. However, since the computers are priced in the currencies of the countries of manufacture, the price comparison is not so straightforward. Your mission—should you decide to accept it—is to determine which of the two models is cheaper.

To complete your assignment you will need two pieces of information: the nominal exchange rate between the dollar and the yen and the prices of the two models in terms of the currencies of their countries of manufacture. Example 17.2 shows how you can use this information to determine which model is cheaper.

**EXAMPLE 17.2** **Comparing prices expressed in different currencies**

A U.S.-made computer costs $2,400, and a similar Japanese-made computer costs 242,000 yen. If the nominal exchange rate is 110 yen per dollar, which computer is the better buy?

To make this price comparison, we must measure the prices of both computers in terms of the same currency. To make the comparison in dollars, we first convert the Japanese computer's price into dollars. The price in terms of Japanese yen is ¥242,000 (the symbol ¥ means "yen"), and we are told that ¥110 = $1. To find the dollar price of the computer, then, we observe that for any good or service,

$$\text{Price in yen} = \text{Price in dollars} \times \text{Value of dollar in terms of yen.}$$

Note that the value of a dollar in terms of yen is just the yen–dollar exchange rate. Making this substitution and solving, we get

$$\begin{aligned}\text{Price in dollars} &= \frac{\text{Price in yen}}{\text{Yen–dollar exchange rate}}\\ &= \frac{¥242{,}000}{¥110/\$1} = \$2{,}200.\end{aligned}$$

Notice that the yen symbol appears in both the numerator and the denominator of the ratio, so it cancels out. Our conclusion is that the Japanese computer is cheaper than the U.S. computer at $2,200, or $200 less than the price of the U.S. computer, $2,400. The Japanese computer is the better deal.

**EXERCISE 17.2**

**Continuing Example 17.2, compare the prices of the Japanese and American computers by expressing both prices in terms of yen.**

In Example 17.2, the fact that the Japanese computer was cheaper implied that your firm would choose it over the U.S.-made computer. In general, a country's ability to compete in international markets depends in part on the prices of its goods and services *relative* to the prices of foreign goods and services, when the prices are measured in a common currency. In the hypothetical example of the Japanese and U.S. computers, the price of the domestic (U.S.) good relative to the price of the foreign (Japanese) good is $2,400/$2,200, or 1.09. So the U.S. computer is 9 percent more expensive than the Japanese computer, putting the U.S. product at a competitive disadvantage.

***real exchange rate*** the price of the average domestic good or service *relative* to the price of the average foreign good or service, when prices are expressed in terms of a common currency

More generally, economists ask whether *on average* the goods and services produced by a particular country are expensive relative to the goods and services produced by other countries. This question can be answered by the country's *real exchange rate*. Specifically, a country's **real exchange rate** is the price of the average domestic good or service *relative* to the price of the average foreign good or service, when prices are expressed in terms of a common currency.

To obtain a formula for the real exchange rate, recall that $e$ equals the nominal exchange rate (the number of units of foreign currency per dollar) and that $P$ equals the domestic price level, as measured, for example, by the consumer price index. We will use $P$ as a measure of the price of the "average" domestic good or service. Similarly, let $P^f$ equal the foreign price level. We will use $P^f$ as the measure of the price of the "average" foreign good or service.

The real exchange rate equals the price of the average domestic good or service relative to the price of the average foreign good or service. It would not be correct, however, to define the real exchange rate as the ratio $P/P^f$, because the two price levels are expressed in different currencies. As we saw in Example 17.2, to convert foreign prices into dollars, we must divide the foreign price by the exchange rate. By this rule, the price in dollars of the average foreign good or service equals $P^f/e$. Now we can write the real exchange rate as

$$\text{Real exchange rate} = \frac{\text{Price of domestic good}}{\text{Price of foreign good, in dollars}} = \frac{P}{P^f/e}.$$

To simplify this expression, multiply the numerator and denominator by $e$ to get

$$\text{Real exchange rate} = \frac{eP}{P^f}, \quad (17.1)$$

which is the formula for the real exchange rate.

To check this formula, let's use it to re-solve the computer example, Example 17.2. (For this exercise, we imagine that computers are the only good produced by the United States and Japan, so the real exchange rate becomes just the price of U.S. computers relative to Japanese computers.) In that example, the nominal exchange rate $e$ was ¥110/$1, the domestic price $P$ (of a computer) was $2,400, and the foreign price $P^f$ was ¥242,000. Applying Equation 17.1, we get

$$\text{Real exchange rate (for computers)} = \frac{(\text{¥}110/\$1) \times \$2{,}400}{\text{¥}242{,}000} = \frac{\text{¥}264{,}000}{\text{¥}242{,}000} = 1.09,$$

which is the same answer we got earlier.

The real exchange rate, an overall measure of the cost of domestic goods relative to foreign goods, is an important economic variable. As Example 17.2 suggests, when the real exchange rate is high, domestic goods are on average more expensive than foreign goods (when priced in the same currency). A high real exchange rate implies that domestic producers will have difficulty exporting to other countries (domestic goods will be "overpriced"), while foreign goods will sell well in the home country (because imported goods are cheap relative to goods produced at home). Since a high real exchange rate tends to reduce exports and increase imports, we conclude that *net exports will tend to be low when the real exchange rate is high*. Conversely, if the real exchange rate is low, then the home country will find it easier to export (because its goods are priced below those of foreign competitors), while domestic residents will buy fewer imports (because imports are expensive relative to domestic goods). *Thus net exports will tend to be high when the real exchange rate is low.*

Equation 17.1 also shows that the real exchange rate tends to move in the same direction as the nominal exchange rate $e$ (since $e$ appears in the numerator of the formula for the real exchange rate). To the extent that real and nominal exchange rates move in the same direction, we can conclude that net exports will be hurt by a high nominal exchange rate and helped by a low nominal exchange rate.

ECONOMIC NATURALIST 17.1

### Does a strong currency imply a strong economy?

Politicians and the public sometimes take pride in the fact that their national currency is "strong," meaning that its value in terms of other currencies is high or rising. Likewise, policymakers sometimes view a depreciating ("weak") currency as a sign of economic failure. Does a strong currency necessarily imply a strong economy?

Contrary to popular impression, there is no simple connection between the strength of a country's currency and the strength of its economy. For example, Figure 17.1 shows that the value of the U.S. dollar relative to other major currencies was greater in the year 1973 than in the year 2000, though U.S. economic performance was considerably better in 2000 than in 1973, a period of deep recession and rising inflation. Indeed, the one period shown in Figure 17.1 during which the dollar rose markedly in value, 1980–1985, was a time of recession and high unemployment in the United States.

One reason a strong currency does not necessarily imply a strong economy is that an appreciating currency (an increase in e) tends to raise the real exchange rate (equal to $eP/P^f$), which may hurt a country's net exports. For example, if the dollar strengthens against the yen (that is, if a dollar buys more yen than before), Japanese goods will become cheaper in terms of dollars. The result may be that Americans prefer to buy Japanese goods rather than goods produced at home. Likewise, a stronger dollar implies that each yen buys fewer dollars, so exported U.S. goods become more expensive to Japanese consumers. As U.S. goods become more expensive in terms of yen, the willingness of Japanese consumers to buy U.S. exports declines. A strong dollar may therefore imply lower sales and profits for U.S. industries that export, as well as for U.S. industries (like automobile manufacturers) that compete with foreign firms for the domestic U.S. market.

## EXCHANGE RATES

- The nominal exchange rate between two currencies is the rate at which the currencies can be traded for each other. More precisely, the nominal exchange rate $e$ for any given country is the number of units of foreign currency that can be bought for one unit of the domestic currency.
- An appreciation is an increase in the value of a currency relative to other currencies (a rise in $e$); a depreciation is a decline in a currency's value (a fall in $e$).
- An exchange rate can be either flexible—meaning that it varies freely according to supply and demand for the currency in the foreign exchange market—or fixed, meaning that its value is established by official government policy.
- The real exchange rate is the price of the average domestic good or service relative to the price of the average foreign good or service, when prices are expressed in terms of a common currency. A useful formula for the real exchange rate is $eP/P^f$, where $e$ is the nominal exchange rate, $P$ is the domestic price level, and $P^f$ is the foreign price level.
- An increase in the real exchange rate implies that domestic goods are becoming more expensive relative to foreign goods, which tends to reduce exports and stimulate imports. Conversely, a decline in the real exchange rate tends to increase net exports.

# THE DETERMINATION OF THE EXCHANGE RATE

Countries that have flexible exchange rates, such as the United States, see the international values of their currencies change continually. What determines the value of the nominal exchange rate at any point in time? In this section we will try to answer this basic economic question. Again, our focus for the moment is on flexible exchange rates, whose values are determined by the foreign exchange market. Later in the chapter we discuss the case of fixed exchange rates.

## A SIMPLE THEORY OF EXCHANGE RATES: PURCHASING POWER PARITY (PPP)

The most basic theory of how nominal exchange rates are determined is called *purchasing power parity,* or PPP. To understand this theory, we must first discuss a fundamental economic concept, called *the law of one price.* The **law of one price** states that if transportation costs are relatively small, the price of an internationally traded commodity must be the same in all locations. For example, if transportation costs are not too large, the price of a bushel of wheat ought to be the same in Bombay, India, and Sydney, Australia. Suppose that were not the case—that the price of wheat in Sydney were only half the price in Bombay. In that case grain merchants would have a strong incentive to buy wheat in Sydney and ship it to Bombay, where it could be sold at double the price of purchase. As wheat left Sydney, reducing the local supply, the price of wheat in Sydney would rise, while the inflow of wheat into Bombay would reduce the price in Bombay. According to the *equilibrium principle* (Chapter 3), the international market for wheat would return to equilibrium only when unexploited opportunities to profit had been eliminated—specifically, only when the prices of wheat in Sydney and in Bombay became equal or nearly equal (with the difference being less than the cost of transporting wheat from Australia to India).

***law of one price*** if transportation costs are relatively small, the price of an internationally traded commodity must be the same in all locations

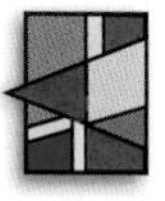

If the law of one price were to hold for all goods and services (which is not a realistic assumption, as we will see shortly), then the value of the nominal exchange rate would be determined as Example 17.3 illustrates.

**EXAMPLE 17.3**

### How many Indian rupees equal one Australian dollar? (1)

Suppose that a bushel of grain costs 5 Australian dollars in Sydney and 150 rupees in Bombay. If the law of one price holds for grain, what is the nominal exchange rate between Australia and India?

Because the market value of a bushel of grain must be the same in both locations, we know that the Australian price of wheat must equal the Indian price of wheat, so that

$$5 \text{ Australian dollars} = 150 \text{ Indian rupees}.$$

Dividing by 5, we get

$$1 \text{ Australian dollar} = 30 \text{ Indian rupees}.$$

Thus the nominal exchange rate between Australia and India should be 30 rupees per Australian dollar.

**EXERCISE 17.3**

**The price of gold is $300 per ounce in New York and 2,500 kronor per ounce in Stockholm, Sweden. If the law of one price holds for gold, what is the nominal exchange rate between the U.S. dollar and the Swedish krona?**

***purchasing power parity (PPP)*** the theory that nominal exchange rates are determined as necessary for the law of one price to hold

Example 17.3 and Exercise 17.3 illustrate the application of the purchasing power parity theory. According to the **purchasing power parity (PPP)** theory, nominal exchange rates are determined as necessary for the law of one price to hold.

A particularly useful prediction of the PPP theory is that in the long run, the *currencies of countries that experience significant inflation will tend to depreciate.* To see why, we will extend the analysis in Example 17.3.

**EXAMPLE 17.4**

**How many Indian rupees equal one Australian dollar? (2)**

Suppose India experiences significant inflation so that the price of a bushel of grain in Bombay rises from 150 to 300 rupees. Australia has no inflation, so the price of grain in Sydney remains unchanged at 5 Australian dollars. If the law of one price holds for grain, what will happen to the nominal exchange rate between Australia and India?

As in Example 17.3, we know that the market value of a bushel of grain must be the same in both locations. Therefore,

$$5 \text{ Australian dollars} = 300 \text{ rupees.}$$

Equivalently,

$$1 \text{ Australian dollar} = 60 \text{ rupees.}$$

The nominal exchange rate is now 60 rupees per Australian dollar. Before India's inflation, the nominal exchange rate was 30 rupees per Australian dollar (Example 17.3). So in this example, inflation has caused the rupee to depreciate against the Australian dollar. Conversely, Australia, with no inflation, has seen its currency appreciate against the rupee.

This link between inflation and depreciation makes economic sense. Inflation implies that a nation's currency is losing purchasing power in the domestic market. Analogously, exchange rate depreciation implies that the nation's currency is losing purchasing power in international markets.

Figure 17.2 shows annual rates of inflation and nominal exchange rate depreciation for the 10 largest South American countries from 1995 to 2001. Inflation is measured as the annual rate of change in the country's consumer price index; depreciation is measured relative to the U.S. dollar. As you can see, inflation varied greatly among South American countries during the period. For example, Argentina's inflation rate was essentially the same as that of the United States, while Venezuela had inflation of 37 percent per year and Ecuador's inflation was 45 percent per year.

Figure 17.2 shows that, as the PPP theory implies, countries with higher inflation during the 1995–2001 period tended to experience the most rapid depreciation of their currencies.

## SHORTCOMINGS OF THE PPP THEORY

Empirical studies have found that the PPP theory is useful for predicting changes in nominal exchange rates over the relatively long run. In particular, this theory helps to explain the tendency of countries with high inflation to experience depreciation of their exchange rates, as shown in Figure 17.2. However, the theory is less successful in predicting short-run movements in exchange rates.

A particularly dramatic failure of the PPP theory occurred in the United States in the early 1980s. As Figure 17.1 indicates, between 1980 and 1985 the value of the U.S. dollar rose nearly 50 percent relative to the currencies of U.S. trading partners. This strong appreciation was followed by an even more rapid depreciation during 1986 and 1987. PPP theory could explain this roller-coaster behavior

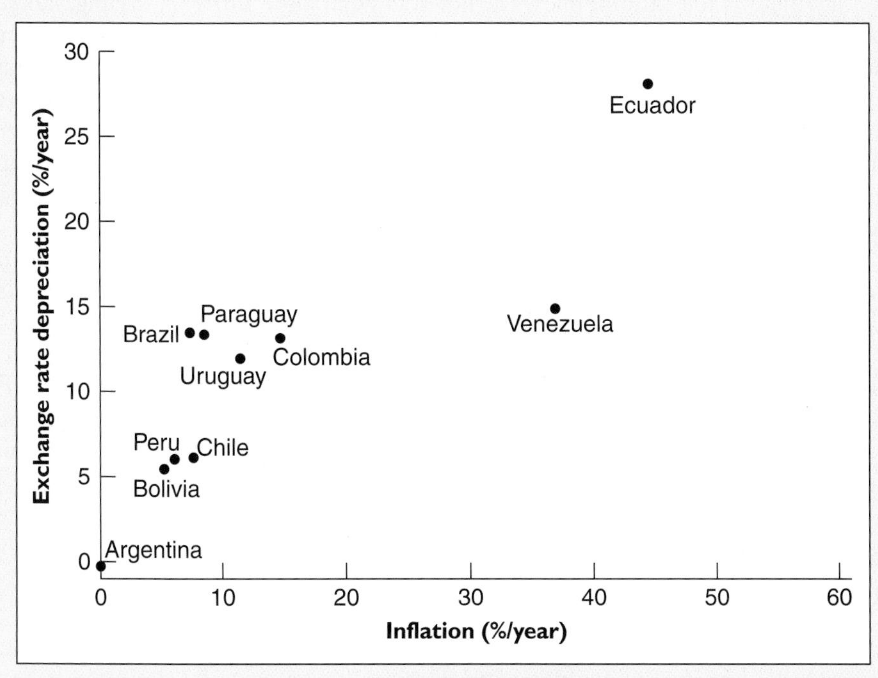


SOURCE: International Monetary Fund, *International Financial Statistics,* and authors' calculations.

**FIGURE 17.2**
**Inflation and Currency Depreciation in South America, 1995–2001.**
The annual rates of inflation and nominal exchange-rate depreciation (relative to the U.S. dollar) in 10 South American countries varied considerably during 1995–2001. High inflation was associated with rapid depreciation of the nominal exchange rate.

only if inflation were far lower in the United States than in U.S. trading partners from 1980 to 1985, and far higher from 1986 to 1987. In fact, inflation was similar in the United States and its trading partners throughout both periods.

Why does the PPP theory work less well in the short run than the long run? Recall that this theory relies on the law of one price, which says that the price of an internationally traded commodity must be the same in all locations. The law of one price works well for goods such as grain or gold, which are standardized commodities that are traded widely. However, *not all goods and services are traded internationally,* and *not all goods are standardized commodities.*

Many goods and services are not traded internationally, because the assumption underlying the law of one price—that transportation costs are relatively small—does not hold for them. For example, for Indians to export haircuts to Australia, they would need to transport an Indian barber to Australia every time a Sydney resident desired a trim. Because transportation costs prevent haircuts from being traded internationally, the law of one price does not apply to them. Thus, even if the price of haircuts in Australia were double the price of haircuts in India, market forces would not necessarily force prices toward equality in the short run. (Over the long run, some Indian barbers might emigrate to Australia.) Other examples of nontraded goods and services are agricultural land, buildings, heavy construction materials (whose value is low relative to their transportation costs), and highly perishable foods. In addition, some products use nontraded goods and services as inputs: A McDonald's hamburger served in Moscow has both a tradable component (frozen hamburger patties) and a nontradable component (the labor of counter workers). In general, the greater the share of nontraded goods and services in a nation's output, the less precisely the PPP theory will apply to the country's exchange rate.[1]

The second reason the law of one price and the PPP theory sometimes fail to apply is that not all internationally traded goods and services are perfectly standardized commodities, like grain or gold. For example, U.S.-made automobiles

[1]Trade barriers, such as tariffs and quotas, also increase the costs associated with shipping goods from one country to another. Thus trade barriers reduce the applicability of the law of one price in much the same way that physical transportation costs do.

and Japanese-made automobiles are not identical; they differ in styling, horsepower, reliability, and other features. As a result, some people strongly prefer one nation's cars to the other's. Thus if Japanese cars cost 10 percent more than American cars, U.S. automobile exports will not necessarily flood the Japanese market, since many Japanese will still prefer Japanese-made cars even at a 10 percent premium. Of course, there are limits to how far prices can diverge before people will switch to the cheaper product. But the law of one price, and hence the PPP theory, will not apply exactly to nonstandardized goods.

To summarize, the PPP theory works reasonably well as an explanation of exchange rate behavior over the long run, but not in the short run. Because transportation costs limit international trade in many goods and services, and because not all goods that are traded are standardized commodities, the law of one price (on which the PPP theory is based) works only imperfectly in the short run. To understand the short-run movements of exchange rates we need to incorporate some additional factors. In the next section we will study a supply and demand framework for the determination of exchange rates.

## THE DETERMINATION OF THE EXCHANGE RATE: A SUPPLY AND DEMAND ANALYSIS

Although the PPP theory helps to explain the long-run behavior of the exchange rate, supply and demand analysis is more useful for studying its short-run behavior. As we will see, dollars are demanded in the foreign exchange market by foreigners who seek to purchase U.S. goods and assets and are supplied by U.S. residents who need foreign currencies to buy foreign goods and assets. The equilibrium exchange rate is the value of the dollar that equates the number of dollars supplied and demanded in the foreign exchange market. In this section we will discuss the factors that affect the supply and demand for dollars, and thus the U.S. exchange rate.

One note before we proceed: In Chapter 14 we described how the supply of money by the Fed and the demand for money by the public help to determine the nominal interest rate. However, the supply and demand for money in the domestic economy, as presented in Chapter 14, are *not* equivalent to the supply and demand for dollars in the foreign exchange market. As mentioned, the foreign exchange market is the market in which the currencies of various nations are traded for one another. The supply of dollars to the foreign exchange market is *not* the same as the money supply set by the Fed; rather, it is the number of dollars U.S. households and firms offer to trade for other currencies. Likewise, the demand for dollars in the foreign exchange market is *not* the same as the domestic demand for money, but the number of dollars holders of foreign currencies seek to buy. To understand the distinction, it may help to keep in mind that while the Fed determines the total supply of dollars in the U.S. economy, a dollar does not "count" as having been supplied to the foreign exchange market until some holder of dollars, such as a household or firm, tries to trade it for a foreign currency.

### The Supply of Dollars

Anyone who holds dollars, from an international bank to a Russian citizen whose dollars are buried in the backyard, is a potential supplier of dollars to the foreign exchange market. In practice, however, the principal suppliers of dollars to the foreign exchange market are U.S. households and firms. Why would a U.S. household or firm want to supply dollars in exchange for foreign currency? There are two major reasons. First, a U.S. household or firm may need foreign currency *to purchase foreign goods or services*. For example, a U.S. automobile importer may need yen to purchase Japanese cars, or an American tourist may need yen to make purchases in Tokyo. Second, a U.S. household or firm may need foreign currency *to purchase foreign assets.* For example, an American mutual fund may wish to acquire stocks issued by Japanese companies, or an individual U.S. saver may want to purchase Japanese government bonds. Because Japanese assets are priced in yen, the U.S. household or firm will need to trade dollars for yen to acquire these assets.

The supply of dollars to the foreign exchange market is illustrated in Figure 17.3. We will focus on the market in which dollars are traded for Japanese yen, but bear in mind that similar markets exist for every other pair of traded currencies. The vertical axis of the figure shows the U.S.–Japanese exchange rate as measured by the number of yen that can be purchased with each dollar. The horizontal axis shows the number of dollars being traded in the yen–dollar market.

Note that the supply curve for dollars is upward-sloping. In other words, the more yen each dollar can buy, the more dollars people are willing to supply to the foreign exchange market. Why? At given prices for Japanese goods, services, and assets, the more yen a dollar can buy, the cheaper those goods, services, and assets will be in dollar terms. For example, if a video game costs 5,000 yen in Japan, and a dollar can buy 100 yen, the dollar price of the video game will be \$50. However, if a dollar can buy 200 yen, then the dollar price of the same video game will be \$25. Assuming that lower dollar prices will induce Americans to increase their expenditures on Japanese goods, services, and assets, a higher yen–dollar exchange rate will increase the supply of dollars to the foreign exchange market. Thus the supply curve for dollars is upward-sloping.

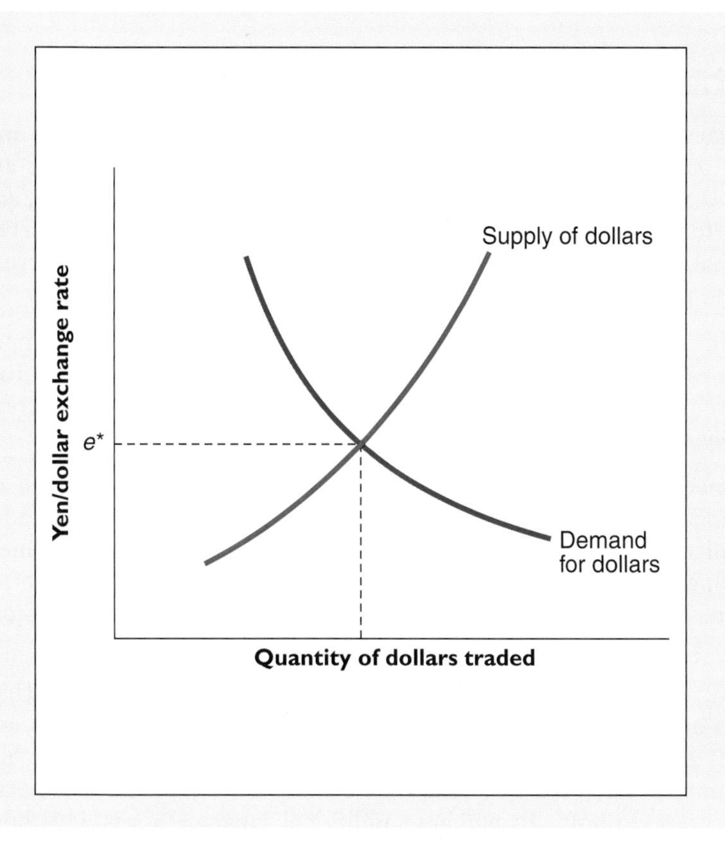


**FIGURE 17.3**
**The Supply and Demand for Dollars in the Yen–Dollar Market.**
The supply of dollars to the foreign exchange market is upward-sloping, because an increase in the number of yen offered for each dollar makes Japanese goods, services, and assets more attractive to U.S. buyers. Similarly, the demand for dollars is downward-sloping, because holders of yen will be less willing to buy dollars the more expensive they are in terms of yen. The equilibrium exchange rate $e^*$, also called the *fundamental value of the exchange rate,* equates the quantities of dollars supplied and demanded.

## The Demand for Dollars

In the yen–dollar foreign exchange market, demanders of dollars are those who wish to acquire dollars in exchange for yen. Most demanders of dollars in the yen–dollar market are Japanese households and firms, although anyone who happens to hold yen is free to trade them for dollars. Why demand dollars? The reasons for acquiring dollars are analogous to those for acquiring yen. First, households and firms that hold yen will demand dollars *so that they can purchase U.S. goods and services.* For example, a Japanese firm that wants to license U.S.-produced software needs dollars to pay the required fees, and a Japanese student studying in an American university must pay tuition in dollars. The firm or the student can acquire the necessary dollars only by offering yen in exchange. Second, households and firms demand dollars *in order to purchase U.S. assets.* The purchase of Hawaiian real estate by a Japanese company or the acquisition of Microsoft stock by a Japanese pension fund are two examples.

The demand for dollars is represented by the downward-sloping curve in Figure 17.3. The curve slopes downward because the more yen a Japanese citizen must pay to acquire a dollar, the less attractive U.S. goods, services, and assets will be. Hence the demand for dollars will be low when dollars are expensive in terms of yen and high when dollars are cheap in terms of yen.

### The Equilibrium Value of the Dollar

As mentioned earlier, the United States maintains a flexible, or floating, exchange rate, which means that the value of the dollar is determined by the forces of supply and demand in the foreign exchange market. In Figure 17.3 the equilibrium value of the dollar is $e^*$, the yen–dollar exchange rate at which the quantity of dollars supplied equals the quantity of dollars demanded. The equilibrium value of the exchange rate is also called the **fundamental value of the exchange rate.** In general, the equilibrium value of the dollar is not constant but changes with shifts in the supply of and demand for dollars in the foreign exchange market.

***fundamental value of the exchange rate (or equilibrium exchange rate)*** the exchange rate that equates the quantities of the currency supplied and demanded in the foreign exchange market

## CHANGES IN THE SUPPLY OF DOLLARS

Recall that people supply dollars to the yen–dollar foreign exchange market in order to purchase Japanese goods, services, and assets. Factors that affect the desire of U.S. households and firms to acquire Japanese goods, services, and assets will therefore affect the supply of dollars to the foreign exchange market. Some factors that will *increase* the supply of dollars, shifting the supply curve for dollars to the right, include:

- An increased preference for Japanese goods. For example, suppose that Japanese firms produce some popular new consumer electronics. To acquire the yen needed to buy these goods, American importers will increase their supply of dollars to the foreign exchange market.
- An increase in U.S. real GDP. An increase in U.S. real GDP will raise the incomes of Americans, allowing them to consume more goods and services (recall the consumption function, introduced in Chapter 13). Some part of this increase in consumption will take the form of goods imported from Japan. To buy more Japanese goods, Americans will supply more dollars to acquire the necessary yen.
- An increase in the real interest rate on Japanese assets. Recall that U.S. households and firms acquire yen in order to purchase Japanese assets as well as goods and services. Other factors, such as risk, held constant, the higher the real interest rate paid by Japanese assets, the more Japanese assets Americans will choose to hold. To purchase additional Japanese assets, U.S. households and firms will supply more dollars to the foreign exchange market.

Conversely, reduced demand for Japanese goods, a lower U.S. GDP, or a lower real interest rate on Japanese assets will *reduce* the number of yen Americans need, in turn reducing their supply of dollars to the foreign exchange market and shifting the supply curve for dollars to the left. Of course, any shift in the supply curve for dollars will affect the equilibrium exchange rate, as Example 17.5 shows.

Supplying dollars, demanding yen.

EXAMPLE 17.5

**Video games, the yen, and the dollar**

Suppose Japanese firms come to dominate the video game market, with games that are more exciting and realistic than those produced in the United States. All else being equal, how will this change affect the relative value of the yen and the dollar?

The increased quality of Japanese video games will increase the demand for the games in the United States. To acquire the yen necessary to buy more Japanese video games, U.S. importers will supply more dollars to the foreign exchange market. As Figure 17.4 shows, the increased supply of dollars will reduce the value of the dollar. In other words, a dollar will buy fewer yen than it did before. At the same time, the yen will increase in value: A given number of yen will buy more dollars than it did before.

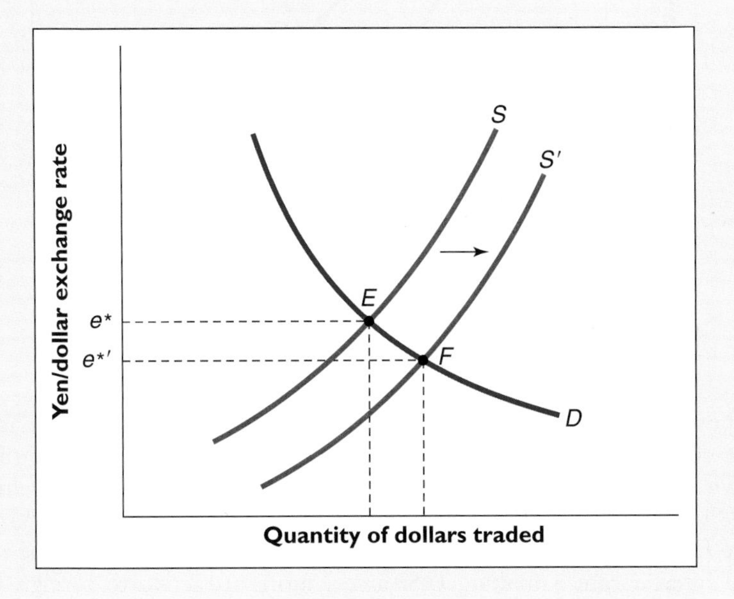


**FIGURE 17.4**
**An Increase in the Supply of Dollars Lowers the Value of the Dollar.**
Increased U.S. demand for Japanese video games forces Americans to supply more dollars to the foreign exchange market to acquire the yen they need to buy the games. The supply curve for dollars shifts from *S* to *S*′, lowering the value of the dollar in terms of yen. The fundamental value of the exchange rate falls from $e^*$ to $e^{*\prime}$.

EXERCISE 17.4

**The U.S. goes into a recession, and real GDP falls. All else equal, how is this economic weakness likely to affect the value of the dollar?**

## CHANGES IN THE DEMAND FOR DOLLARS

The factors that can cause a change in the demand for dollars in the foreign exchange market, and thus a shift of the dollar demand curve, are analogous to the factors that affect the supply of dollars. Factors that will *increase* the demand for dollars include:

- An increased preference for U.S. goods. For example, Japanese airlines might find that U.S.-built aircraft are superior to others, and decide to expand the number of American-made planes in their fleets. To buy the American planes, Japanese airlines would demand more dollars on the foreign exchange market.
- An increase in real GDP abroad, which implies higher incomes abroad, and thus more demand for imports from the United States.
- An increase in the real interest rate on U.S. assets, which would make those assets more attractive to foreign savers. To acquire U.S. assets, Japanese savers would demand more dollars.

## MONETARY POLICY AND THE EXCHANGE RATE

Of the many factors that could influence a country's exchange rate, among the most important is the monetary policy of the country's central bank. As we will see, monetary policy affects the exchange rate primarily through its effect on the real interest rate.

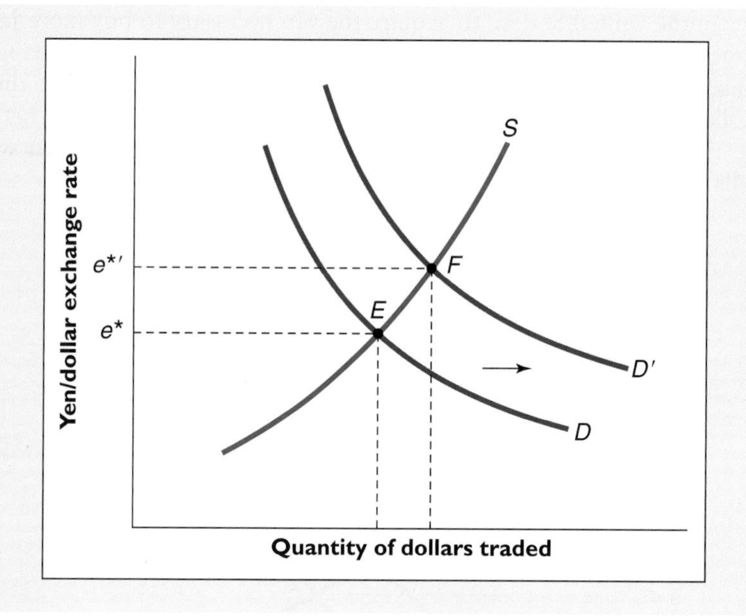


**FIGURE 17.5**
**A Tightening of Monetary Policy Strengthens the Dollar.**
Tighter monetary policy in the United States raises the domestic real interest rate, increasing the demand for U.S. assets by foreign savers. An increased demand for U.S. assets in turn increases the demand for dollars. The demand curve shifts from $D$ to $D'$, leading the exchange rate to appreciate from $e^*$ to $e^{*\prime}$.

Suppose the Fed is concerned about inflation and tightens U.S. monetary policy in response. The effects of this policy change on the value of the dollar are shown in Figure 17.5. Before the policy change, the equilibrium value of the exchange rate is $e^*$, at the intersection of supply curve $S$ and the demand curve $D$ (point $E$ in the figure). The tightening of monetary policy raises the domestic U.S. real interest rate $r$, making U.S. assets more attractive to foreign financial investors. The increased willingness of foreign investors to buy U.S. assets increases the demand for dollars, shifting the demand curve rightward from $D$ to $D'$ and the equilibrium point from $E$ to $F$. As a result of this increase in demand, the equilibrium value of the dollar rises from $e^*$ to $e^{*\prime}$.

In short, a tightening of monetary policy by the Fed raises the demand for dollars, causing the dollar to appreciate. By similar logic, an easing of monetary policy, which reduces the real interest rate, would weaken the demand for the dollar, causing it to depreciate.

**Why did the dollar appreciate nearly 50 percent in the first half of the 1980s?**

Figure 17.1 showed the strong appreciation of the U. S. dollar in 1980–1985, followed by a sharp depreciation in 1986–1987. We saw earlier that the PPP theory cannot explain this roller-coaster behavior. What *can* explain it?

Tight monetary policy, and the associated high real interest rate, were important causes of the dollar's remarkable appreciation during 1980–1985. As we saw in Economic Naturalist 15.4, U.S. inflation peaked at 13.5 percent in 1980. Under the leadership of Chairman Paul Volcker, the Fed responded to the surge in inflation by raising the real interest rate sharply in hopes of reducing aggregate demand and inflationary pressures. As a result, the real interest rate in the United States rose from negative values in 1979 and 1980 to more than 7 percent in 1983 and 1984 (see Table 15.2). Attracted by these high real returns, foreign savers rushed to buy U.S. assets, driving the value of the dollar up significantly.

The Fed's attempt to bring down inflation was successful. By the middle of the 1980s the Fed was able to ease U.S. monetary policy. The resulting decline in the real interest rate reduced the demand for U.S. assets, and thus for dollars, at which point the dollar fell back almost to its 1980 level.

## THE EXCHANGE RATE AS A TOOL OF MONETARY POLICY

In a closed economy, monetary policy affects aggregate demand solely through the real interest rate. For example, by raising the real interest rate, a tight monetary policy reduces consumption and investment spending. We will see next that in an open economy with a flexible exchange rate, the exchange rate serves as another channel for monetary policy, one that reinforces the effects of the real interest rate.

To illustrate, suppose that policymakers are concerned about inflation and decide to restrain aggregate demand. To do so, they increase the real interest rate, reducing consumption and investment spending. But, as Figure 17.5 shows, the higher real interest rate also increases the demand for dollars, causing the dollar to appreciate. The stronger dollar, in turn, further reduces aggregate demand. Why? As we saw in discussing the real exchange rate, a stronger dollar reduces the cost of imported goods, increasing imports. It also makes U.S. exports more costly to foreign buyers, which tends to reduce exports. Recall that net exports—or exports minus imports—is one of the four components of aggregate demand. Thus, by reducing exports and increasing imports, a stronger dollar (more precisely, a higher real exchange rate) reduces aggregate demand.

In sum, when the exchange rate is flexible, a tighter monetary policy reduces net exports (through a stronger dollar) as well as consumption and investment spending (through a higher real interest rate). Conversely, an easier monetary policy weakens the dollar and stimulates net exports, reinforcing the effect of the lower real interest rate on consumption and investment spending. Thus, relative to the case of a closed economy we studied earlier, *monetary policy is more effective in an open economy with a flexible exchange rate.*

The tightening of monetary policy under Fed Chairman Volcker in the early 1980s illustrates the effect of monetary policy on net exports (the trade balance). As we saw in Economic Naturalist 17.2, Volcker's tight-money policies were a major reason for the 50 percent appreciation of the dollar during 1980–1985. In 1980 and 1981, the United States enjoyed a trade surplus, with exports that modestly exceeded imports. Largely in response to a stronger dollar, the U.S. trade balance fell into deficit after 1981. By the end of 1985 the U.S. trade deficit was about 3 percent of GDP, a substantial shift in less than half a decade.

### RECAP DETERMINING THE EXCHANGE RATE

- The most basic theory of nominal exchange rate determination, purchasing power parity (PPP), is based on the law of one price. The law of one price states that if transportation costs are relatively small, the price of an internationally traded commodity must be the same in all locations. According to the PPP theory, the nominal exchange rate between two currencies can be found by setting the price of a traded commodity in one currency equal to the price of the same commodity expressed in the second currency.
- A useful prediction of the PPP theory is that the currencies of countries that experience significant inflation will tend to depreciate over the long run. However, the PPP theory does not work well in the short run. The fact that many goods and services are nontraded, and that not all traded goods are standardized, reduces the applicability of the law of one price, and hence of the PPP theory.

- Supply and demand analysis is a useful tool for studying the short-run determination of the exchange rate. U.S. households and firms supply dollars to the foreign exchange market to acquire foreign currencies, which they need to purchase foreign goods, services, and assets. Foreigners demand dollars in the foreign exchange market to purchase U.S. goods, services, and assets. The equilibrium exchange rate, also called the fundamental value of the exchange rate, equates the quantities of dollars supplied and demanded in the foreign exchange market.
- An increased preference for foreign goods, an increase in U.S. real GDP, or an increase in the real interest rate on foreign assets will increase the supply of dollars on the foreign exchange market, lowering the value of the dollar. An increased preference for U.S. goods by foreigners, an increase in real GDP abroad, or an increase in the real interest rate on U.S. assets will increase the demand for dollars, raising the value of the dollar.
- A tight monetary policy raises the real interest rate, increasing the demand for dollars and strengthening the dollar. A stronger dollar reinforces the effects of tight monetary policy on aggregate spending by reducing net exports, a component of aggregate demand. Conversely, an easy monetary policy lowers the real interest rate, weakening the dollar.

## FIXED EXCHANGE RATES

So far we have focused on the case of flexible exchange rates, the relevant case for most large industrial countries like the United States. However, the alternative approach, fixing the exchange rate, has been quite important historically and is still used in many countries, especially small or developing nations. In this section we will see how our conclusions change when the nominal exchange rate is fixed rather than flexible. One important difference is that when a country maintains a fixed exchange rate, its ability to use monetary policy as a stabilization tool is greatly reduced.

### HOW TO FIX AN EXCHANGE RATE

In contrast to a flexible exchange rate, whose value is determined solely by supply and demand in the foreign exchange market, the value of a fixed exchange rate is determined by the government (in practice, usually the finance ministry or treasury department, with the cooperation of the central bank). Today, the value of a fixed exchange rate is usually set in terms of a major currency (for instance, Argentina pegs its currency one-for-one to the dollar), or relative to a "basket" of currencies, typically those of the country's trading partners. Historically, currency values were often fixed in terms of gold or other precious metals, but in recent years precious metals have rarely if ever been used for that purpose.

***devaluation*** a reduction in the official value of a currency (in a fixed-exchange-rate system)

Once an exchange rate has been fixed, the government usually attempts to keep it unchanged for some time.[2] However, sometimes economic circumstances force the government to change the value of the exchange rate. A reduction in the official value of a currency is called a **devaluation**; an increase in the official

[2]There are exceptions to this statement. Some countries employ a *crawling peg* system, under which the exchange rate is fixed at a value that changes in a preannounced way over time. For example, the government may announce that the value of the fixed exchange rate will fall 2 percent each year. Other countries use a *target zone* system, in which the exchange rate is allowed to deviate by a small amount from its fixed value. To focus on the key issues, we will assume that the exchange rate is fixed at a single value for a protracted period.

value is called a **revaluation.** The devaluation of a fixed exchange rate is analogous to the depreciation of a flexible exchange rate; both involve a reduction in the currency's value. Conversely, a revaluation is analogous to an appreciation.

***revaluation*** an increase in the official value of a currency (in a fixed-exchange-rate system)

The supply and demand diagram we used to study flexible exchange rates can be adapted to analyze fixed exchange rates. Let's consider the case of a country called Latinia, whose currency is called the peso. Figure 17.6 shows the supply and demand for the Latinian peso in the foreign exchange market. Pesos are *supplied* to the foreign exchange market by Latinian households and firms who want to acquire foreign currencies to purchase foreign goods and assets. Pesos are *demanded* by holders of foreign currencies who need pesos to purchase Latinian goods and assets. Figure 17.6 shows that the quantities of pesos supplied and demanded in the foreign exchange market are equal when a peso equals 0.1 dollars (10 pesos to the dollar). Hence 0.1 dollars per peso is the *fundamental value* of the peso. If Latinia had a flexible-exchange-rate system, the peso would trade at 10 pesos to the dollar in the foreign exchange market.

But let's suppose that Latinia has a fixed exchange rate and that the government has decreed the value of the Latinian peso to be 8 pesos to the dollar, or 0.125 dollars per peso. This official value of the peso, 0.125 dollars, is indicated by the solid horizontal line in Figure 17.6. Notice that it is greater than the fundamental value, corresponding to the intersection of the supply and demand curves. When the officially fixed value of an exchange rate is greater than its fundamental value, the exchange rate is said to be **overvalued.** The official value of an exchange rate can also be lower than its fundamental value, in which case the exchange rate is said to be **undervalued.**

***overvalued exchange rate*** an exchange rate that has an officially fixed value greater than its fundamental value

***undervalued exchange rate*** an exchange rate that has an officially fixed value less than its fundamental value

In this example, Latinia's commitment to hold the peso at 8 to the dollar is inconsistent with the fundamental value of 10 to the dollar, as determined by supply and demand in the foreign exchange market (the Latinian peso is overvalued). How could the Latinian government deal with this inconsistency? There are several possibilities. First, Latinia could simply devalue its currency, from 0.125 dollars per peso to 0.10 dollars per peso, which would bring the peso's official value into line with its fundamental value. As we will see, devaluation is often the ultimate result of an overvaluation of a currency. However, a country with a fixed exchange rate will be reluctant to change the official value of its exchange rate every time the fundamental value changes. If a country must continuously adjust its exchange rate to market conditions, it might as well switch to a flexible exchange rate.

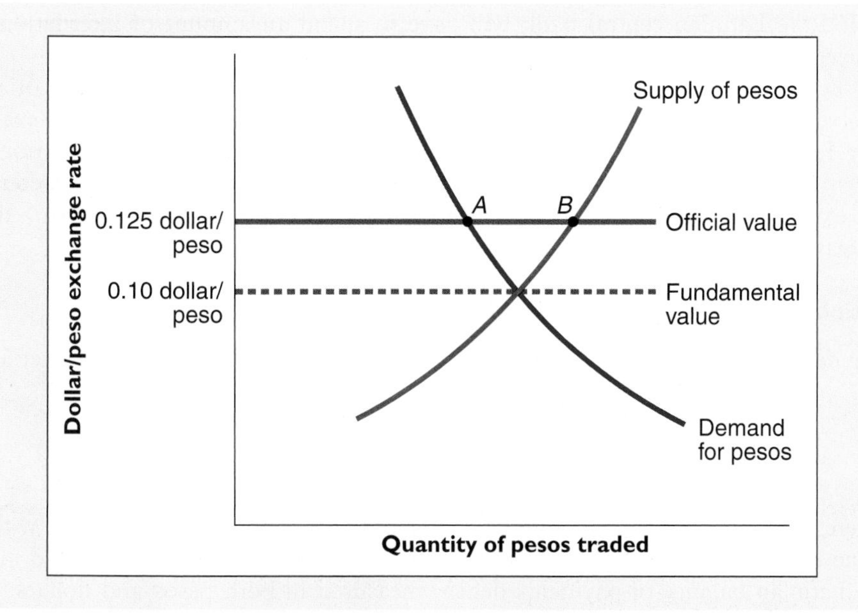


**FIGURE 17.6**
**An Overvalued Exchange Rate.**
The peso's official value (0.125 dollars) is shown as greater than its fundamental value (0.10 dollars), as determined by supply and demand in the foreign exchange market. Thus the peso is overvalued. To maintain the fixed value, the government must purchase pesos in the quantity *AB* each period.

As a second alternative, Latinia could try to maintain its overvalued exchange rate by restricting international transactions. Imposing quotas on imports and prohibiting domestic households and firms from acquiring foreign assets would effectively reduce the supply of pesos to the foreign exchange market, raising the fundamental value of the currency. An even more extreme action would be to prohibit Latinians from exchanging the peso for other currencies without government approval, a policy that would effectively allow the government to determine directly the supply of pesos to the foreign exchange market. Such measures might help to maintain the official value of the peso. However, restrictions on trade and capital flows are extremely costly to the economy, because they reduce the gains from specialization and trade and deny domestic households and firms access to foreign capital markets. Thus, a policy of restricting international transactions to maintain a fixed exchange rate is likely to do more harm than good.

The third and most widely used approach to maintaining an overvalued exchange rate is for the government to become a demander of its own currency in the foreign exchange market. Figure 17.6 shows that at the official exchange rate of 0.125 dollars per peso, the private sector supply of pesos (point $B$) exceeds the private sector demand for pesos (point $A$). To keep the peso from falling below its official value, in each period the Latinian government could purchase a quantity of pesos in the foreign exchange market equal to the length of the line segment $AB$ in Figure 17.6. If the government followed this strategy, then at the official exchange rate of 0.125 dollars per peso, the total demand for pesos (private demand at point $A$ plus government demand $AB$) would equal the private supply of pesos (point $B$). This situation is analogous to government attempts to keep the price of a commodity, like grain or milk, above its market level. To maintain an official price of grain that is above the market-clearing price, the government must stand ready to purchase the excess supply of grain forthcoming at the official price. In the same way, to keep the "price" of its currency above the market-clearing level, the government must buy the excess pesos supplied at the official price.

***international reserves*** foreign currency assets held by a government for the purpose of purchasing the domestic currency in the foreign exchange market

***balance-of-payments deficit*** the net decline in a country's stock of international reserves over a year

***balance-of-payments surplus*** the net increase in a country's stock of international reserves over a year

To be able to purchase its own currency and maintain an overvalued exchange rate, the government (usually the central bank) must hold foreign currency assets, called **international reserves,** or simply *reserves*. For example, the Latinian central bank may hold dollar deposits in U.S. banks or U.S. government debt, which it can trade for pesos in the foreign exchange market as needed. In the situation shown in Figure 17.6, to keep the peso at its official value, in each period the Latinian central bank will have to spend an amount of international reserves equal to the length of the line segment $AB$.

Because a country with an overvalued exchange rate must use part of its reserves to support the value of its currency in each period, over time its available reserves will decline. The net decline in a country's stock of international reserves over a year is called its **balance-of-payments deficit.** Conversely, if a country experiences a net increase in its international reserves over the year, the increase is called its **balance-of-payments surplus.**

**EXAMPLE 17.6** **Latinia's balance-of-payments deficit**

The demand for and supply of Latinian pesos in the foreign exchange market are

$$\text{Demand} = 25{,}000 - 50{,}000e,$$
$$\text{Supply} = 17{,}600 + 24{,}000e,$$

where the Latinian exchange rate $e$ is measured in dollars per peso. Officially, the value of the peso is 0.125 dollars. Find the fundamental value of the peso and the Latinian balance-of-payments deficit, measured in both pesos and dollars.

To find the fundamental value of the peso, equate the demand and supply for pesos:

$$25{,}000 - 50{,}000e = 17{,}600 + 24{,}000e.$$

Solving for $e$, we get

$$7{,}400 = 74{,}000e$$
$$e = 0.10.$$

So the fundamental value of the exchange rate is 0.10 dollars per peso, as in Figure 17.6.

At the official exchange rate, 0.125 dollars per peso, the demand for pesos is $25{,}000 - 50{,}000(0.125) = 18{,}750$, and the supply of pesos is $17{,}600 + 24{,}000(0.125) = 20{,}600$. Thus the quantity of pesos supplied to the foreign exchange market exceeds the quantity of pesos demanded by $20{,}600 - 18{,}750 = 1{,}850$ pesos. To maintain the fixed rate, the Latinian government must purchase 1,850 pesos per period, which is the Latinian balance-of-payments deficit. Since pesos are purchased at the official rate of 8 pesos to the dollar, the balance-of-payments deficit in dollars is $(1{,}850 \text{ pesos}) \times (0.125 \text{ dollars/peso}) = \$(1{,}850/8) = \$231.25$.

**EXERCISE 17.5**

**Repeat Example 17.6 under the assumption that the fixed value of the peso is 0.15 dollars per peso. What do you conclude about the relationship between the degree of currency overvaluation and the resulting balance-of-payments deficit?**

Although a government can maintain an overvalued exchange rate for a time by offering to buy back its own currency at the official price, there is a limit to this strategy, since no government's stock of international reserves is infinite. Eventually the government will run out of reserves, and the fixed exchange rate will collapse. As we will see next, the collapse of a fixed exchange rate can be quite sudden and dramatic.

**EXERCISE 17.6**

**Diagram a case in which a fixed exchange rate is *undervalued* rather than overvalued. Show that, to maintain the fixed exchange rate, the central bank must use domestic currency to purchase foreign currency in the foreign exchange market. With an undervalued exchange rate, is the country's central bank in danger of running out of international reserves? (*Hint:* Keep in mind that a central bank is always free to print more of its own currency.)**

## SPECULATIVE ATTACKS

A government's attempt to maintain an overvalued exchange rate can be ended quickly and unexpectedly by the onset of a *speculative attack*. A **speculative attack** involves massive selling of domestic currency assets by both domestic and foreign financial investors. For example, in a speculative attack on the Latinian peso, financial investors would attempt to get rid of any financial assets—stocks, bonds, deposits in banks—denominated in pesos. A speculative attack is most likely to occur when financial investors fear that an overvalued currency will soon be devalued, since in a devaluation, financial assets denominated in the domestic currency suddenly become worth much less in terms of other currencies. Ironically,

***speculative attack*** a massive selling of domestic currency assets by financial investors speculative attacks, which are usually prompted by *fear* of devaluation, may turn out to be the *cause* of devaluation. Thus a speculative attack may actually be a self-fulfilling prophecy.

The effects of a speculative attack on the market for pesos are shown in Figure 17.7. At first, the situation is the same as in Figure 17.6: The supply and demand for Latinian pesos are indicated by the curves marked *S* and *D*, implying a fundamental value of the peso of 0.10 dollars per peso. As before, the official value of the peso is 0.125 dollars per peso—greater than the fundamental value—so the peso is overvalued. To maintain the fixed value of the peso, each period the Latinian central bank must use its international reserves to buy back pesos, in the amount corresponding to the line segment *AB* in the figure.

Suppose, though, that financial investors fear that Latinia may soon devalue its currency, perhaps because the central bank's reserves are getting low. If the peso were to be devalued from its official value of 8 pesos to the dollar to its fundamental value of 10 pesos per dollar, then a 1 million peso investment, worth $125,000 at the fixed exchange rate, would suddenly be worth only $100,000. To try to avoid these losses, financial investors will sell their peso-denominated assets and offer pesos on the foreign exchange market. The resulting flood of pesos into the market will shift the supply curve of pesos to the right, from *S* to *S′* in Figure 17.7.

This speculative attack creates a serious problem for the Latinian central bank. Prior to the attack, maintaining the value of the peso required the central bank to spend each period an amount of international reserves corresponding to the line segment *AB*. Now suddenly the central bank must spend a larger quantity of reserves, equal to the distance *AC* in Figure 17.7, to maintain the fixed

**FIGURE 17.7**
**A Speculative Attack on the Peso.**
Initially, the peso is overvalued at 0.125 dollars per peso. To maintain the official rate, the central bank must buy pesos in the amount *AB* each period. Fearful of possible devaluation, financial investors launch a speculative attack, selling peso-denominated assets and supplying pesos to the foreign exchange market. As a result, the supply of pesos shifts from *S* to *S′*, lowering the fundamental value of the currency still further and forcing the central bank to buy pesos in the amount *AC* to maintain the official exchange rate. This more rapid loss of reserves may lead the central bank to devalue the peso, confirming financial investors' fears.

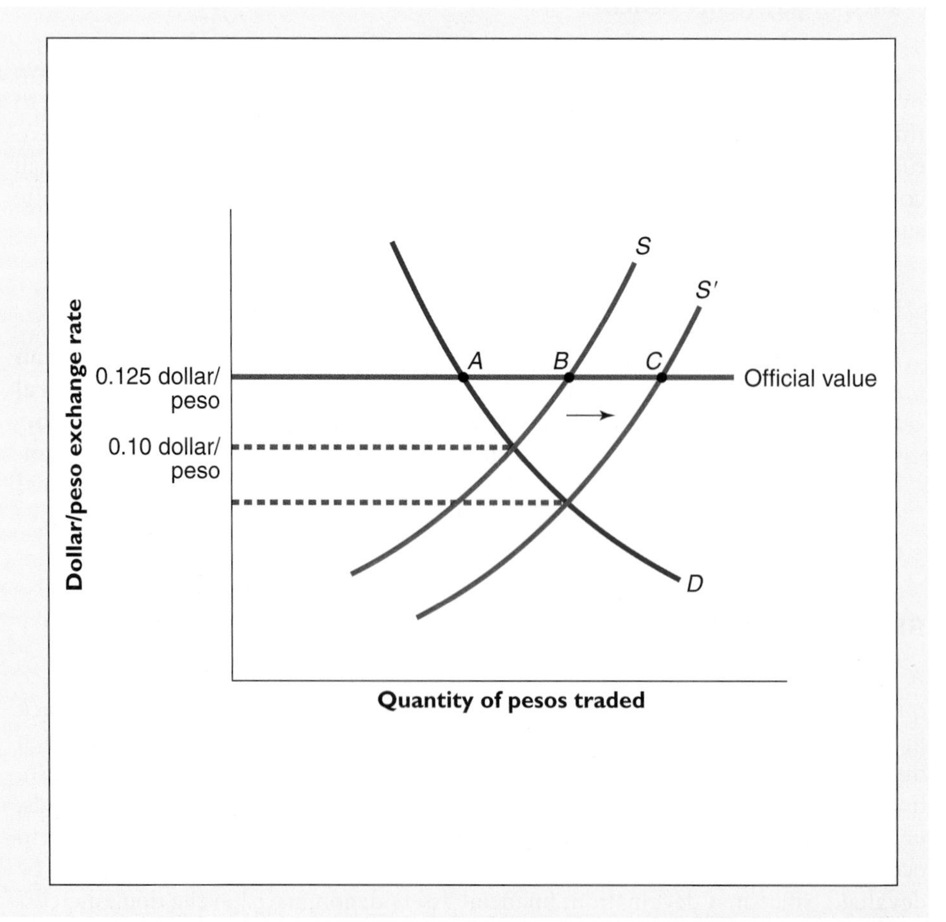

exchange rate. These extra reserves are needed to purchase the pesos being sold by panicky financial investors. In practice, such speculative attacks often force a devaluation by reducing the central bank's reserves to the point where further defense of the fixed exchange rate is considered hopeless. Thus a speculative attack ignited by fears of devaluation may actually end up producing the very devaluation that was feared.

### Can a speculative attack occur under flexible exchange rates?

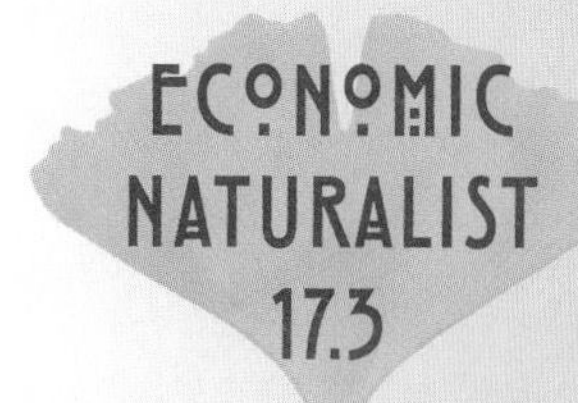


The last section described the self-fulfilling nature of speculative attacks in a fixed-exchange rate system: fears of a currency devaluation often lead to actual currency devaluation. Can a speculative attack occur in a flexible exchange rate system?

As noted earlier a speculative attack on a currency involves large-scale selling of a country's domestic currency assets by both domestic and foreign financial investors based on fears of a future decline in the value of the country's currency. The economic effects of a speculative attack are illustrated in Figure 17.7: an increase in the supply of the domestic currency that lowers the fundamental value of the currency, often dramatically. Brazil experienced just this sort of deterioration in the value of its currency, the real (pronounced ray-al′), during 2002, in anticipation of a presidential victory by Brazilian Worker's Party candidate, Luis Inacio da Silva, who goes by the nickname "Lula." Investors were concerned that da Silva would break from the economic policies of outgoing president Fernando Henrique Cardoso, who helped to stabilize Brazil's economy, South America's largest, since his first presidential election in 1994.

In the early 1990s, just prior to Mr. Cardoso's election, high inflation stunted Brazil's economic performance and reduced foreign investment, weakening future economic growth prospects. Inflation averaged over 20 percent per month since the late 1980s and reached a monthly rate of nearly 50 percent (over 10,000 percent per year!) by June 1994, when newly elected president Cardoso instituted a bold economic plan that included the creation of a new currency—the real—that was directly linked to the value of the U.S. dollar. Immediately the inflation rate dropped to less than 10 percent per month and by 1997 the *annual* inflation rate was in the single digits. However, by 1998, increasing Brazilian budget and current-account deficits led to a speculative attack that led to a dramatic devaluation of the real in January 1999 and an uncoupling of the real from the U.S. dollar. Since 1999 the Brazilian currency has been allowed to fluctuate freely but has remained relatively stable, along with the Brazilian economy: inflation remains in the single digits, poverty has been reduced, and a variety of social indicators have improved. Progress has come with a cost, however: rising government debt to pay for Cardoso's social programs.

Concerns about the growing government debt and the increasing popularity of opposition party candidate da Silva led to another round of downward pressure on the real during 2002 as investors began to sell off financial assets denominated in reals. Investors were concerned that da Silva, if elected president, would be unwilling to make the tough government spending and taxing decisions that would be necessary to stabilize the government's debt, raising concerns of a government default, higher inflation, and even further downward pressure on the real. As a result, by the end of September the real had reached record lows and inflation was on the rise.

However, following da Silva's landslide election victory in early October, pledges that he would continue Cardoso's fight against inflation and promote fiscal policy discipline along with a new international loan package from the International Monetary Fund, helped to calm currency and financial markets, sparking hopes for continued economic growth with low inflation, rising currency values, and falling interest rates. Whether da Silva will be able to avoid further downward pressure on the real will depend upon investors' confidence that he will carry out his presidential pledges.

## MONETARY POLICY AND THE FIXED EXCHANGE RATE

We have seen that there is no really satisfactory way of maintaining a fixed exchange rate above its fundamental value for an extended period. A central bank can maintain an overvalued exchange rate for a time by using international reserves to buy up the excess supply of its currency in the foreign exchange market. But a country's international reserves are limited and may eventually be exhausted by the attempt to keep the exchange rate artificially high. Moreover, speculative attacks often hasten the collapse of an overvalued exchange rate.

An alternative to trying to maintain an overvalued exchange rate is to take actions that increase the fundamental value of the exchange rate. If the exchange rate's fundamental value can be raised enough to equal its official value, then the overvaluation problem will be eliminated. The most effective way to change the exchange rate's fundamental value is through monetary policy. As we saw earlier in the chapter, a tight monetary policy that raises the real interest rate will increase the demand for the domestic currency, as domestic assets become more attractive to foreign financial investors. Increased demand for the currency will in turn raise its fundamental value.

The use of monetary policy to support a fixed exchange rate is shown in Figure 17.8. At first, the demand and supply of the Latinian peso in the foreign exchange market are given by the curves *D* and *S*, so the fundamental value of the peso equals 0.10 dollars per peso—less than the official value of 0.125 dollars per peso. Just as before, the peso is overvalued. This time, however, the Latinian central bank uses monetary policy to eliminate the overvaluation problem. To do so, the central bank increases the domestic real interest rate, making Latinian assets more attractive to foreign financial investors and raising the demand for pesos from *D* to *D′*. After this increase in the demand for pesos, the fundamental value of the peso equals the officially fixed value, as can be seen in

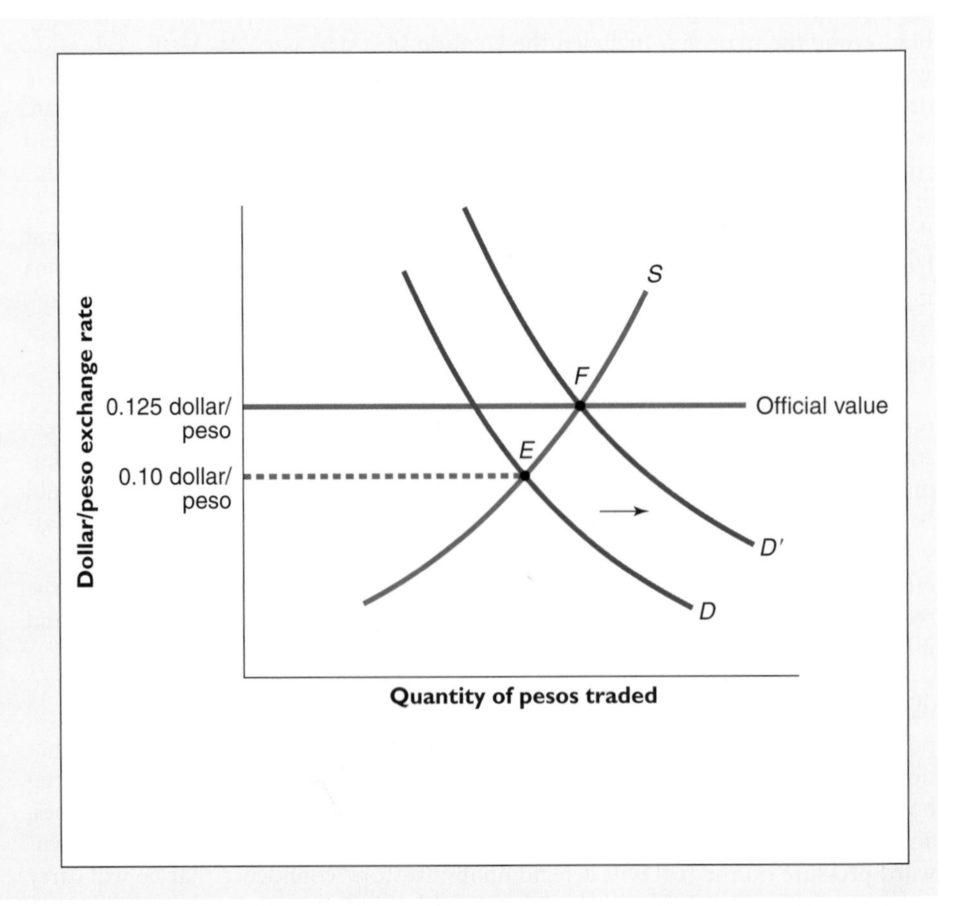


**FIGURE 17.8**
**A Tightening of Monetary Policy Eliminates an Overvaluation.**
With the demand for the peso given by *D* and the supply given by *S*, equilibrium occurs at point *E* and the fundamental value of the peso equals 0.10 dollars per peso—below the official value of 0.125 dollars per peso. The overvaluation of the peso can be eliminated by tighter monetary policy, which raises the domestic real interest rate, making domestic assets more attractive to foreign financial investors. The resulting increase in demand for the peso, from *D* to *D′*, raises the peso's fundamental value to 0.125 dollars per peso, the official value. The peso is no longer overvalued.

Figure 17.8. Because the peso is no longer overvalued, it can be maintained at its fixed value without loss of international reserves or fear of speculative attack. Conversely, an easing of monetary policy (a lower real interest rate) could be used to remedy an undervaluation, in which the official exchange rate is below the fundamental value.

Although monetary policy can be used to keep the fundamental value of the exchange rate equal to the official value, using monetary policy in this way has some drawbacks. In particular, *if monetary policy is used to set the fundamental value of the exchange rate equal to the official value, it is no longer available for stabilizing the domestic economy.* Suppose, for example, that the Latinian economy were suffering a recession due to insufficient aggregate demand at the same time that its exchange rate is overvalued. The Latinian central bank could lower the real interest rate to increase spending and output, or it could raise the real interest rate to eliminate overvaluation of the exchange rate, *but it cannot do both.* Hence, if Latinian officials decide to maintain the fixed exchange rate, they must give up any hope of fighting the recession using monetary policy. The fact that a fixed exchange rate limits or eliminates the use of monetary policy for the purpose of stabilizing aggregate demand is one of the most important features of a fixed-exchange-rate system.

The conflict monetary policymakers face, between stabilizing the exchange rate and stabilizing the domestic economy, is most severe when the exchange rate is under a speculative attack. A speculative attack lowers the fundamental value of the exchange rate still further, by increasing the supply of the currency in the foreign exchange market (see Figure 17.7). To stop a speculative attack, the central bank must raise the fundamental value of the currency a great deal, which requires a large increase in the real interest rate. (In a famous episode in 1992, the Swedish central bank responded to an attack on its currency by raising the short-term interest rate to 500 percent!) However, because the increase in the real interest rate that is necessary to stop a speculative attack reduces aggregate demand, it can cause a severe economic slowdown. Economic Naturalist 17.4 describes a real-world example of this phenomenon.

*"It's just a flesh wound. I got it defending the dollar."*

**What were the causes and consequences of the East Asian crisis of 1997–1998?**

During the past three decades the countries of East Asia have enjoyed impressive economic growth and stability. But the "East Asian miracle" seemed to end in 1997, when a wave of speculative attacks hit the region's currencies. Thailand, which had kept a constant value for its currency in terms of the U.S. dollar for more than a decade, was the first to come under attack, but the crisis spread to other countries, including South Korea, Indonesia, and Malaysia. Each of these countries was ultimately forced to devalue its currency. What caused this crisis, and what were its consequences?

Because of the impressive economic record of the East Asian countries, the speculative attacks on their currencies were unexpected by most policymakers, economists, and financial investors. With the benefit of hindsight, however, we can identify some problems in the East Asian economies that contributed to the crisis. Perhaps the most serious problems concerned their banking systems. In the decade prior to the crisis, East Asian banks received large inflows of capital from foreign financial investors hoping to profit from the East Asian miracle. Those inflows would have been a boon if they had been well invested, but unfortunately, many bankers used the funds to make loans to family members, friends, or the politically well-connected—a phenomenon that became known as *crony capitalism.* The results were poor returns on investment and defaults by many borrowers. Ultimately, foreign investors realized that the returns to investing in East Asia would be much lower than expected. When they began to sell off their assets, the process snowballed into a full-fledged speculative attack on the East Asian currencies.

Despite assistance by international lenders such as the International Monetary Fund (see Box 17.1), the effects of the speculative attacks on the East Asian economies were severe. The prices of assets such as stocks and land plummeted, and there were banking panics in several nations. (See Chapter 10 for a discussion of banking panics.) In an attempt to raise the fundamental values of their exchange rates and stave off additional devaluation, several of the countries increased their real interest rates sharply. However, the rise in real interest rates depressed aggregate demand, contributing to sharp declines in output and rising unemployment.

Fortunately, by 1999 most East Asian economies had begun to recover. Still, the crisis impressed the potential dangers of fixed exchange rates quite sharply in the minds of policymakers in developing countries. Another lesson from the crisis is that banking regulations need to be structured so as to promote economically sound lending rather than crony capitalism.

**BOX 17.1: THE INTERNATIONAL MONETARY FUND**

The International Monetary Fund (IMF) was established after World War II. An international agency, the IMF is controlled by a 24-member Executive Board. Eight Executive Board members represent individual countries (China, France, Germany, Japan, Russia, Saudi Arabia, the United Kingdom, and the United States); the other 16 members each represent a group of countries. A managing director oversees the IMF's operations and its approximately 2,600 employees.

The original purpose of the IMF was to help manage the system of fixed exchange rates, called the Bretton Woods system, put in place after the war. Under Bretton Woods, the IMF's principal role was to lend international reserves to member countries who needed them so that those countries could maintain their exchange rates at the official values. However, by 1973 the United States, the United Kingdom, Germany, and most other industrial nations had abandoned fixed exchange rates for flexible rates, leaving the IMF to find a new mission. Since 1973 the IMF has been involved primarily in lending to developing countries. It lent heavily to Mexico when that country experienced speculative attacks in 1994, and it made loans to East Asian countries during the 1997–1998 crisis. Other countries that received large IMF loans in recent years include Russia and Brazil.

The IMF's performance in recent crises has been controversial. Many observers credit the IMF with helping Mexico, the East Asian nations, and others to recover quickly from the effects of speculative attacks and contend that the IMF plays a vital role in maintaining international economic stability. However, some critics have charged that the IMF has required recipients of its loans to follow economic policies—such as tight monetary policies and fiscal cutbacks—that have turned out to be ill-advised. Others have claimed that the IMF's loans help foreign financial investors and the richest people in the countries receiving loans, rather than the average person. (The IMF has been severely embarrassed by reports that much of the nearly $5 billion it lent to Russia in 1998 has disappeared into the bank accounts of unscrupulous citizens, including gangsters.)

The IMF has also come into conflict with the World Bank, a separate international institution that was set up at about the same time as the IMF. The World Bank, whose mission is to provide long-term loans to help poor nations develop their economies, has complained that IMF interventions in poor countries interfered with World Bank programs and objectives. In 2000, a report commissioned by the U.S. Congress recommended reducing the IMF's powers (as well as, incidentally, those of the World Bank). The debate over the IMF's proper role will no doubt continue.

**How did policy mistakes contribute to the Great Depression?**

ECONOMIC NATURALIST 17.5

Chapter 4 introduced the study of macroeconomics with the claim that policy mistakes played a major role in causing the Great Depression. Now that we are close to completing our study of macroeconomics, we can be more specific about that claim. How did policy mistakes contribute to the Great Depression?

Many policy mistakes (as well as a great deal of bad luck) contributed to the severity of the Depression. For example, U.S. policymakers, in an attempt to protect domestic industries, imposed the infamous Hawley-Smoot tariff in 1930. Other countries quickly retaliated with their own tariffs, leading to the virtual collapse of international trade.

However, the most serious mistakes by far were in the realm of monetary policy.[3] As we saw in Chapter 10, the U.S. money supply contracted by one-third between 1929 and 1933 (Table 10.7). Associated with this unprecedented decline in the money supply were sharply falling output and prices and surging unemployment.

At least three separate policy errors were responsible for the collapse of the U.S. money supply between 1929 and 1933. First, the Federal Reserve tightened monetary policy significantly in 1928 and 1929, despite the absence of inflation. Fed officials took this action primarily in an attempt to "rein in" the booming stock market, which they feared was rising too quickly. Their "success" in dampening stock market speculation was more than they bargained for, however, as rising interest rates and a slowing economy contributed to a crash in stock prices that began in October 1929.

The second critical policy error was allowing thousands of U.S. banks to fail during the banking panics of 1930 to 1933. Apparently officials believed that the failures would eliminate only the weakest banks, strengthening the banking system

[3]A classic 1963 book by Milton Friedman and Anna Schwartz, *A Monetary History of the United States: 1867–1960* (Princeton University Press), was the first to provide detailed support for the view that poor monetary policy helped to cause the Depression.

overall. However, the banking panics sharply reduced bank deposits and the overall money supply, for reasons discussed in Economic Naturalist 10.2.

The third policy error, related to the subject of this chapter, arose from the U.S. government's exchange rate policies. When the Depression began, the United States, like most other major countries, was on the gold standard, with the value of the dollar officially set in terms of gold.[4] By establishing a fixed value for the dollar, the United States effectively created a fixed exchange rate between the dollar and other currencies whose values were set in terms of gold. As the Depression worsened, Fed officials were urged by Congress to ease monetary policy to stop the fall in output and prices. However, as we saw earlier, under a fixed exchange rate monetary policy cannot be used to stabilize the domestic economy. Specifically, policymakers of the early 1930s feared that if they eased monetary policy, foreign financial investors might perceive the dollar to be overvalued and launch a speculative attack, forcing a devaluation of the dollar or even the abandonment of the gold standard altogether. The Fed therefore made no serious attempt to arrest the collapse of the money supply.

With hindsight, we can see that the Fed's decision to put a higher priority on remaining on the gold standard than on stimulating the economy was a major error. Indeed, countries that abandoned the gold standard in favor of a floating exchange rate, such as Great Britain and Sweden, or which had never been on the gold standard (Spain and China), were able to increase their money supplies and to recover much more quickly from the Depression than the United States did. The Fed evidently believed, erroneously as it turned out, that stability of the exchange rate would somehow translate into overall economic stability.

Upon taking office in March 1933, Franklin D. Roosevelt reversed several of these policy errors. He took active measures to restore the health of the banking system, and he suspended the gold standard. The money supply stopped falling and began to grow rapidly. Output, prices, and stock prices recovered rapidly during 1933 to 1937, although unemployment remained high. However, ultimate recovery from the Depression was interrupted by another recession in 1937–1938.

## RECAP FIXED EXCHANGE RATES

- The value of a fixed exchange rate is set by the government. The official value of a fixed exchange rate may differ from its fundamental value, as determined by supply and demand in the foreign exchange market. An exchange rate whose officially fixed value exceeds its fundamental value is overvalued; an exchange rate whose officially fixed value is below its fundamental value is undervalued.
- For an overvalued exchange rate, the quantity of the currency supplied to the foreign exchange market at the official exchange rate exceeds the quantity demanded. The government can maintain an overvalued exchange rate for a time by using its international reserves (foreign currency assets) to purchase the excess supply of its currency. The net decline in a country's stock of international reserves during the year is its balance-of-payments deficit.
- Because a country's international reserves are limited, it cannot maintain an overvalued exchange rate indefinitely. Moreover, if financial investors fear an impending devaluation of the exchange rate, they may launch a speculative attack, selling domestic currency assets and supplying large amounts of the country's currency to the foreign exchange market—an

[4]The value of the dollar in 1929 was such that the price of 1 ounce of gold was fixed at $20.67.

action that exhausts the country's reserves even more quickly. Because rapid loss of reserves may force a devaluation, financial investors' fear of devaluation may prove a self-fulfilling prophecy.

- A tight monetary policy, which increases the real interest rate, raises the demand for the currency and hence its fundamental value. By raising a currency's fundamental value to its official value, tight monetary policies can eliminate the problem of overvaluation and stabilize the exchange rate. However, if monetary policy is used to set the fundamental value of the exchange rate, it is no longer available for stabilizing the domestic economy.

## SHOULD EXCHANGE RATES BE FIXED OR FLEXIBLE?

Should countries adopt fixed or flexible exchange rates? In briefly comparing the two systems, we will focus on two major issues: (1) the effects of the exchange rate system on monetary policy and (2) the effects of the exchange rate system on trade and economic integration.

On the issue of monetary policy, we have seen that the type of exchange rate a country has strongly affects the central bank's ability to use monetary policy to stabilize the economy. A flexible exchange rate actually strengthens the impact of monetary policy on aggregate demand. But a fixed exchange rate prevents policymakers from using monetary policy to stabilize the economy, because they must instead use it to keep the exchange rate's fundamental value at its official value (or else risk speculative attack).

In large economies like that of the United States, giving up the power to stabilize the domestic economy via monetary policy makes little sense. Thus large economies should nearly always employ a flexible exchange rate. However, in small economies, giving up this power may have some benefits. An interesting case is that of Argentina, which for the period 1991–2001 maintained a one-to-one exchange rate between its peso and the U.S. dollar. Although prior to 1991 Argentina had suffered periods of hyperinflation, while the peso was pegged to the dollar Argentina's inflation rate essentially equaled that of the United States (see Figure 17.2). By tying its currency to the dollar and giving up the freedom to set its monetary policy, Argentina attempted to commit itself to avoiding the inflationary policies of the past, and instead placed itself under the "umbrella" of the Federal Reserve. Unfortunately, early in 2002 investors' fears that Argentina would not be able to repay its international debts led to a speculative attack on the Argentine peso. The fixed exchange rate collapsed, the peso depreciated, and Argentina experienced an economic crisis. The lesson is that a fixed exchange rate alone cannot stop inflation in a small economy, if other policies are not sound as well. Large fiscal deficits, which were financed by foreign borrowing, ultimately pushed Argentina into crisis.

The second important issue is the effect of the exchange rate on trade and economic integration. Proponents of fixed exchange rates argue that fixed rates promote international trade and cross-border economic cooperation by reducing uncertainty about future exchange rates. For example, a firm that is considering building up its export business knows that its potential profits will depend on the future value of its own country's currency relative to the currencies of the countries to which it exports. Under a flexible-exchange-rate regime, the value of the home currency fluctuates with changes in supply and demand and is therefore difficult to predict far in advance. Such uncertainty may make the firm reluctant to expand its export business. Supporters of fixed exchange rates argue that if the exchange rate is officially fixed, uncertainty about the future exchange rate is reduced or eliminated.

One problem with this argument, which has been underscored by episodes like the East Asian crisis and the Argentine crisis, is that fixed exchange rates are not guaranteed to remain fixed forever. Although they do not fluctuate from day to day as flexible rates do, a speculative attack on a fixed exchange rate may lead suddenly and unpredictably to a large devaluation. Thus a firm that is trying to forecast the exchange rate 10 years into the future may face as much uncertainty if the exchange rate is fixed as if it is flexible.

The potential instability of fixed exchange rates caused by speculative attacks has led some countries to try a more radical solution to the problem of uncertainty about exchange rates: the adoption of a common currency. Economic Naturalist 17.6 describes an important instance of this strategy.

ECONOMIC NATURALIST 17.6

**Why have 11 European countries adopted a common currency?**

Effective January 1, 1999, eleven western European nations, including France, Germany, and Italy, adopted a common currency, called the euro. In several stages the euro replaced the French franc, the German mark, the Italian lira, and other national currencies. The process was completed in early 2002 when the old currencies were completely eliminated and replaced by euros. Why have these nations adopted a common currency?

For some decades the nations of western Europe have worked to increase economic cooperation and trade among themselves. European leaders recognized that a unified and integrated European economy would be more productive and perhaps more competitive with the U.S. economy than a fragmented one. As part of this effort, these countries established fixed exchange rates under the auspices of a system called the European Monetary System (EMS). Unfortunately, the EMS did not prove stable. Numerous devaluations of the various currencies occurred, and in 1992 severe speculative attacks forced several nations, including Great Britain, to abandon the fixed-exchange-rate system.

In December 1991, in Maastricht in the Netherlands, the member countries of the European Community (EC) adopted a treaty popularly known as the Maastricht Treaty. One of the major provisions of the treaty, which took effect in November 1993, was that member countries would strive to adopt a common currency. This common currency, known as the euro, was formally adopted on January 1, 1999. The advent of the euro means that Europeans will no longer have to change currencies when trading with other European countries, much as Americans from different states can trade with each other without worrying that a "New York dollar" will change in value relative to a "California dollar." The euro should help to promote European trade and cooperation while eliminating the problem of speculative attacks on the currencies of individual countries.

Because western Europe now has a single currency, it also must have a common monetary policy. The EC members agreed that European monetary policy would be put under the control of a new European Central Bank (ECB), a multinational institution located in Frankfurt, Germany. The ECB has in effect become "Europe's Fed." One potential problem with having a single monetary policy for 11 different countries is that different countries may face different economic conditions, so a single monetary policy cannot respond to all of them. What will the ECB do, for example, if Italy is suffering from a recession (which requires an easing of monetary policy) while Germany is worried about inflation (which requires a tightening)? Whether the requirement of a single monetary policy will create conflicts of interest among the member nations of the European Community remains to be seen.

# ■ SUMMARY ■

- The *nominal exchange rate* between two currencies is the rate at which the currencies can be traded for each other. A rise in the value of a currency relative to other currencies is called an *appreciation;* a decline in the value of a currency is called a *depreciation.*

- Exchange rates can be flexible or fixed. The value of a *flexible exchange rate* is determined by the supply and demand for the currency in the *foreign exchange market,* the market on which currencies of various nations are traded for one another. The government sets the value of a *fixed exchange rate.*

- The *real exchange rate* is the price of the average domestic good or service *relative* to the price of the average foreign good or service, when prices are expressed in terms of a common currency. An increase in the real exchange rate implies that domestic goods and services are becoming more expensive relative to foreign goods and services, which tends to reduce exports and increase imports. Conversely, a decline in the real exchange rate tends to increase net exports.

- A basic theory of nominal exchange rate determination, the *purchasing power parity* (PPP) theory, is based on the law of one price. The *law of one price* states that if transportation costs are relatively small, the price of an internationally traded commodity must be the same in all locations. According to the PPP theory, we can find the nominal exchange rate between two currencies by setting the price of a commodity in one of the currencies equal to the price of the commodity in the second currency. The PPP theory correctly predicts that the currencies of countries that experience significant inflation will tend to depreciate in the long run. However, the fact that many goods and services are not traded internationally, and that not all traded goods are standardized, makes the PPP theory less useful for explaining short-run changes in exchange rates.

- Supply and demand analysis is a useful tool for studying the determination of exchange rates in the short run. The equilibrium exchange rate, also called the *fundamental value of the exchange rate,* equates the quantities of the currency supplied and demanded in the foreign exchange market. A currency is supplied by domestic residents who wish to acquire foreign currencies to purchase foreign goods, services, and assets. An increased preference for foreign goods, an increase in the domestic GDP, or an increase in the real interest rate on foreign assets will all increase the supply of a currency on the foreign exchange market and thus lower its value. A currency is demanded by foreigners who wish to purchase domestic goods, services, and assets. An increased preference for domestic goods by foreigners, an increase in real GDP abroad, or an increase in the domestic real interest rate will all increase the demand for the currency on the foreign exchange market and thus increase its value.

- If the exchange rate is flexible, a tight monetary policy (by raising the real interest rate) increases the demand for the currency and causes it to appreciate. The stronger currency reinforces the effects of the tight monetary policy on aggregate demand by reducing net exports. Conversely, easy monetary policy lowers the real interest rate and weakens the currency, which in turn stimulates net exports.

- The value of a fixed exchange rate is officially established by the government. A fixed exchange rate whose official value exceeds its fundamental value in the foreign exchange market is said to be *overvalued.* An exchange rate whose official value is below its fundamental value is *undervalued.* A reduction in the official value of a fixed exchange rate is called a *devaluation;* an increase in its official value is called a *revaluation.*

- For an overvalued exchange rate, the quantity of the currency supplied at the official exchange rate exceeds the quantity demanded. To maintain the official rate, the country's central bank must use its *international reserves* (foreign currency assets) to purchase the excess supply of its currency in the foreign exchange market. Because a country's international reserves are limited, it cannot maintain an overvalued exchange rate indefinitely. Moreover, if financial investors fear an impending devaluation of the exchange rate, they may launch a *speculative attack,* selling their domestic currency assets and supplying large quantities of the currency to the foreign exchange market. Because speculative attacks cause a country's central bank to spend its international reserves even more quickly, they often force a devaluation.

- A tight monetary policy, by raising the fundamental value of the exchange rate, can eliminate the problem of overvaluation. However, if monetary policy is used to set the fundamental value of the exchange rate equal to the official value, it is no longer available for stabilizing the domestic economy. Thus under fixed exchange rates, monetary policy has little or no power to affect domestic output and employment.

- Because a fixed exchange rate implies that monetary policy can no longer be used for domestic stabilization, most large countries employ a flexible exchange rate. A fixed exchange rate may benefit a small country by forcing its central bank to follow the monetary policies of the country to which it has tied its rate. Advocates of fixed exchange rates argue that they increase trade and economic integration by making the exchange rate more predictable. However, the threat of speculative attacks greatly reduces the long-term predictability of a fixed exchange rate.

## KEY TERMS

appreciation (472)
balance-of-payments deficit (488)
balance-of-payment surplus (488)
depreciation (472)
devaluation (486)
fixed exchange rate (473)
flexible exchange rate (473)
foreign exchange market (473)
fundamental value of the exchange rate (482)
international reserves (488)
law of one price (477)
nominal exchange rate (471)
overvalued exchange rate (487)
purchasing power parity (PPP) (478)
real exchange rate (474)
revaluation (487)
speculative attack (489)
undervalued exchange rate (487)

## REVIEW QUESTIONS

1. Japanese yen trade at 110 yen per dollar and Mexico pesos trade at 10 pesos per dollar. What is the nominal exchange rate between the yen and the peso? Express in two ways.
2. Define *nominal exchange rate* and *real exchange rate*. How are the two concepts related? Which type of exchange rate most directly affects a country's ability to export its goods and services?
3. Would you expect the law of one price to apply to crude oil? To fresh milk? To taxi rides? To compact discs produced in different countries by local recording artists? Explain your answer in each case.
4. Why do U.S. households and firms supply dollars to the foreign exchange market? Why do foreigners demand dollars in the foreign exchange market?
5. Under a flexible exchange rate, how does an easing of monetary policy (a lower real interest rate) affect the value of the exchange rate? Does this change in the exchange rate tend to weaken or strengthen the effect of the monetary ease on output and employment? Explain.
6. Define *overvalued exchange rate*. Discuss four ways in which government policymakers can respond to an overvaluation. What are the drawbacks of each approach?
7. Use a supply and demand diagram to illustrate the effects of a speculative attack on an overvalued exchange rate. Why do speculative attacks often result in a devaluation?
8. Contrast fixed and flexible exchange rates in terms of how they affect (a) the ability of monetary policy to stabilize domestic output and (b) the predictability of future exchange rates.

## PROBLEMS

1. Using the data in Table 17.1, find the nominal exchange rate between the Mexican peso and the Japanese yen. Express in two ways. How do your answers change if the peso appreciates by 10 percent against the dollar while the value of the yen against the dollar remains unchanged?
2. A British-made automobile is priced at £20,000 (20,000 British pounds). A comparable U.S.-made car costs $26,000. One pound trades for $1.50 in the foreign exchange market. Find the real exchange rate for automobiles from the perspective of the United States and from the perspective of Great Britain. Which country's cars are more competitively priced?
3. Between last year and this year, the CPI in Blueland rose from 100 to 110 and the CPI in Redland rose from 100 to 105. Blueland's currency unit, the blue, was worth $1 (U.S.) last year and is worth 90 cents (U.S.) this year. Redland's currency unit, the red, was worth 50 cents (U.S.) last year and is worth 45 cents (U.S.) this year.

   Find the percentage change from last year to this year in Blueland's *nominal* exchange rate with Redland and in Blueland's *real* exchange rate with Redland. (Treat Blueland as the home country.) Relative to Redland, do you expect Blueland's exports to be helped or hurt by these changes in exchange rates?
4. The demand for U.S.-made cars in Japan is given by

$$\text{Japanese demand} = 10{,}000 - 0.001(\text{Price of U.S. cars in yen}).$$

Similarly, the demand for Japanese-made cars in the United States is

U.S. demand = 30,000 − 0.2(Price of Japanese cars in dollars).

The domestic price of a U.S.-made car is $20,000, and the domestic price of a Japanese-made car is ¥2,500,000. From the perspective of the United States, find the real exchange rate in terms of cars and net exports of cars to Japan, if:

a. The nominal exchange rate is 100 yen per dollar.
b. The nominal exchange rate is 125 yen per dollar.

How does an appreciation of the dollar affect U.S. net exports of automobiles (considering only the Japanese market)?

5. a. Gold is $350 per ounce in the United States and 2,800 pesos per ounce in Mexico. What nominal exchange rate between U.S. dollars and Mexican pesos is implied by the PPP theory?
   b. Mexico experiences inflation so that the price of gold rises to 4,200 pesos per ounce. Gold remains $350 per ounce in the United States. According to the PPP theory, what happens to the exchange rate? What general principle does this example illustrate?
   c. Gold is $350 per ounce in the United States and 4,200 pesos per ounce in Mexico. Crude oil (excluding taxes and transportation costs) is $30 per barrel in the United States. According to the PPP theory, what should a barrel of crude oil cost in Mexico?
   d. Gold is $350 per ounce in the United States. The exchange rate between the United States and Canada is 0.70 U.S. dollars per Canadian dollar. How much does an ounce of gold cost in Canada?

6. How would each of the following be likely to affect the value of the dollar, all else being equal? Explain.
   a. U.S. stocks are perceived as having become much riskier financial investments.
   b. European computer firms switch from U.S.-produced software to software produced in India, Israel, and other nations.
   c. As East Asian economies recover, international financial investors become aware of many new, high-return investment opportunities in the region.
   d. The U.S. government imposes a large tariff on imported automobiles.
   e. The Federal Reserve reports that it is less concerned about inflation and more concerned about an impending recession in the United States.
   f. U.S. consumers increase their spending on imported goods.

7. The demand for and supply of shekels in the foreign exchange market is

$$\text{Demand} = 30{,}000 - 8{,}000e,$$
$$\text{Supply} = 25{,}000 + 12{,}000e,$$

   where the nominal exchange rate is expressed as U.S. dollars per shekel.
   a. What is the fundamental value of the shekel?
   b. The shekel is fixed at 0.30 U.S. dollars. Is the shekel overvalued, undervalued, or neither? Find the balance-of-payments deficit or surplus in both shekels and dollars. What happens to the country's international reserves over time?
   c. Repeat part b for the case in which the shekel is fixed at 0.20 U.S. dollars.

8. The annual demand for and supply of shekels in the foreign exchange market is as given in Problem 7. The shekel is fixed at 0.30 dollars per shekel. The country's international reserves are $600. Foreign financial investors hold checking accounts in the country in the amount of 5,000 shekels.
   a. Suppose that foreign financial investors do not fear a devaluation of the shekel, and thus do not convert their shekel checking accounts into dollars. Can the shekel be maintained at its fixed value of 0.30 U.S. dollars for the next year?
   b. Now suppose that foreign financial investors come to expect a possible devaluation of the shekel to 0.25 U.S. dollars. Why should this possibility worry them?
   c. In response to their concern about devaluation, foreign financial investors withdraw all funds from their checking accounts and attempt to convert those shekels into dollars. What happens?
   d. Discuss why the foreign investors' forecast of devaluation can be considered a "self-fulfilling prophecy."

9. Eastland's currency is called the eastmark, and Westland's currency is called the westmark. In the market in which eastmarks and westmarks are traded for each other, the supply of and demand for eastmarks is given by

$$\text{Demand} = 25{,}000 - 5{,}000e - 50{,}000(r_E - r_W),$$
$$\text{Supply} = 18{,}500 + 8{,}000e - 50{,}000(r_E - r_W).$$

The nominal exchange rate $e$ is measured as westmarks per eastmark, and $r_E$ and $r_W$ are the real interest rates prevailing in Eastland and Westland, respectively.

a. Explain why it makes economic sense for the two real interest rates to appear in the demand and supply equations in the way they do.
b. Initially, $r_E = r_W = 0.10$, or 10 percent. Find the fundamental value of the eastmark.
c. The Westlandian central bank grows concerned about inflation and raises Westland's real interest rate to 12 percent. What happens to the fundamental value of the eastmark?
d. Assume that the exchange rate is flexible and that Eastland does not change its real interest rate following the increase in Westland's real interest rate. Is the action of the Westlandian central bank likely to increase or reduce aggregate demand in Eastland? Discuss.
e. Now suppose that the exchange rate is fixed at the value you found in part b. After the action by the Westlandian central bank, what will the Eastlandian central bank have to do to keep its exchange rate from being overvalued? What effect will this action have on the Eastlandian economy?
f. In the context of this example, discuss the effect of fixed exchange rates on the ability of a country to run an independent monetary policy.

## ANSWERS TO IN-CHAPTER EXERCISES

17.1 Answers will vary, depending on when the data is obtained.

17.2 The dollar price of the U.S. computer is $2,400, and each dollar is equal to 110 yen. Therefore the yen price of the U.S. computer is (110 yen/dollar) × ($2,400), or 264,000 yen. The price of the Japanese computer is 242,000 yen. Thus the conclusion that the Japanese model is cheaper does not depend on the currency in which the comparison is made.

17.3 Since the law of one price holds for gold, its price per ounce must be the same in New York and Stockholm:

$$\$300 = 2{,}500 \text{ kronor.}$$

Dividing both sides by 300, we get

$$\$1 = 8.33 \text{ kronor.}$$

So the exchange rate is 8.33 kronor per dollar.

17.4 A decline in U.S. GDP reduces consumer incomes and hence imports. As Americans are purchasing fewer imports, they supply fewer dollars to the foreign exchange market, so the supply curve for dollars shifts to the left. Reduced supply raises the equilibrium value of the dollar.

17.5 At a fixed value for the peso of 0.15 dollars, the demand for the peso equals $25{,}000 - 50{,}000(0.15) = 17{,}500$. The supply of the peso equals $17{,}600 + 24{,}000(0.15) = 21{,}200$. The quantity supplied at the official rate exceeds the quantity demanded by 3,700. Latinia will have to purchase 3,700 pesos each period, so its balance-of-payments deficit will equal 3,700 pesos, or $3{,}700 \times 0.15 = 555$ dollars. This balance-of-payments deficit is larger than we found in Example 17.6. We conclude that the greater the degree of overvaluation, the larger the country's balance-of-payments deficit is likely to be.

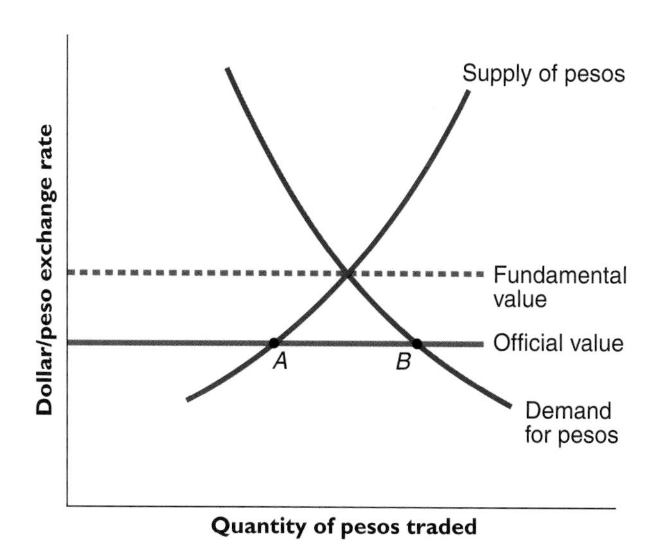


17.6 The figure shows a situation in which the official value of the currency is *below* the fundamental value, as determined by the supply of and demand for the currency in the foreign exchange market, so the currency is undervalued. At the official value of the exchange rate, the quantity demanded of the domestic currency (point *B*) exceeds the quantity supplied (point *A*). To maintain the official value, the central bank must supply domestic currency to the foreign exchange market each period in the amount *AB*. In contrast to the case of an overvalued exchange rate, here the central bank is providing its own currency to the foreign exchange market and receiving foreign currencies in return.

The central bank can print as much of its own currency as it likes, and so with an undervalued currency there is no danger of running out of international reserves. Indeed, the central bank's stock of international reserves increases in the amount *AB* each period, as it receives foreign currencies in exchange for the domestic currency it supplies.

# Monetary Policy and the Equilibrium Price Level and Inflation Rate

An alternate chapter by Norman P. Obst

To begin the analysis of determining both the equilibrium price level path and the steady-state (long run equilibrium) inflation rate, we return to the key assumption of the basic Keynesian model. Firms do not respond, in the short run, by changing price from what it otherwise would have been in response to changes in the demand for their products. Instead, they meet the demand at that price by adjusting production. In this chapter, we allow for the possibility of a positive long run equilibrium inflation rate so prices are automatically changing in a way consistent with that inflation rate. Our assumption, therefore, that firms do not respond, in the short run, by changing price from what it otherwise would be still means firms meet the demand for their products at preset prices, but those preset prices are moving according to what the equilibrium inflation rate determines.

## THE OUTPUT GAP AND INFLATION RATE CHANGES

For a number of reasons, the rate of inflation tends to be relatively constant over time. Expectations of steady-state inflation play a key role in determining how rapidly the inflation rate itself changes, but such expectations tend not to change rapidly without clear evidence of a change, or expected change, in monetary policy. Also important to the slow adjustment of inflation is the existence of long term wage and price contracts that have a built-in rate of nominal increase based on inflation expectations. The tendency of inflation to continue at or near its existing rate is sometimes referred to as *inflation inertia*. In the absence of shocks to the economy, if the output gap is zero so there is no shortage or surplus of goods and services at full-employment output, the rate of inflation will tend to remain constant as prices increase at the same rate to cover cost increases, which are also increasing at that rate.

However, if an expansionary gap exists, so inventories would be dropping unintentionally if production were at full-employment output because that output is not high enough to meet planned purchases, then as we move into the long run, firms will adjust prices by more than their costs are rising so the inflation rate will increase. *When an expansionary gap exists, the inflation rate will increase in the long run.* Similarly, if inventories would be increasing unintentionally if production were at full-employment output due to a surplus of full-employment output over planned expenditures on that output, firms will tend to raise their prices more slowly in the long run as they have an incentive to cut prices from what they otherwise would have been in order to increase their sales. *When a recessionary gap exists, therefore, the inflation rate will decrease in the long run.*

To summarize our discussion, in the short run firms do not change price from what it otherwise would have been in response to output gaps, but instead meet demand by adjusting output. In the long run, firms increase the inflation rate (the rate at which they are increasing prices) in the presence of an expansionary gap and firms decrease the inflation rate (reduce the rate at which they are increasing prices) in response to a recessionary gap. In the absence of a gap, the inflation rate does not change.

# THE STEADY-STATE INFLATION RATE

The quantity equation (which is sometimes called the equation of exchange) states nominal GDP demanded, $P \times Y$, is determined by the product of the money supply, M, and the speed at which money circulates, V, called the income velocity of money.

$$M \times V = P \times Y \qquad (15.1)$$

This equation defines velocity and provides a method of determining the long run inflation rate.

Because the inflation rate is the same as the growth rate of prices, we need a method of examining the growth rates of the variables in the quantity equation. To do this, we shall use an approximation developed from an equation that many students were required to learn in their high school math courses. The equation states the logarithm of the product of two variables is equal to the sum of the logarithms of each of the variables ($\log(AB) = \log(A)+\log(B)$). Since the growth rate of a variable is equal to the slope of the corresponding logarithm function, the same rule applies to growth rates. Therefore, the growth rate of the left side of 15.1 is equal to the sum of the growth rates of M and V and the growth rate of the right side is equal to the sum of the growth rates of P and Y.

$$Gr(M) + Gr(V) = Gr(P) + Gr(Y) \qquad (15.2)$$

The operator, "Gr", means growth rate. For the long run, we assume output is at potential output, so Y is equal to Y*. The growth rate of price is the inflation rate symbolized as $\pi$. If we assume the velocity of money, the speed at which money circulates, is constant in the long run so its growth rate is zero, then equation 15.2 can be written in a way that illustrates the value of the equilibrium inflation rate.

$$Gr(M) = \pi + Gr(Y^*) \tag{15.3}$$

The long-run equilibrium inflation rate, $\pi$, is equal to the growth rate of the money supply less the growth rate of potential output as can be observed by solving equation 15.3 for $\pi$. For example, if potential output is growing at a rate of 3% per year and the money supply is growing at a rate of 5% per year, then the steady-state inflation rate is 2% per year.

We represent the steady-state inflation rate on a graph with the inflation rate on the vertical axis and the ratio of output to potential output on the horizontal axis. At

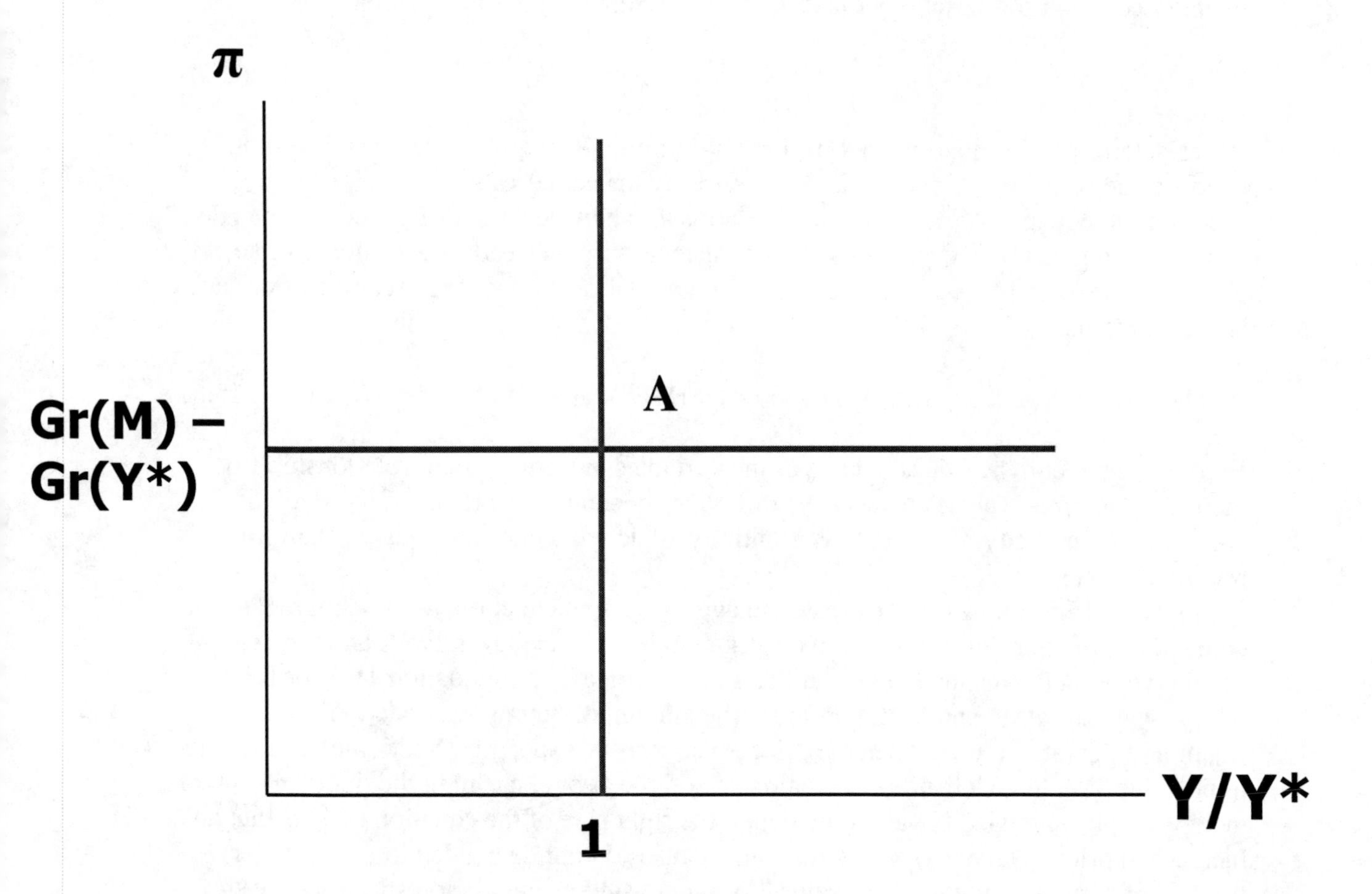


**Figure 15.1**

**Figure 15.1 illustrates the value of the steady-state inflation rate is equal to the growth rate of the money supply minus the growth rate of potential output. At intersection point A, the inflation rate is at its steady-state value and, since the ratio of actual output to potential output is one, actual output must be at its long run equilibrium value of potential output.**

intersection point A in figure 15.1, actual output, Y, is at potential output Y*, and the inflation rate is at its steady-state value.

## THE AGGREGATE DEMAND CURVE

Now that we have found the long run equilibrium inflation rate, we can specify the equilibrium price level that is consistent with that inflation rate. If we use M* to represent the long run trend value of the money supply, V* to represent the long run trend value of velocity, and P* to represent the long run trend value of price, then the quantity equation for these long run trend values will be similar to equation 15.1.

$$M^* \times V^* = P^* \times Y^* \tag{15.4}$$

P*, as determined in equation 15.4, is the long run moving equilibrium price level. In contrast, the actual price level, P, is determined by the actual values of M, V, and Y according to equation 15.1. To obtain a relationship between aggregate output demanded and the aggregate price level that is also consistent with the steady-state inflation rate, we divide equation 15.1 by equation 15.4 to obtain a relationship between relative prices and relative outputs.

$$(M/M^*) \times (V/V^*) = (P/P^*) \times (Y/Y^*) \tag{15.5}$$

Expressing the current values of each of the variables in terms of their relationship to their long run trend values creates a type of quantity equation in terms of these relative values. For simplicity, we assume that initially all actual values are equal to their long run trend values.

Equation 15.5 is now used to create the aggregate demand curve (AD). Aggregate demand is a relationship between the aggregate relative price level, P/P*, and the amount of relative output demanded, Y/Y*, at that relative price level. Equation 15.5 for fixed relative values, M/M* and V/V*, provides the amount of output demanded, Y, as a fraction of potential output when the price level, P, is at a specified fraction of its moving trend value. With the left side of equation 15.5 fixed at one, meaning the trend and actual money supply and velocity values are equal, the right side of the equation tells us that any increase in price relative to its long run value causes output demanded to be a smaller fraction of potential output. With values for money and its circulation speed relative to long run moving equilibrium values fixed, the fraction current price is of trend price multiplied by the fraction output is of potential output must also be fixed. (Furthermore, since actual values are beginning at long run trend values, the left side of equation 15.5 is equal to one.) Any increase in current price relative to trend price would necessarily cause a decrease in current output demanded relative to potential output since these relative money supplies and circulation speeds are all fixed. With the higher relative price level, less relative output would have to be demanded since only that smaller amount can be demanded with the relative amounts of money and of its circulation speed both fixed. The aggregate demand curve is illustrated in figure 15.2. Its equation is that the product of the variables on the axes must equal one.

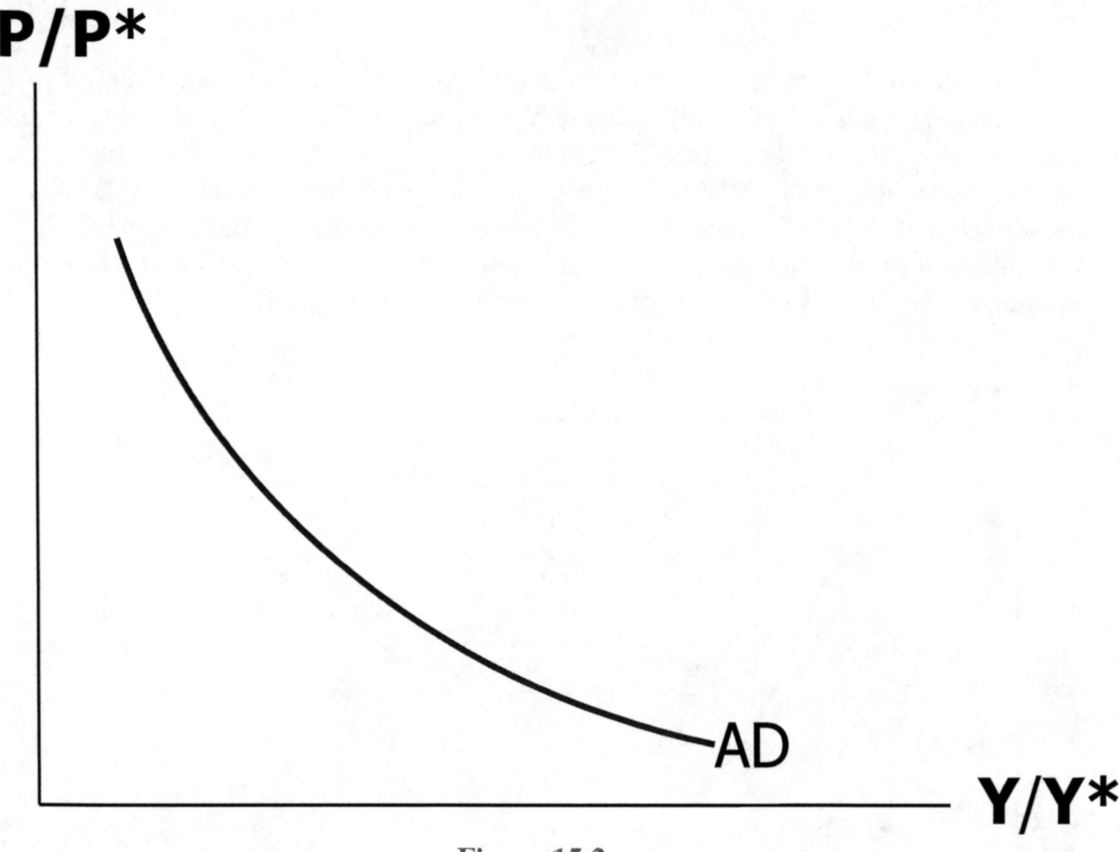


Figure 15.2

## SHIFTS OF THE AGGREGATE DEMAND CURVE

The aggregate demand curve is drawn for fixed values of M/M* and of V/V*. An increase in the current money supply, M, relative to its trend value, M*, would shift the curve to the right. The left side of equation 15.5 would increase above one meaning that for any value of relative price, P/P*, a higher value of Y/Y* would now be consistent with the increased value on the left side of the equation. Similarly, a decrease in the current money supply, M, relative to its trend value, M*, to put it below its long run trend value, would shift the aggregate demand curve to the left. It is also true that autonomous changes in planned aggregate expenditures, PAE, relative to its trend value, will also shift the aggregate demand curve. Autonomous increases in any component of planned aggregate expenditures that increase the overall value of Y relative to Y* will increase the velocity of money, V, relative to its trend value, V*, and, therefore, also shift the aggregate demand curve to the right. To see this, note Y in equation 15.1 will be equal to planned aggregate expenditures in short-run equilibrium so V must increase for the equation to remain valid.

$$M \times V = P \times (C+I^{p}+G+NX) \quad (15.6)$$

For any value of P/P* in equation 15.5, therefore, an increase in any component of PAE above its long run trend value will increase V and increase (V/V*) if trend value V* is not changing. (Even if the value of V* is changing, it would still be true that the increase in (Y/Y*) would increase (V/V*) if there are no changes in (P/P*) and (M/M*)). The higher value of (Y/Y*) for each value of (P/P*) means that the aggregate demand curve will shift to the right. Similarly any decrease in components of PAE relative to their long run trend values would shift the aggregate demand curve to the left.

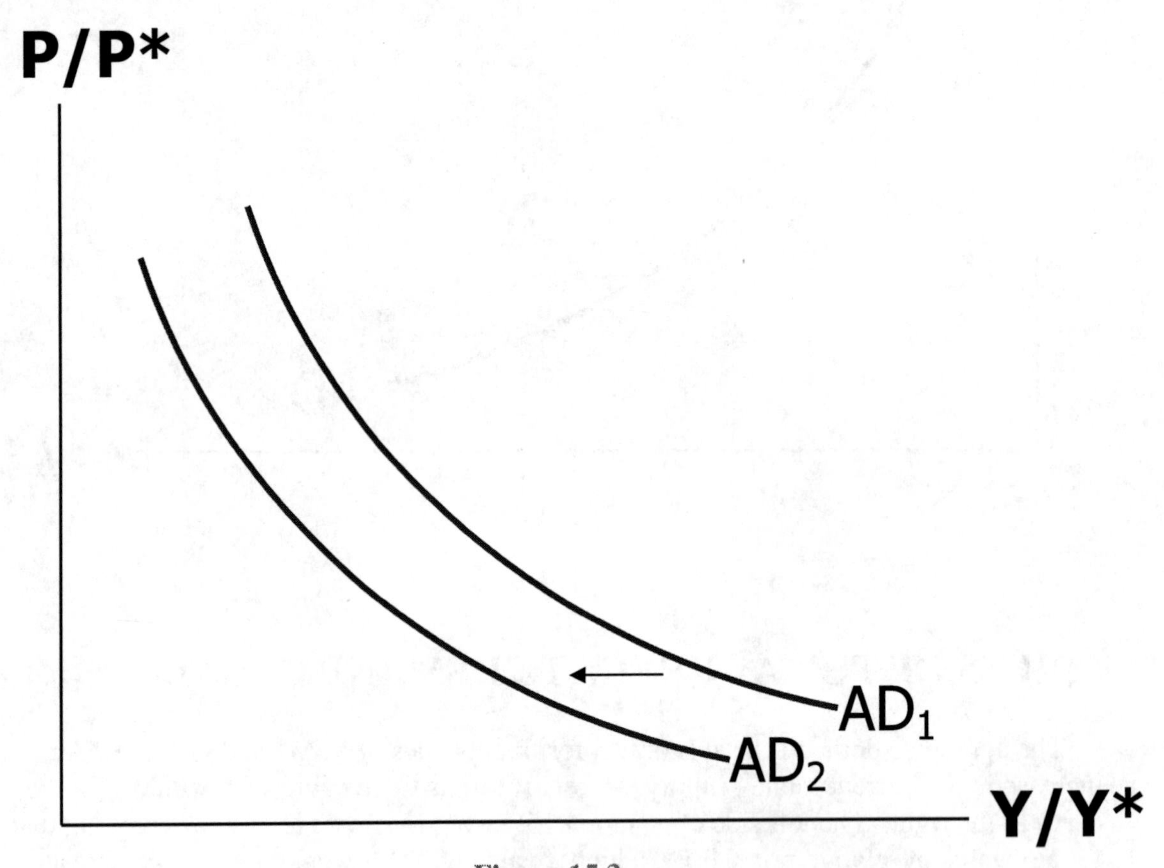


**Figure 15.3**

**A shift in the aggregate demand curve to the left can be caused by a decrease in the money supply relative to its long term trend value. The same is true for decreases in autonomous components of planned expenditures relative to their long term trend value, which cause a decrease in velocity relative to its trend value.**

## SHORT RUN AGGREGATE SUPPLY

We return to the short run assumption that firms meet a change in demand for their product by not changing the price level from what it otherwise would have been, but by adjusting output to what is demanded. We can represent this short run behavior by indicating the relationship between price and its long-run equilibrium existing trend value is not changed in the short run when demand changes. In our graph, with the ratio of price to its existing long run trend value on the vertical axis and the ratio of output to potential output on the horizontal axis, an unchanged price level path in the short run is illustrated by a horizontal short run aggregate supply line.

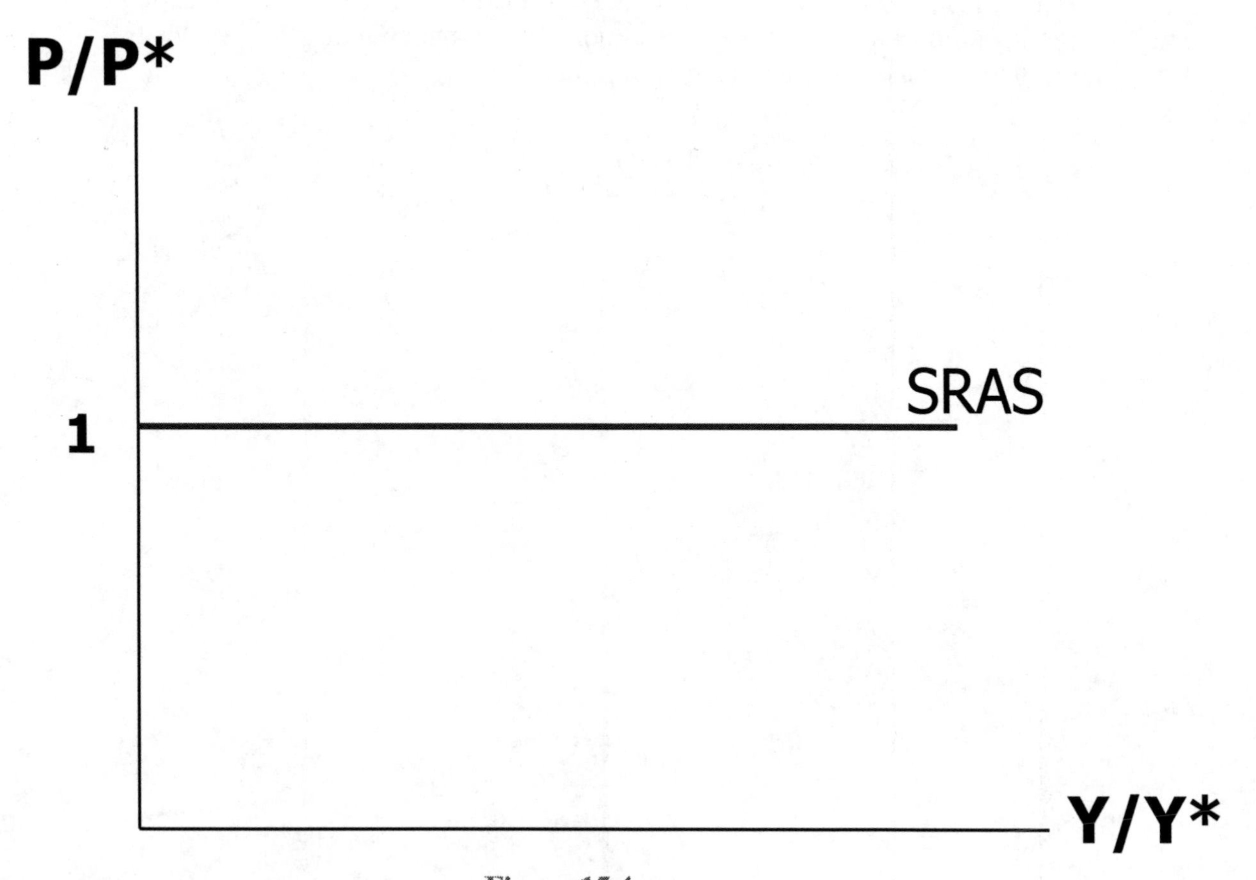


**Figure 15.4**

**The short run aggregate supply line is horizontal indicating firms do not adjust the relationship between current price and its existing long run trend value, but instead adjust output to demand. In the short run, firms do not adjust the price level from what it otherwise would have been when demand changes. Since we have assumed price is initially at its long run trend value, the short run aggregate supply line will be horizontal at current price, P, equal to its existing long run trend value, P*, so that (P/P*) = 1.**

## LONG RUN AGGREGATE SUPPLY

Before considering how the economy behaves in response to supply and demand changes, we determine the long run aggregate supply line. Our long run assumption is that price will adjust according to how the inflation rate changes in response to expansionary and recessionary gaps. Of course, we know the long run equilibrium inflation rate is determined by the long run growth rate of the money supply and the growth rate of potential output. Thus, changes in the inflation rate in response to gaps must eventually bring the inflation rate into conformity with its long run equilibrium value.

The long run aggregate supply line is a vertical line at (Y/Y*) = 1, indicating that at long run equilibrium, actual output is at potential output and the price level is free to adjust to any value relative to its long run trend value.

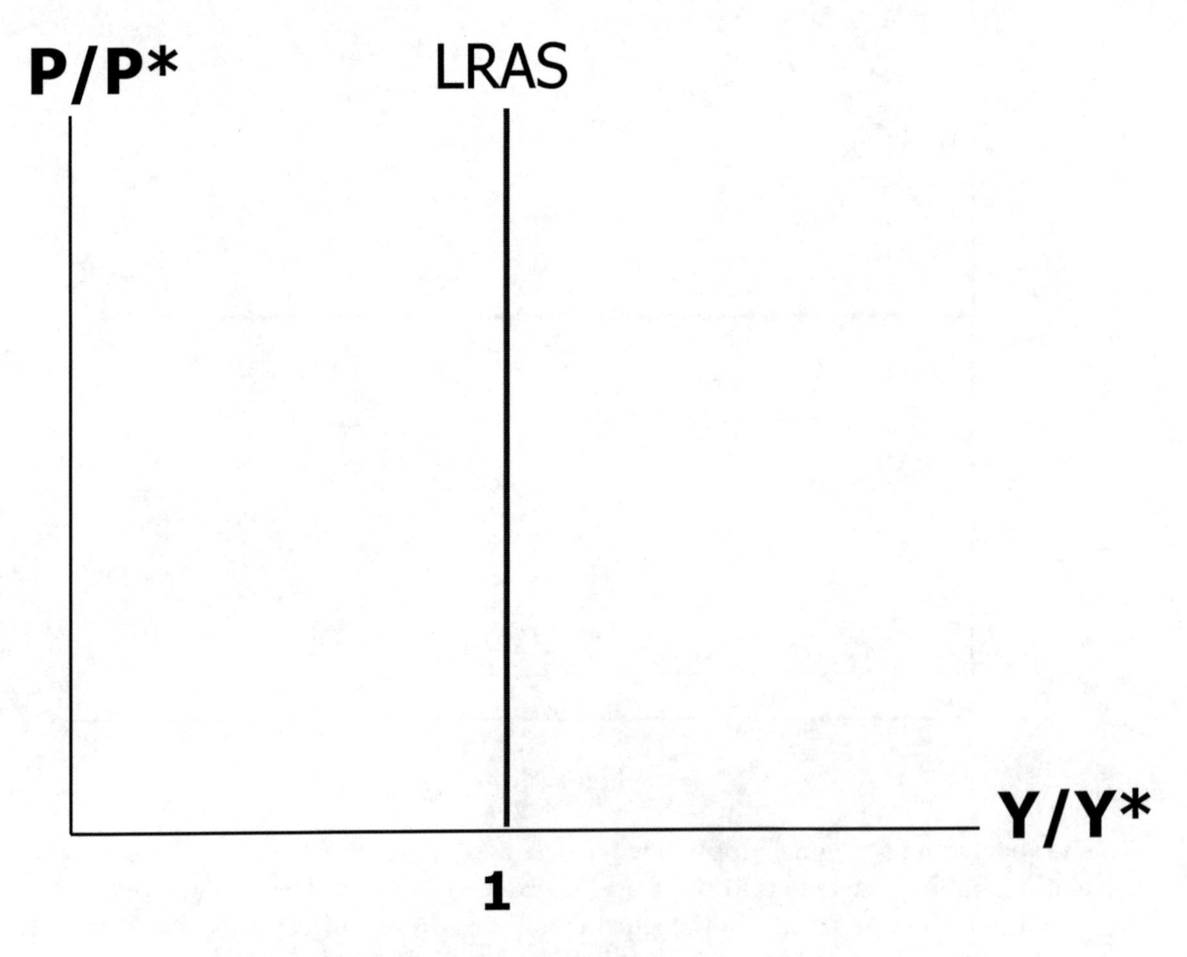


**Figure 15.5**

**The long run aggregate supply curve is vertical at one indicating output is at potential output in the long run and the price level is free to adjust in the process of actual output moving to potential output.**

## THE PRICE ADJUSTMENT TO A DECREASE IN DEMAND

We can use our aggregate demand and supply graph to demonstrate how price, output, and the inflation rate adjust to a decrease in demand. Recall that demand can decrease either because of a change in the money supply to put it below its long run trend value or because of a change in any autonomous component of planned expenditures which puts it below its long run trend values. We shall examine first either a money supply decrease, or an increase to a value less than M* so, in both cases, (M/M*)<1. We assume the money supply growth rate continues unchanged from its preceding value after the one-time money supply change.

Since we have assumed initially V equals V* and V remains equal to long run trend value V*, the left side of equation 15.5 becomes equal to (M/M*). For example, if M changes to become only nineteen-twentieths as large as M would have been had it still been equal to the trend value of M, M*, the left side of equation 15.5 would be equal to .95. Thus, the aggregate demand curve would shift down (and to the left) in such a way that at (Y/Y*) equal to 1, (P/P*) would equal .95. That must be true to satisfy equation 15.5. To generalize, to whatever value (M/M*) drops, that is exactly the aggregate demand shift down at (Y/Y*) equal to 1; that is, (P/P*) will drop to the same value as

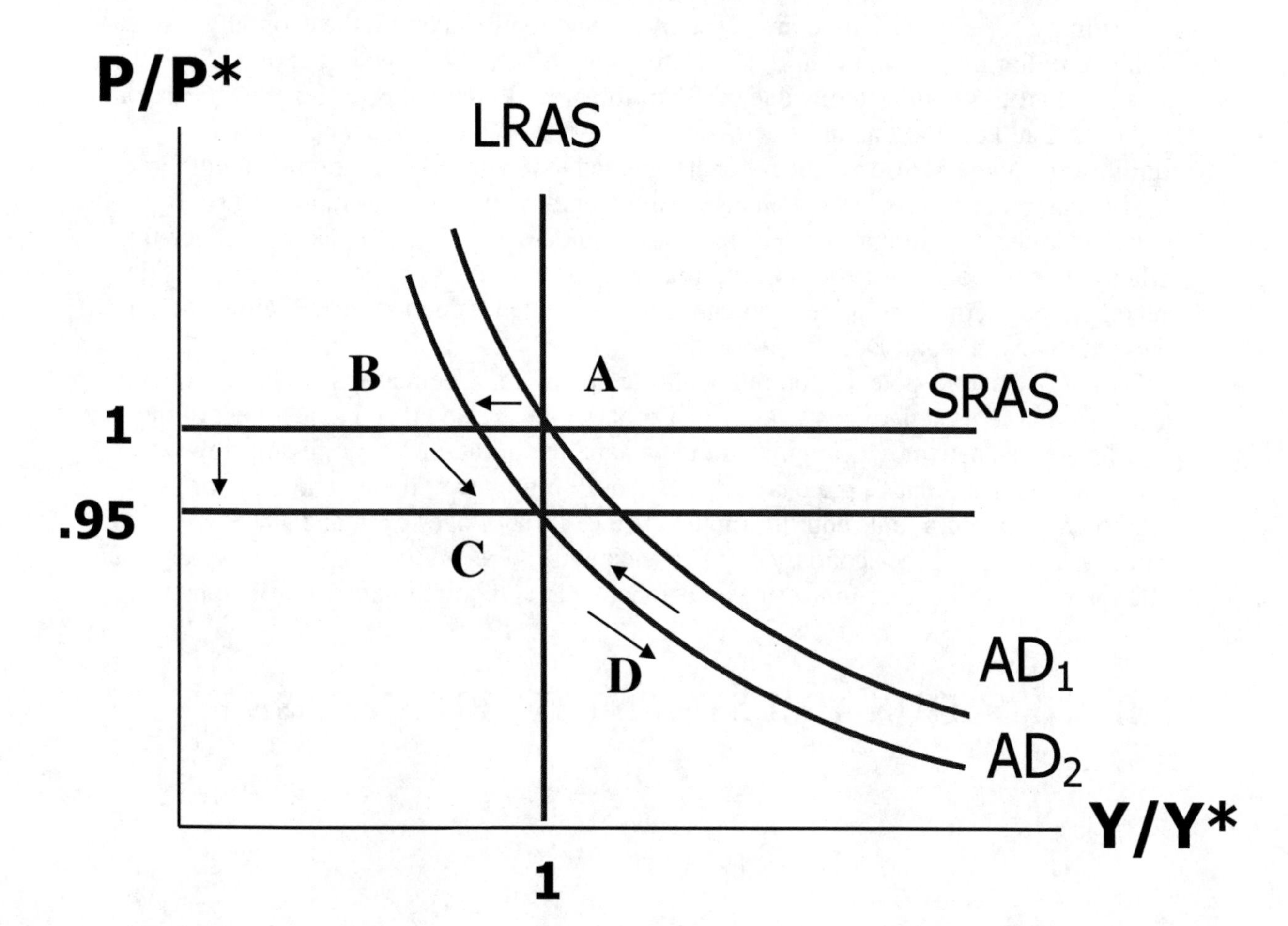

**Figure 15.6**

**Figure 15.6 illustrates the effect of a one-time change in the money supply from equal to the long-run trend value to equal to .95 of the trend value of the money supply. The money supply continues to grow at its preceding rate after the one-time change to the level of the money supply. The short run effect is a movement from intersection point A to intersection point B. In the short run, price is not adjusted away from its existing long term trend value, which means price continues to rise according to the equilibrium inflation rate. Output is adjusted to the lower value of demand indicating that output falls below potential output and a recessionary gap is created. In response to the recessionary gap, the inflation rate falls in the long run. The economy moves to intersection point C where the price level is now equal to .95 of its preceding long term trend value (which allows for the possibility of the actual price level never falling, so there is disinflation, or for the possibility that deflation occurs so the price level does fall).**

(M/M*) at (Y/Y*) equal to 1. The economy moves from A to B in the short run and from B to C in the long run

# OVERSHOOTING

Interestingly, after the one-time money supply reduction from trend, the economy is not back in equilibrium at point C in figure 15.6 although output is at potential output and the price level is at its equilibrium value of .95 of the preceding trend price level. The reason for this is that the inflation rate is below its equilibrium value. It must be below its equilibrium value because the recessionary gap led to a reduced inflation rate, but the steady-state inflation rate has not changed (the money supply is continuing to be increased at the preceding rate after the one-time money supply level change). Since the inflation rate is too low for equilibrium, the price level will not be rising as rapidly as the preceding long-term trend price level causing a continued reduction in P/P' along $AD_2$ in figure 15.6 toward point D.

Output overshoots potential output resulting in a cycle. The reasons why the economy moves from a recessionary gap to an expansionary gap are the lagged adjustment of the price level to its new moving equilibrium level and the inflation rate adjusting slowly to gaps. Although at point C, price is at its new equilibrium level, there is no way for the inflation rate to return to its equilibrium level since no shortages exist and since, without perfect knowledge of the economy by all participants, there are no incentives to return the inflation rate to its equilibrium. An expansionary gap is required before that will happen.

# THE INFLATION ADJUSTMENT TO A DECREASE IN DEMAND

In the above illustration of a decrease in demand, we have assumed the money supply drops below its preceding long term trend value and then remains at that same relative value to trend. Thus, we have been assuming an adjustment in the money supply which

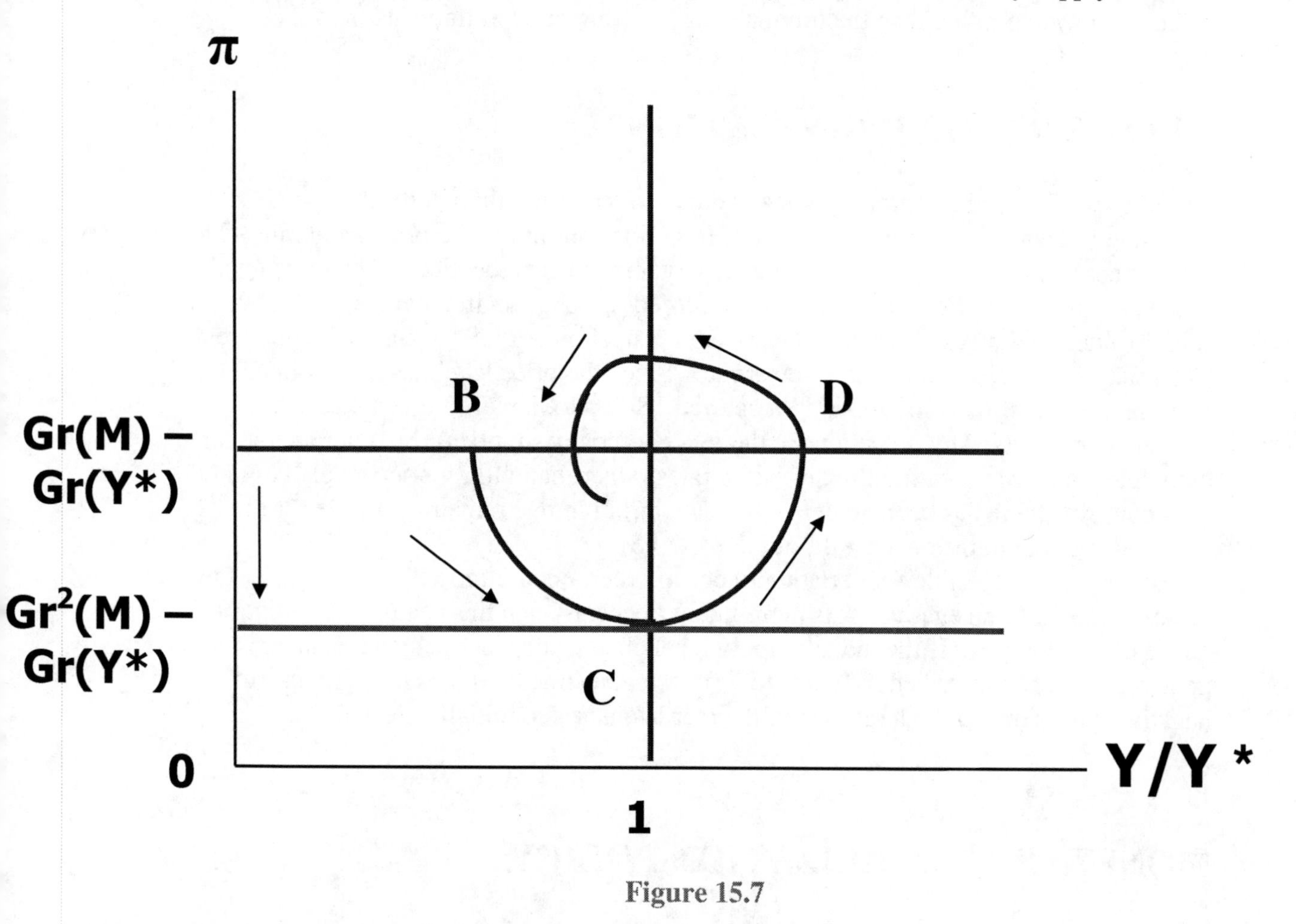


Figure 15.7

**The inflation rate adjusts in the long run to a decrease in demand.**

changes its value, but then continues to raise that value at the same rate the money supply had been increasing in its preceding long run equilibrium trend. Consequently, although the price level adjusts to a value below what it would have been had the money supply change not occurred, the long run equilibrium inflation rate does not change. It is valuable to demonstrate inflation rate changes on a graph. Therefore, we shall modify figure 15.1 to display the inflation rate path.

Notice in figure 15.7 the inflation rate cycles about its steady-state value. The movement of relative price in figure 15.6 is occurring while the inflation rate is moving from B to C in figure 15.7. The inflation rate also overshoots its equilibrium value. When the inflation rate is rising in the expansionary gap region from C to D in figure 15.6, it is nevertheless below the equilibrium inflation rate so the expansionary gap is increasing. It is only when the inflation rate rises above its equilibrium value at point D in figure 15.7 that the movement back towards C and B from D occurs in figure 15.6.

The analysis of a reduction in the relative money supply together with a change in the money growth rate, rather than leaving the growth rate the same, is mostly similar to the analysis above. The major exception is that the inflation rate will cycle about its new steady-state value rather than the original value as illustrated in figure 15.7.

## DEFLATION OR DISINFLATION?

Note the shift in the aggregate demand curve to the left, although resulting in a movement down the vertical axis in figure 15.6, does not imply the price level falls. The ratio of actual price to the level it would have otherwise been declines. The price level moves to the level indicated at C, below its preceding long run trend value, but above its initial value. In fact, as described, the price level itself never falls during the adjustment. The price level increases, but at a slower rate. Since the price level has not declined, disinflation, rather than deflation, has occurred. Shifts to the left of the aggregate demand curve caused by reductions in the relative money supply, or by other factors, are completely consistent with falling inflation rates rather than falling price levels. Figure 15.7 confirms there has been no deflation. The inflation rate remains positive for the entire adjustment to the new equilibrium.

However, it is possible for deflation to occur in certain circumstances; that is, for the inflation rate to turn negative. It is more likely to occur when the steady-state inflation rate is close to zero, or if the money supply level change, or other demand change, is so great that the path described in figure 15.7 for the adjustment crosses the zero axis to negative values due to the large size of the recessionary gap initially created.

## MONETARY STABILIZATION POLICY

An objective of monetary policy is to prevent the cycle from occurring. The Federal Reserve will want to prevent the economy from moving away from point C in figure 15.6 to prevent overshooting. It can achieve that objective by measuring the size of the output gap and by observing inflation rate changes. When output returns to full employment and no gap exists, the Federal Reserve will want to adjust the money growth rate so it is consistent with the lower inflation rate at intersection point C in figure 15.7. The lower inflation rate will be accommodated by reducing the money growth rate. The Federal Reserve will be able to tell if it is successful by monitoring inflation rate changes. Should the inflation rate turn upwards on a sustained basis, the central bank will know that it has not succeeded in reducing the money growth rate sufficiently.

By reducing the money growth rate to $Gr^2(M)$ at point C in figure 5.7, the Federal Reserve will establish that point as the new equilibrium and prevent the cycle. It will attempt to assure success by checking for the formation of an expansionary gap and by monitoring inflation rate changes and the inflation rate level. Should changes start to become consistently positive with a rise in inflation expectations accompanying the inflation rate increase, the central bank will need to adjust the money growth rate further to assure consistency with point C and to assure market participants that it is committed to the lower equilibrium inflation rate. To achieve success, the central bank must have market participants believe the higher steady-state inflation rate will not return and that the central bank is committed to a permanently lower equilibrium inflation rate.

## THE PRICE ADJUSTMENT TO AN INCREASE IN DEMAND

Appropriate monetary policy will be necessary as well if a cycle is to be avoided when there is an increase in aggregate demand, but policy changes must be put in place promptly when the shock occurs. As we shall see, the Federal Reserve will need to act aggressively. Otherwise, either a recession or a higher steady-state inflation rate will be inevitable.

As noted above, either a money supply change to increase its relative level, or autonomous increases in components of planned aggregate expenditures, PAE, relative to their trend values will shift the aggregate demand curve to the right. Again, starting at long run equilibrium trend values, an increase in government expenditures, for example, to above its trend value, must result in a relative velocity increase. For purposes of this

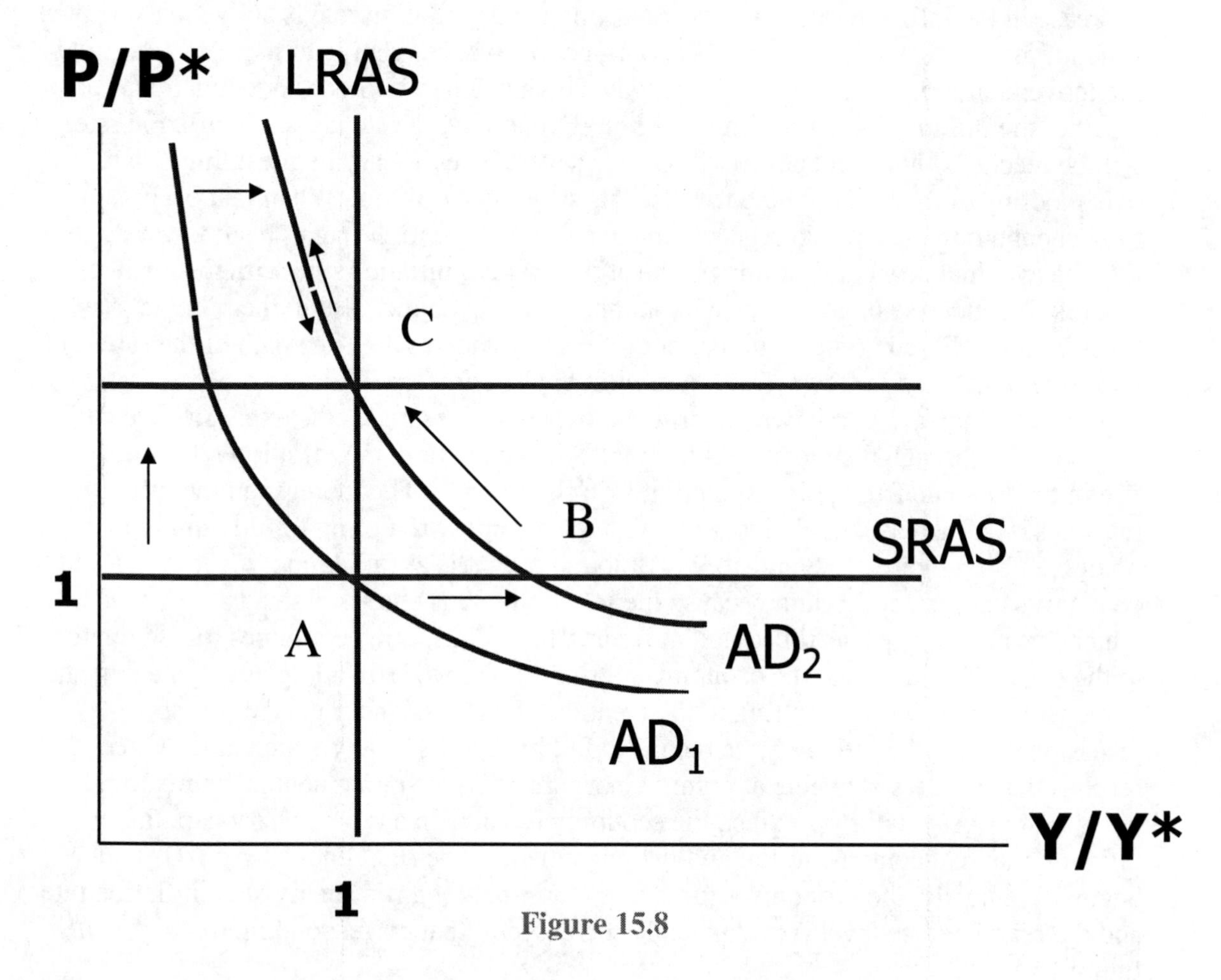


**Figure 15.8**

**An increase in demand leads to a higher relative price level.**

example, we shall assume an increase in military expenditures relative to trend with the economy beginning at full employment, at the steady-state inflation rate and at the equilibrium price level, which is the trend price level consistent with the steady-state inflation rate and equation 15.5. With the increase in actual velocity relative to trend caused by the increase in government expenditures relative to its trend value, the aggregate demand curve will shift to the right indicating that at each relative price level, more relative output will be demanded.

In the short run, firms will meet the increase in demand with an increase in relative output as demonstrated in the figure 15.8 with a movement from intersection point A toward intersection point B. As Y/Y* moves above one, output moves above full employment output and an expansionary gap is created. In the long run, the inflation rate will increase causing the price level to rise relative to the trend price level, demonstrated in figure 15.8 as a movement from B towards C along the new aggregate demand curve. The economy will not be in equilibrium at point C, however, since the actual inflation rate is now too high. With the steady-state inflation rate unchanged (determined by the difference between the growth rates of the money supply and of potential output), the price level will continue to advance relative to its trend value with the economy continuing to move left along the new aggregate demand past intersection point C. As that movement occurs, the economy will experience a recessionary gap resulting in a decrease in the inflation rate. However, as long as the inflation rate is above the steady-state inflation rate, the price level will continue to advance relative to its trend value and the movement to the left along the aggregate demand curve will continue regardless of whether the inflation rate is declining or not. Eventually, the steady-state inflation rate will be reached, but the economy will be in a severe recessionary gap resulting in an overshooting of the inflation rate to below its steady-state value. When that occurs, the movement along the new aggregate demand curve will be to the right, back toward point C. The eventual new equilibrium will be at the original inflation rate with the result of this relative increase in government expenditures being an increase in the relative price level (at point C) rather than a higher equilibrium inflation rate. The path of the inflation rate is demonstrated in figure 15.9. A similar explanation applies.

Once the increase in relative government expenditures creates the expansionary gap in the short run through the movement to point B in figure 15.8, the inflation rate starts to rise in the long run from point B to point C in figure 15.9. The increase in the inflation rate causes an increase in relative price with the economy returning to full employment at point C in both figures 15.8 and 15.9. Although output is at full employment output, the economy is not in equilibrium because the inflation rate is above its steady-state value, which has not changed, as illustrated in figure 15.9. Thus, price continues to rise relative to the trend price level and the economy continues to move left and up aggregate demand curve two in figure 15.8 resulting in the creation of a recessionary gap. With the recessionary gap, the inflation rate begins to fall back to its steady-state value. Once it reaches that steady-state value at point D in figure 15.9, the movement in figure 15.8 to the left along $AD_2$ will stop. Since the economy remains in a recessionary gap, the inflation rate will continue to fall and the movement to the right back along $AD_2$ will begin. Eventually, the economy returns to the original long run steady-state inflation rate and the relative price level is permanently increased to that corresponding to point C in figure 15.8.

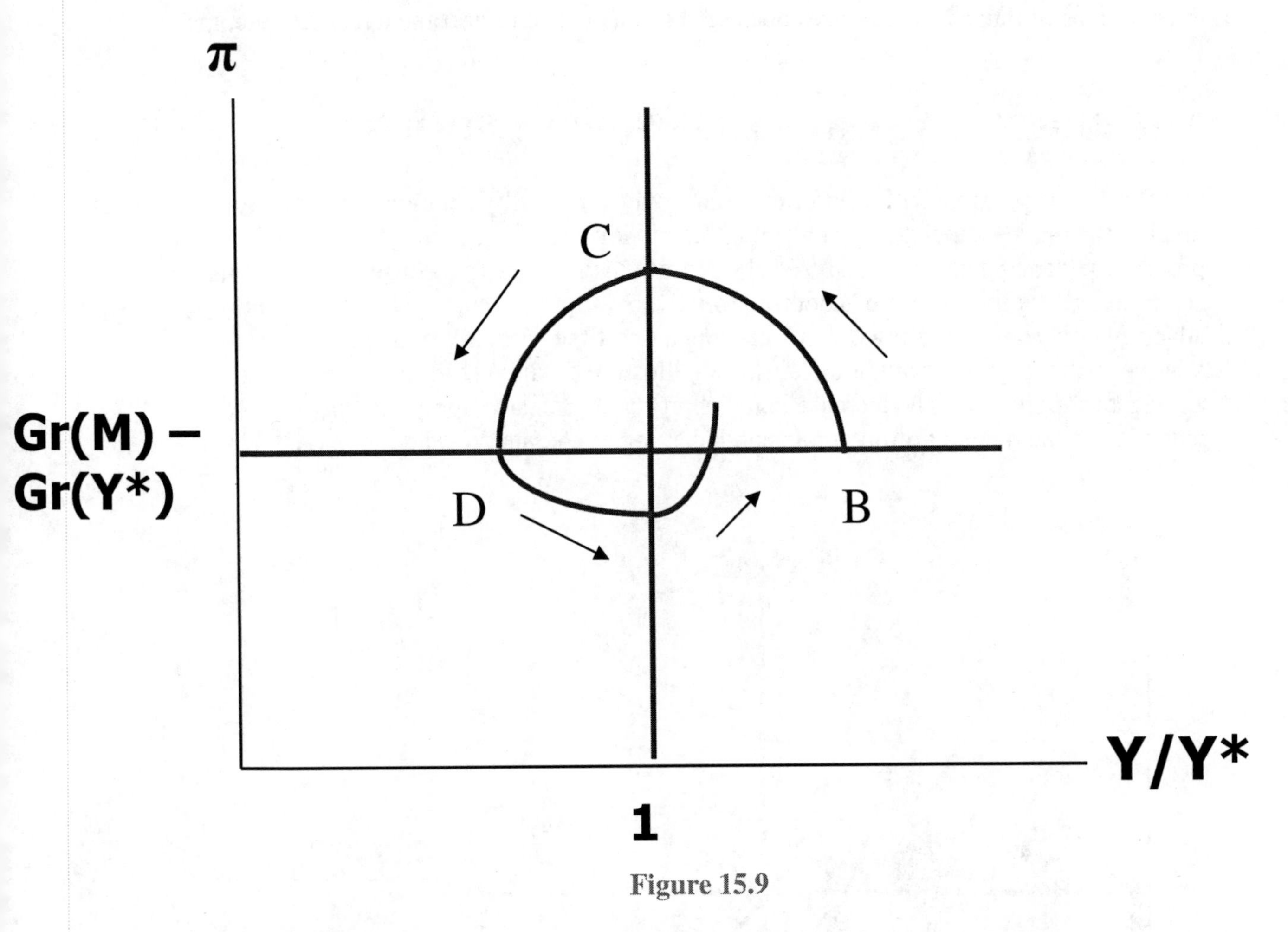


**Figure 15.9**

**Starting with an expansionary gap at B, the inflation rate rises above its long run steady state value. Although the expansionary gap is gone at C, the inflation rate is too high. Price continues to rise above that level consistent with full employment causing a recessionary gap and a consequent reduction in the inflation rate.**

The increase in relative government expenditures has the long run consequence, therefore, of not affecting the steady-state inflation rate, but of permanently increasing the price level compared to what it would have otherwise been. Since government expenditures have increased relative to what it would have been and since output is back at potential output, it must be true that another component of planned expenditures is less than what it would have otherwise been. That reduction is referred to as *crowding out.* Typically, crowding out is of investment expenditures and is the direct result of the rise in interest rates that accompanies the drop in the relative real money supply caused by the rise in the relative price ratio.

The only way monetary policy can prevent the inflation rate increase and the resulting recessionary gap from forming is if the relative money supply is reduced initially to prevent the increase in relative government expenditures from forming an expansionary gap. The Federal Reserve needs to react at once so a rise in inflation rates and in expectations of those inflation rates does not have a chance to become a market outcome.

## THE PRICE LEVEL ADVERSE SUPPLY SHOCK

While it is possible with correct monetary policy, either by changing the relative level of the money supply, or by changing the money growth rate, or by changing both, to prevent the demand changes analyzed above from having negative consequences on the economy, generally, the same is not true for adverse supply shocks. The price level adverse supply shock is a price shock causing a one-time increase in the price level relative to the long run trend price level. It is illustrated with a shift up in the short run aggregate supply line. The increase in the short run aggregate supply line results in a reduction in the quantity of output demanded along aggregate demand curve ($AD_1$) as

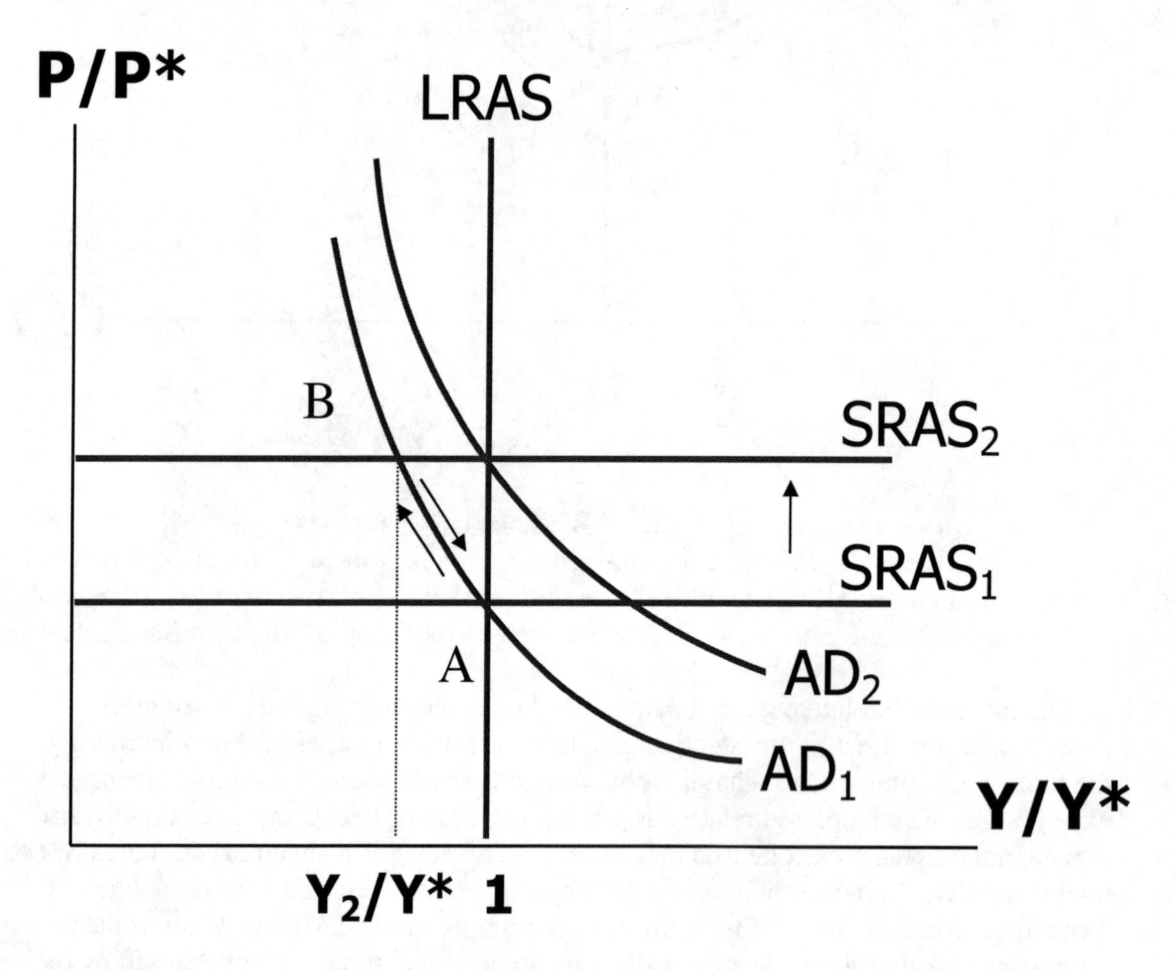


**Figure 15.10**

**The price level adverse supply shock leads to stagflation.**

illustrated in figure 15.10. In the short run, firms will decrease relative output to meet the lower demand moving output from potential output to output $Y_2$. This decline in relative output along with the increase in the relative price level is known as *stagflation*; that is, a combination of stagnation and price increases. While the price level is indeed rising from the shock, the steady state inflation rate has not changed, however. Once output responds to the reduction in aggregate demand, the resulting recessionary gap would lead to a reduction in the inflation rate reducing the relative price level along the aggregate demand curve. The relative price level would return to its initial value while the inflation rate declines below its unchanged steady-state value. An adjustment similar to that described with figure 15.9 would occur.

It is tempting to conclude that an increase in the relative money supply level, shifting the aggregate demand curve to the right to $AD_2$, would prevent the recession, but if the adverse shock was the result of input price increases such as wage increases or oil price increases, the recession may be postponed and exacerbated by such action. With input price increases, generally, time is needed for resource market quantities demanded to decline and quantities supplied to increase. If the economy is moved back to full employment before it has had time to complete those quantity adjustments, then resource owners may perceive no reaction to the prior price changes and put into effect additional price changes in the same direction. In figure 15.10, the short run aggregate supply line would move up again resulting in a new recessionary gap. If the Federal Reserve were to continue to increase the relative money supply in response, the risk would be a higher steady-state inflation rate as the money supply increases lead to a high money supply growth rate. The Federal Reserve needs to be careful not to accommodate adverse price level supply shocks for concern that what should be only a price level adjustment turns out to be an eventual increase in the steady-state inflation rate.

## THE INFLATION RATE ADVERSE SUPPLY SHOCK

The inflation rate adverse supply shock involves not a one time price level increase, but a one-time increase in the rate at which prices are changing. The actual inflation rate would jump higher from the shock. Since we assume the shock involves neither a change in the money supply growth rate nor in the potential output growth rate, the steady state inflation rate would not change.

The inflation rate adverse supply shock is illustrated in figure 15.9 as an instant increase in the inflation rate from its equilibrium to the level associated with, say, point C. An adjustment similar to that exhibited in figure 15.9 would occur as a recessionary gap is created from the high inflation rate resulting in the price level advancing relative to the trend value. The only way recession can be avoided is if the Federal Reserve accommodates the inflation rate increase with an equal increase in the money supply growth rate raising the steady state inflation rate to that higher value resulting from the shock. Either recession or a permanently higher inflation rate is a necessary consequence of the shock.

## THE OUTPUT ADVERSE SUPPLY SHOCK

In addition to supply shocks that affect prices immediately, we analyze supply shocks that affect quantities initially. First, we shall examine a shock that reduces potential output to a lower level, but the shock permits potential output to continue to grow at its preceding rate nevertheless. It becomes useful to examine actual economy values relative to their new long term trend values rather than to the existing trend values. The drop in potential output, therefore, would result in a movement to the right in figure 15.11 along the AD curve as actual output relative to the now lower level of potential output is higher. To better understand this, return to quantity equation 15.4 for long run equilibrium values. For the product of M* and V* to equal the product of P* and Y* with a reduction in Y* and no change in M* and V*, the value of P* must increase. Thus, the same shock that decreases the level of potential output also increases the long run equilibrium trend price level above what it otherwise would have been. In figure 15.11, the changes in Y* and P* are represented by a movement to the right along the AD curve since (Y/Y*) is increasing with the decrease in Y* and (P/P*) is decreasing with the increase in P*. The inflationary gap created by the movement to the right, starts the inflation rate increasing

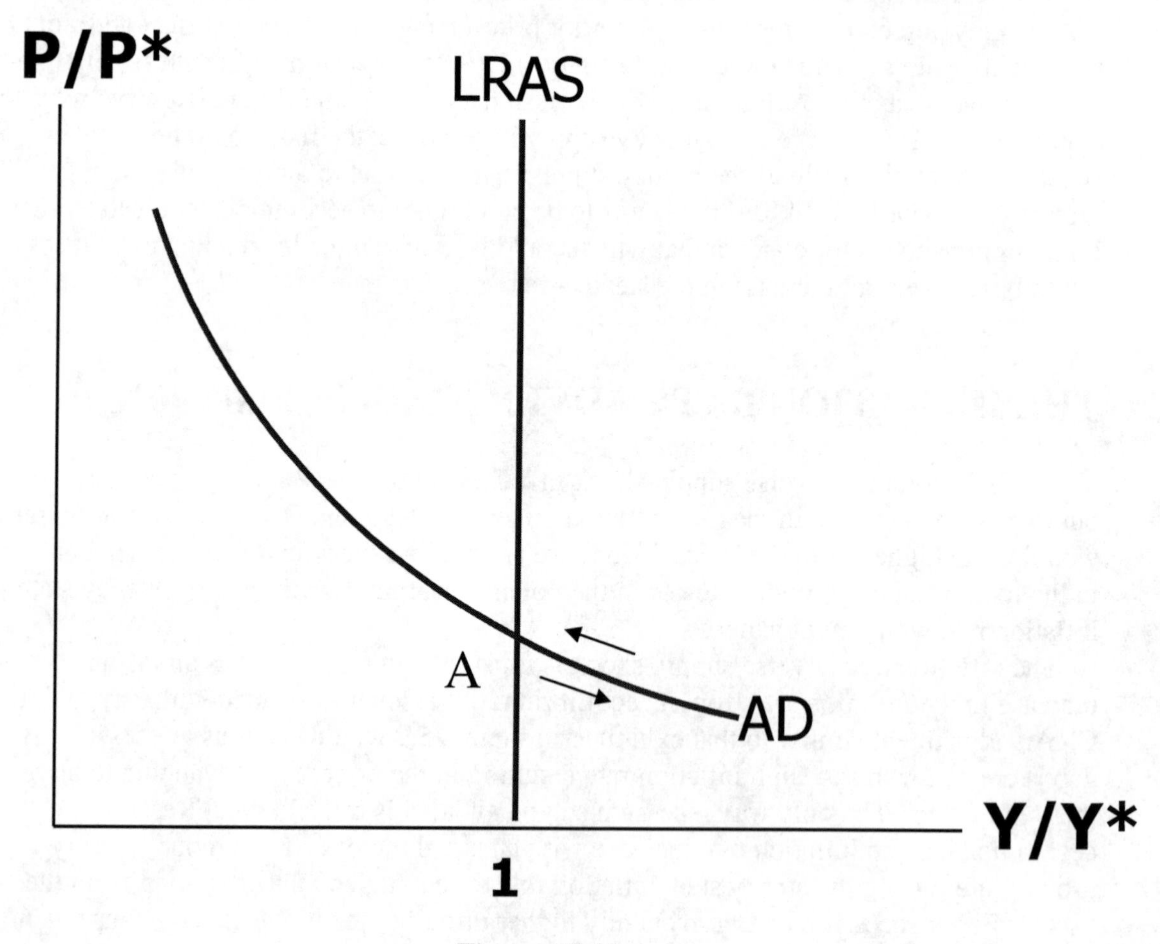


Figure 15.11

resulting in a movement back along the AD curve to point A as the price level begins rising relative to its new higher value of P*. When A is reached, output is at its new lower level of potential output and price is at its new higher long run level. The steady state inflation rate will be unchanged. (Of course, the adjustment to that inflation rate will continue as overshooting occurs similar to what is displayed in figure 15.9.)

## THE OUTPUT GROWTH RATE ADVERSE SUPPLY SHOCK

The potential output growth rate adverse supply shock involves a decrease in the long run growth rate of potential output without any initial change in the level of potential output. Equation 15.3 tells us the steady-state inflation rate will increase as a result. If we use the new steady state values for our long run trend values satisfying equation 15.4, the actual inflation rate relative to the now higher steady state inflation rate results in a decline in the actual price level relative to the new long run trend price level. As a consequence, the economy moves down and to the right along the existing aggregate demand curve creating an expansionary gap as illustrated in figure 15.11 (although for different reasons that those arising from a potential output level shock). Federal Reserve monetary policy can easily offset the inflation effects, if it realizes the shock is occurring, by reducing the money growth rate by the same amount as the potential output growth rate declines. Actual and long run trend money growth rates would adjust according to the decline in the potential output growth rate decline. Of course, monetary policy cannot change the lower growth rate of potential output. The adjustment of actual output would be in each period as output demanded would exactly equal the level of potential output each period if the money supply growth rate is adjusted. Actual output would adjust to demand; that is, to potential output each period.

## RESTATEMENT

The long run inflation rate is determined by the difference between the money supply growth rate and the growth rate of potential output. One time only decreases or increases in the level of the money supply from its long run path cause a change in the equilibrium price level from what it otherwise would have been, but the steady-state inflation rate does not change. The same is true for changes in planned expenditures relative to their preceding long term path values. Generally, reductions in aggregate demand do not involve deflation. They do involve disinflation with downward movements along the vertical axis of the aggregate demand-aggregate supply graph illustrating price level increases that are, at the same time, a reduction from the equilibrium price level path that would have prevailed had aggregate demand not changed.

One time only changes in the price level, the inflation rate, or in the level of potential output have no influence on the steady-state inflation rate, but may have an impact on the equilibrium price level relative to what it otherwise would have been.

Changes in the money supply growth rate or in the potential output growth rate will have an effect on the steady-state inflation rate. Only changes in either of these two factors can have an effect on the steady-state inflation rate. (Of course, money supply growth rate changes, in turn, can be the result of government's inability to otherwise finance its expenditures with increased taxes or increased debt.) If the Federal Reserve does not permit an increase in the money supply growth rate and if the potential output growth rate does not fall, the steady-state inflation rate will not increase.

If monetary policy is to minimize the likelihood of recessions, it needs to have as its paramount objective the preventing of the inflation rate from rising to unacceptable levels. Monetary policy can offset the effects of demand shocks by adjusting the level of the money supply to compensate for the shock. Monetary policy can compensate for potential output growth rate shocks to prevent recession. Other supply shocks, however, can have effects on the price level that monetary policy cannot change without risking higher steady-state inflation rates or severe recession as well as have effects on potential output that monetary policy cannot change. Adjustments to supply shocks can create recessions that cannot be offset with monetary policy without causing a sustained higher price level than what would have otherwise occurred and possibly resulting in higher inflation rates from repeated money supply adjustments as a consequence.

When changes in the money supply or in the growth rate of the money supply are needed, the Federal Reserve implements those changes with its control over the Federal Funds rate. In an intermediate macroeconomics course, the effect of interest rate changes on velocity can be explored directly. That can include analyzing the implications of velocity varying when the money supply changes.

## SUMMARY

- This chapter provided a framework for analyzing monetary policy while determining both the equilibrium price level path and the steady-state inflation rate. By using the quantity equation to link current values to long run trend values, we created an aggregate demand curve measuring the actual price level relative to the long run trend price level on the vertical axis and the ratio of current output to potential output on the vertical axis. We maintained the Keynesian model assumptions relating to short run and long run adjustment. In the short run the inflation rate does not change, but in the long run the inflation rate adjusts according to whether an expansionary or recessionary gap exists.

- The aggregate demand curve is downward sloping because an increase in price relative to the long run trend price level reduces the fraction output can be of potential output considering expenditures are fixed determined by the money supply and the turnover rate of the money supply, both assumed equal to their long term trend values. A movement down the vertical axis of the aggregate demand curve can be consistent with either deflation or disinflation as only the ratio of the price level to the long run trend price level is declining.

- A shift to the left of the aggregate demand curve can be caused either by a decline in the current money supply level or a decline in velocity resulting from a reduction in any autonomous component of planned expenditures. In the short run, output will adjust to demand forming a recessionary gap. In the long run, the inflation rate will fall causing price to fall below trend price. The economy moves back to potential output along the new aggregate demand curve. However, the actual inflation rate is below its unchanged steady-state value leading to overshooting.

- The steady-state inflation rate is equal to the money supply growth rate minus the growth rate of potential output. Therefore, long run inflation can only be the result of an increase in the money supply growth rate or a reduction in the potential output growth rate. Only if government budget deficits are financed with money creation, can budget deficits be the direct cause of higher inflation rates. Generally, higher government expenditures and reduced taxes lead to a higher price level than would otherwise have occurred, but not to a higher steady-state inflation rate.

- Without a change in regime (in the structure of the economy suspending prior expectations), the only way monetary policy can reduce the steady-state inflation rate is through recession. (An opportune increase in the growth rate of potential output (an advantageous supply shock) could lead to a lower inflation rate without recession.) Therefore, a primary objective of the Federal Reserve is to prevent the inflation rate from increasing to higher sustained levels, if it is at all possible.

- Adverse supply shocks can cause stagflation in which output is falling and the price level or the inflation rate is rising. Generally, the price level, not the steady-state inflation rate, will be affected by adverse supply shocks unless the shock is to the growth rate of potential output and the Federal Reserve takes no action to adjust the money growth rate in response.

## KEY TERMS

adverse supply shock (16-19)
aggregate demand (AD) curve (4)
crowding out (15)
deflation (12)
disinflation (12)
inflation inertia (1)
long run aggregate supply (LRAS) line (8)
overshooting (10)
short run aggregate supply (SRAS) line (7)
steady-state inflation rate (3)

## REVIEW QUESTIONS

1. Discuss the relationship between the quantity equation and the aggregate demand curve (AD). Explain why a movement down the vertical axis of the curve is consistent with rising prices.

2. Explain how each of the following affects the aggregate demand curve.
   a. A decrease in government purchases
   b. An increase in the level of the money supply.
   c. An increase in planned investment spending.
3. Explain why a policy change of increasing the level of the money supply while not changing its growth rate leads to the economy remaining in disequilibrium even after the return of output to potential output.
4. Discuss why it is important the Federal Reserve prevents sustained increases in the inflation rate.
5. Discuss which adverse supply shocks are difficult problems for monetary policy and explain why.
6. Discuss the different outcomes arising from a change in the level of the money supply and from a change in the growth rate of the money supply.
7. Explain when deflation, rather than disinflation, is likely to occur.

"Monetary Policy and the Equilibrium Price Level and Inflation Rate", an alternate for chapter 15 of Principles of Macroeconomics by R. H. Frank and B. S. Bernanke, second edition.

Author: Norman P. Obst
Department of Economics
Michigan State University
East Lansing, Michigan 48824-1038

# GLOSSARY

## A

**Absolute advantage.** One person has an absolute advantage over another if he or she takes fewer hours to perform a task than the other person.

**Aggregate demand (*AD*).** Total planned spending on final goods and services.

**Aggregate demand (*AD*) curve.** Shows the relationship between aggregate demand and inflation; because short-run equilibrium output equals aggregate demand, the aggregate demand curve also shows the relationship between short-run equilibrium output and inflation; increases in inflation reduce aggregate demand and short-run equilibrium output, so the aggregate demand curve is downward-sloping.

**Aggregate supply shock.** Either an inflation shock or a shock to potential output; adverse aggregate supply shocks of both types reduce output and increase inflation.

**Aggregation.** The adding up of individual economic variables to obtain economywide totals.

**Appreciation.** An increase in the value of a currency relative to other currencies.

**Assets.** Anything of value that one owns.

**Attainable point.** Any combination of goods that can be produced using currently available resources.

**Autarky.** A situation in which a country is economically self-sufficient.

**Automatic stabilizers.** Provisions in the law that imply automatic increases in government spending or decreases in taxes when real output declines.

**Autonomous expenditure.** The portion of planned aggregate expenditure that is independent of output.

**Average benefit.** Total benefit of undertaking $n$ units of an activity divided by $n$.

**Average cost.** Total cost of undertaking $n$ units of an activity divided by $n$.

**Average labor productivity.** Output per employed worker.

## B

**Balance-of-payments deficit.** The net decline in a country's stock of international reserves over a year.

**Balance-of-payments surplus.** The net increase in a country's stock of international reserves over a year.

**Bank reserves.** Cash or similar assets held by commercial banks for the purpose of meeting depositor withdrawals and payments.

**Banking panic.** An episode in which depositors, spurred by news or rumors of the imminent bankruptcy of one or more banks, rush to withdraw their deposits from the banking system.

**Barter.** The direct trade of goods or services for other goods or services.

**Bequest saving.** Saving done for the purpose of leaving an inheritance.

**Board of Governors.** The leadership of the Fed, consisting of seven governors appointed by the President to staggered 14-year terms.

**Bond.** A legal promise to repay a debt, usually including both the principal amount and regular interest payments.

**Boom.** A particularly strong and protracted expansion.

**Buyer's reservation price.** The largest dollar amount the buyer would be willing to pay for a good.

**Buyer's surplus.** The difference between the buyer's reservation price and the price he or she actually pays.

## C

**Capital gains.** Increases in the value of existing assets.

**Capital good.** A long-lived good, which is itself produced and used to produce other goods and services.

**Capital inflows.** Purchases of domestic assets by foreign households and firms.

**Capital losses.** Decreases in the value of existing assets.

**Capital outflows.** Purchases of foreign assets by domestic households and firms.

**Cash on the table.** Economic metaphor for unexploited gain from exchange.

**Change in demand.** A shift of the entire demand curve.

**Change in supply.** A shift of the entire supply curve.

**Change in the quantity demanded.** A movement along the demand curve that occurs in response to a change in price.

**Change in the quantity supplied.** A movement along the supply curve that occurs in response to a change in price.

**Closed economy.** An economy that does not trade with the rest of the world.

**Comparative advantage.** One person has a comparative advantage over another if his or her opportunity cost of performing a task is lower than the other person's opportunity cost.

**Complements.** Two goods are complements in consumption if an increase in the price of one causes a leftward shift in the demand curve for the other.

**Compound interest.** The payment of interest not only on the original deposit but on all previously accumulated interest.

**Constant.** A quantity that is fixed in value.

**Consumer price index (CPI).** For any period, measures the cost in that period of a standard basket of goods and services relative to the cost of the same basket of goods and services in a fixed year, called the *base year.*

**Consumption expenditure.** Spending by households on goods and services, such as food, clothing, and entertainment.
**Consumption function.** The relationship between consumption spending and its determinants, such as disposable (after-tax) income.
**Consumption possibilities.** The combinations of goods and services that a country's citizens might feasibly consume.
**Contraction.** *See* **Recession.**
**Contractionary monetary policy.** An increase in interest rates by the Fed, made with the intention of reducing an expansionary gap; also known as *monetary tightening*.
**Contractionary policies.** Government policy actions designed to reduce planned spending and output.
**Coupon payments.** Regular interest payments made to the bondholder.
**Coupon rate.** The interest rate promised when a bond is issued.
**Crowding out.** The tendency of increased government deficits to reduce investment spending.
**Cyclical unemployment.** The extra unemployment that occurs during periods of recession.

## D

**Deflating (a nominal quantity).** The process of dividing a nominal quantity by a price index (such as the CPI) to express the quantity in real terms.
**Deflation.** A situation in which the prices of most goods and services are falling over time so that inflation is negative.
**Demand curve.** A curve or schedule showing the total quantity of a good that buyers wish to buy at each price.
**Demand for money.** The amount of wealth an individual or firm chooses to hold in the form of money.
**Dependent variable.** A variable in an equation whose value is determined by the value taken by another variable in the equation.
**Deposit insurance.** A system under which the government guarantees that depositors will not lose any money even if their bank goes bankrupt.
**Depreciation.** A decrease in the value of a currency relative to other currencies.
**Depression.** A particularly severe or protracted recession.
**Devaluation.** A reduction in the official value of a currency (in a fixed-exchange-rate system).
**Diminishing returns to capital.** If the amount of labor and other inputs employed is held constant, then the greater the amount of capital already in use, the less an additional unit of capital adds to production.
**Diminishing returns to labor.** If the amount of capital and other inputs in use is held constant, then the greater the quantity of labor already employed, the less each additional worker adds to production.
**Discount rate.** The interest rate that the Fed charges commercial banks to borrow reserves.
**Discount window lending.** The lending of reserves by the Federal Reserve to commercial banks.
**Discouraged workers.** People who say they would like to have a job but have not made an effort to find one in the past four weeks.
**Disinflation.** A substantial reduction in the rate of inflation.
**Diversification.** The practice of spreading one's wealth over a variety of different financial investments to reduce overall risk.
**Dividend.** A regular payment received by stockholders for each share that they own.
**Duration.** The length of an unemployment spell.

## E

**Economic surplus.** The economic surplus from taking any action is the benefit of taking the action minus its cost.
**Economics.** The study of how people make choices under conditions of scarcity and of the results of those choices for society.
**Efficiency (or economic efficiency).** Condition that occurs when all goods and services are produced and consumed at their respective socially optimal levels.
**Efficient point.** Any combination of goods for which currently available resources do not allow an increase in the production of one good without a reduction in the production of the other.
**Entrepreneurs.** People who create new economic enterprises.
**Equation.** A mathematical expression that describes the relationship between two or more variables.
**Equilibrium.** A stable, balanced, or unchanging situation in which all forces at work within a system are canceled by others.
**Equilibrium exchange rate.** *See* **Fundamental value of the exchange rate.**
**Equilibrium price and equilibrium quantity.** The price and quantity of a good at the intersection of the supply and demand curves for the good.
**Excess demand (or shortage).** The difference between the quantity supplied and the quantity demanded when the price of a good lies below the equilibrium price; buyers are dissatisfied when there is excess demand.
**Excess supply (or surplus).** The difference between the quantity supplied and the quantity demanded when the price of a good exceeds the equilibrium price; sellers are dissatisfied when there is excess supply.
**Expansion.** A period in which the economy is growing at a rate significantly above normal.
**Expansionary gap.** A negative output gap, which occurs when actual output is higher than potential output.
**Expansionary monetary policy.** A reduction in interest rates by the Fed, made with the intention of reducing a recessionary gap; also known as *monetary easing*.
**Expansionary policies.** Government policy actions intended to increase planned spending and output.

## F

**Federal funds rate.** The interest rate that commercial banks charge each other for very short-term (usually overnight) loans; because the Fed frequently sets its policy in the form of a target for the federal funds rate, this rate is closely watched in financial markets.
**Federal Open Market Committee (or FOMC).** The committee that makes decisions concerning monetary policy.
**Federal Reserve System (or Fed).** The central bank of the United States; also called the *Fed*.
**Final goods or services.** Goods or services consumed by the ultimate user; because they are the end products of the production process, they are counted as part of GDP.

**Financial intermediaries.** Firms that extend credit to borrowers using funds raised from savers.

**Fiscal policy.** Decisions that determine the government's budget, including the amount and composition of government expenditures and government revenues.

**Fisher effect.** The tendency for nominal interest rates to be high when inflation is high and low when inflation is low.

**Fixed cost.** A cost that does not vary with the level of an activity.

**Fixed exchange rate.** An exchange rate whose value is set by official government policy.

**Flexible exchange rate.** An exchange rate whose value is not officially fixed but varies according to the supply and demand for the currency in the foreign exchange market.

**Flow.** A measure that is defined per unit of time.

**Foreign exchange market.** The market on which currencies of various nations are traded for one another.

**Fractional-reserve banking system.** A banking system in which bank reserves are less than deposits so that the reserve-deposit ratio is less than 100 percent.

**Frictional unemployment.** The short-term unemployment associated with the process of matching workers with jobs.

**Fundamental value of the exchange rate (or equilibrium exchange rate).** The exchange rate that equates the quantities of the currency supplied and demanded in the foreign exchange market.

## G

**Government budget deficit.** The excess of government spending over tax collections ($G - T$).

**Government budget surplus.** The excess of government tax collections over government spending ($T - G$); the government budget surplus equals public saving.

**Government purchases.** Purchases by federal, state, and local governments of final goods and services; government purchases do *not* include *transfer payments,* which are payments made by the government in return for which no current goods or services are received, nor do they include interest paid on the government debt.

**Gross domestic product (GDP).** The market value of the final goods and services produced in a country during a given period.

## H

**Hyperinflation.** A situation in which the inflation rate is extremely high.

## I

**Income effect.** The change in the quantity demanded of a good that results because a change in the price of a good changes the buyer's purchasing power.

**Income-expenditure multiplier.** The effect of a one-unit increase in autonomous aggregate demand on short-run equilibrium output.

**Independent variable.** A variable in an equation whose value determines the value taken by another variable in the equation.

**Indexing.** The practice of increasing a nominal quantity each period by an amount equal to the percentage increase in a specified price index. Indexing prevents the purchasing power of the nominal quantity from being eroded by inflation.

**Induced aggregate demand.** The portion of aggregate demand that is determined within the model.

**Induced expenditure.** The portion of planned aggregate expenditure that depends on output $Y$.

**Inefficient point.** Any combination of goods for which currently available resources enable an increase in the production of one good without a reduction in the production of the other.

**Inferior good.** A good whose demand curve shifts leftward when the incomes of buyers increase.

**Inflation shock.** A sudden change in the normal behavior of inflation, unrelated to the nation's output gap.

**Intermediate goods or services.** Goods or services used up in the production of final goods and services and therefore not counted as part of GDP.

**International capital flows.** Purchases or sales of real and financial assets across international borders.

**International reserves.** Foreign currency assets held by a government for the purpose of purchasing the domestic currency in the foreign exchange market.

**Investment.** Spending by firms on final goods and services, primarily capital goods and housing.

## L

**Labor force.** The total number of employed and unemployed people in the economy.

**Law of one price.** If transportation costs are relatively small, the price of an internationally traded commodity must be the same in all locations.

**Liabilities.** The debts one owes.

**Life-cycle saving.** Saving to meet long-term objectives, such as retirement, college attendance, or the purchase of a home.

**Long-run aggregate supply (*LRAS*) line.** A vertical line showing the economy's potential output $Y^*$.

**Long-run equilibrium.** A situation in which actual output equals potential output and the inflation rate is stable; graphically, long-run equilibrium occurs when the *AD* curve, the *SRAS* line, and the *LRAS* line all intersect at a single point.

## M

**M1.** Sum of currency outstanding and balances held in checking accounts.

**M2.** All the assets in M1 plus some additional assets that are usable in making payments but at greater cost or inconvenience than currency or checks.

**Macroeconomic policies.** Government actions designed to affect the performance of the economy as a whole.

**Macroeconomics.** The study of the performance of national economies and the policies that governments use to try to improve that performance.

**Marginal benefit.** The marginal benefit of an activity is the increase in total benefit that results from carrying out one additional unit of the activity.

**Marginal cost.** The marginal cost of an activity is the increase in total cost that results from carrying out one additional unit of the activity.

**Marginal propensity to consume (*MPC*).** The amount by which consumption rises when disposable income rises by \$1; we assume that $0 < MPC < 1$.

**Market.** The market for any good consists of all buyers or sellers of that good.

**Market equilibrium.** Occurs when all buyers and sellers are satisfied with their respective quantities at the market price.

**Medium of exchange.** An asset used in purchasing goods and services.

**Menu costs.** The costs of changing prices.

**Microeconomics.** The study of individual choice under scarcity and its implications for the behavior of prices and quantities in individual markets.

**Monetary policy.** Determination of the nation's money supply.

**Money.** Any asset that can be used in making purchases.

**Money demand curve.** Shows the relationship between the aggregate quantity of money demanded $M$ and the nominal interest rate $i$; because an increase in the nominal interest rate increases the opportunity cost of holding money, which reduces the quantity of money demanded, the money demand curve slopes down.

**Multiplier.** *See* **Income-expenditure multiplier.**

**Mutual fund.** A financial intermediary that sells shares in itself to the public, then uses the funds raised to buy a wide variety of financial assets.

## N

**National saving.** The saving of the entire economy, equal to GDP less consumption expenditures and government purchases of goods and services, or $Y - C - G$.

**Natural rate of unemployment, $u^*$.** The part of the total unemployment rate that is attributable to frictional and structural unemployment; equivalently, the unemployment rate that prevails when cyclical unemployment is zero, so the economy has neither a recessionary nor an expansionary output gap.

**Net exports.** Exports minus imports.

**Nominal exchange rate.** The rate at which two currencies can be traded for each other.

**Nominal GDP.** A measure of GDP in which the quantities produced are valued at current-year prices; nominal GDP measures the *current dollar value* of production.

**Nominal interest rate.** The annual percentage increase in the nominal value of a financial asset; also known as the *market interest rate.*

**Nominal quantity.** A quantity that is measured in terms of its current dollar value.

**Normal good.** A good whose demand curve shifts rightward when the incomes of buyers increase.

**Normative analysis.** Addresses the question of whether a policy *should* be used; normative analysis inevitably involves the values of the person doing the analysis.

## O

**Okun's law.** States that each extra percentage point of cyclical unemployment is associated with about a 2 percentage point increase in the output gap, measured in relation to potential output.

**100 percent reserve banking.** A situation in which banks' reserves equal 100 percent of their deposits.

**Open economy.** An economy that trades with other countries.

**Open-market operations.** Open-market purchases and open-market sales.

**Open-market purchase.** The purchase of government bonds from the public by the Fed for the purpose of increasing the supply of bank reserves and the money supply.

**Open-market sale.** The sale by the Fed of government bonds to the public for the purpose of reducing bank reserves and the money supply.

**Opportunity cost.** The opportunity cost of an activity is the value of the next-best alternative that must be forgone to undertake the activity.

**Output gap, $Y^* - Y$.** The difference between the economy's potential output and its actual output at a point in time.

**Overvalued exchange rate.** An exchange rate that has an officially fixed value greater than its fundamental value.

## P

**Participation rate.** The percentage of the working-age population in the labor force (that is, the percentage that is either employed or looking for work).

**Peak.** The beginning of a recession, the high point of economic activity prior to a downturn.

**Planned aggregate expenditure (*PAE*).** Total planned spending on goods and services.

**Policy reaction function.** Describes how the action a policymaker takes depends on the state of the economy.

**Portfolio allocation decision.** The decision about the forms in which to hold one's wealth.

**Positive analysis.** Addresses the economic consequences of a particular event or policy, not whether those consequences are desirable.

**Potential output, $Y^*$.** The amount of output (real GDP) that an economy can produce when using its resources, such as capital and labor, at normal rates; also known as *potential GDP* or *full-employment output.*

**Precautionary saving.** Saving for protection against unexpected setbacks, such as the loss of a job or a medical emergency.

**Price ceiling.** A maximum allowable price, specified by law.

**Price index.** A measure of the average price of a given class of goods or services relative to the price of the same goods and services in a base year.

**Price level.** A measure of the overall level of prices at a particular point in time as measured by a price index such as the CPI.

**Principal amount.** The amount originally lent.

**Private saving.** The saving of the private sector of the economy is equal to the after-tax income of the private sector minus consumption expenditures $(Y - T - C)$; private saving can be further broken down into household saving and business saving.

**Production possibilities curve.** A graph that describes the maximum amount of one good that can be produced for every possible level of production of the other good.

**Protectionism.** The view that free trade is injurious and should be restricted.

**Public saving.** The saving of the government sector is equal to net tax payments minus government purchases $(T - G)$.

**Purchasing power parity (PPP).** The theory that nominal exchange rates are determined as necessary for the law of one price to hold.

## Q

**Quantity equation.** Money times velocity equals nominal GDP: $M \times V = P \times Y$.

**Quota.** A legal limit on the quantity of a good that may be imported.

## R

**Rate of inflation.** The annual percentage rate of change in the price level, as measured, for example, by the CPI.

**Rational person.** Someone with well-defined goals who tries to fulfill those goals as best he or she can.

**Real exchange rate.** The price of the average domestic good or service *relative* to the price of the average foreign good or service, when prices are expressed in terms of a common currency.

**Real GDP.** A measure of GDP in which the quantities produced are valued at the prices in a base year rather than at current prices; real GDP measures the actual *physical volume* of production.

**Real interest rate.** The annual percentage increase in the purchasing power of a financial asset; the real interest rate on any asset equals the nominal interest rate on that asset minus the inflation rate.

**Real quantity.** A quantity that is measured in physical terms—for example, in terms of quantities of goods and services.

**Real wage.** The wage paid to workers measured in terms of real purchasing power; the real wage for any given period is calculated by dividing the nominal (dollar) wage by the CPI for that period.

**Recession (or contraction).** A period in which the economy is growing at a rate significantly below normal.

**Recessionary gap.** A positive output gap, which occurs when potential output exceeds actual output ($Y^* > Y$).

**Relative price.** The price of a specific good or service *in comparison to* the prices of other goods and services.

**Reservation price.** The highest price someone is willing to pay to obtain any good or service, or the lowest payment someone would accept for giving up a good or performing a service.

**Reserve requirements.** Set by the Fed, the minimum values of the ratio of bank reserves to bank deposits that commercial banks are allowed to maintain.

**Reserve-deposit ratio.** Bank reserves divided by deposits.

**Revaluation.** An increase in the official value of a currency (in a fixed-exchange-rate system).

**Risk premium.** The rate of return that financial investors require to hold risky assets minus the rate of return on safe assets.

## S

**Saving.** Current income minus spending on current needs.

**Saving rate.** Saving divided by income.

**Seller's reservation price.** The smallest dollar amount for which a seller would be willing to sell an additional unit, generally equal to marginal cost.

**Seller's surplus.** The difference between the price received by the seller and his or her reservation price.

**Shortage.** *See* **Excess demand.**

**Short-run aggregate supply (*SRAS*) line.** A horizontal line showing the current rate of inflation, as determined by past expectations and pricing decisions.

**Short-run equilibrium.** A situation in which inflation equals the value determined by past expectations and pricing decisions, and output equals the level of short-run equilibrium output that is consistent with that inflation rate; graphically, short-run equilibrium occurs at the intersection of the *AD* curve and the *SRAS* line.

**Short-run equilibrium output.** The level of output at which output *Y* equals aggregate demand *AD*; the level of output that prevails during the period in which prices are predetermined.

**Skill-biased technological change.** Technological change that affects the marginal products of higher-skilled workers differently from those of lower-skilled workers.

**Slope.** In a straight line, the ratio of the vertical distance the straight line travels between any two points (*rise*) to the corresponding horizontal distance (*run*).

**Socially optimal quantity.** The quantity of a good that results in the maximum possible economic surplus from producing and consuming the good.

**Speculative attack.** A massive selling of domestic currency assets by financial investors.

**Stabilization policies.** Government policies that are used to affect aggregate demand, with the objective of eliminating output gaps.

**Stock.** A measure that is defined *at a point in time.*

**Stock (or equity).** A claim to partial ownership of a firm.

**Store of value.** An asset that serves as a means of holding wealth.

**Structural policy.** Government policies aimed at changing the underlying structure, or institutions, of the nation's economy.

**Structural unemployment.** The long-term and chronic unemployment that exists even when the economy is producing at a normal rate.

**Substitutes.** Two goods are substitutes in consumption if an increase in the price of one causes a rightward shift in the demand curve for the other.

**Substitution effect.** The change in the quantity demanded of a good that results because buyers switch to substitutes when the price of a good changes.

**Sunk cost.** A cost that is beyond recovery at the moment a decision must be made.

**Supply curve.** A curve or schedule showing the total quantity of a good that sellers wish to sell at each price.

**Surplus.** *See* **Excess supply.**

## T

**Tariff.** A tax imposed on an imported good.

**Time value of money.** The fact that a given dollar amount today is equivalent to a larger dollar amount in the future, because the money can be invested in an interest-bearing account in the meantime.

**Total surplus.** The difference between the buyer's reservation price and the seller's reservation price.

**Trade balance (or net exports).** The value of a country's exports less the value of its imports in a particular period (quarter or year).

**Trade deficit.** When imports exceed exports, the difference between the value of a country's imports and the value of its exports in a given period.

**Trade surplus.** When exports exceed imports, the difference between the value of a country's exports and the value of its imports in a given period.

**Transfer payments.** Payments the government makes to the public for which it receives no current goods or services in return.

**Trough.** The end of a recession, the low point of economic activity prior to a recovery.

## U

**Unattainable point.** Any combination of goods that cannot be produced using currently available resources.

**Undervalued exchange rate.** An exchange rate that has an officially fixed value less than its fundamental value.

**Unemployment rate.** The number of unemployed people divided by the labor force.

**Unemployment spell.** A period during which an individual is continuously unemployed.

**Unit of account.** A basic measure of economic value.

## V

**Value added.** For any firm, the market value of its product or service minus the cost of inputs purchased from other firms.

**Variable.** A quantity that is free to take a range of different values.

**Variable cost.** A cost that varies with the level of an activity.

**Velocity.** A measure of the speed at which money circulates, or equivalently, the value of a transaction completed in a period of time divided by the stock of money required to make those transactions; numerically, $V = (P \times Y)/M$, where $V$ is velocity, $P \times Y$ is nominal GDP, and $M$ is the money supply whose velocity is being measured.

**Vertical intercept.** The value taken by the dependent variable when the independent variable equals zero.

## W

**Wealth.** The value of assets minus liabilities.

**Wealth effect.** The tendency of changes in asset prices to affect households' wealth and thus their spending on consumer goods.

**Worker mobility.** The movement of workers between jobs, firms, and industries.

**World price.** The price at which a good or service is traded on international markets.

# INDEX

## C

## F

## G

## J

## N



## O

## S

## T



## U

## V



## W



## Y



## Z